AF588238

OXFORD SERIES ON MATERIALS MODELLING

Series Editors

Adrian P. Sutton, FRS
Department of Physics, Imperial College London

Robert E. Rudd
Lawrence Livermore National Laboratory

Oxford Series on Materials Modelling

Materials modelling is one of the fastest growing areas in the science and engineering of materials, both in academe and in industry. It is a very wide field covering materials phenomena and processes that span ten orders of magnitude in length and more than twenty in time. A broad range of models and computational techniques has been developed to model separately atomistic, microstructural and continuum processes. A new field of multi-scale modelling has also emerged in which two or more length scales are modelled sequentially or concurrently. The aim of this series is to provide a pedagogical set of texts spanning the atomistic and microstructural scales of materials modelling, written by acknowledged experts. Each book will assume at most a rudimentary knowledge of the field it covers and it will bring the reader to the frontiers of current research. It is hoped that the series will be useful for teaching materials modelling at the postgraduate level.

APS, London
RER, Livermore, California

1. M. W. Finnis: *Interatomic forces in condensed matter*
2. K. Bhattacharya: *Microstructure of martensite – Why it forms and how it gives rise to the shape-memory effects*
3. V. V. Bulatov, W. Cai: *Computer simulations of dislocations*
4. A. S. Argon: *Strengthening mechanisms in crystal plasticity*

Forthcoming:
L. P. Kubin, B. Devincre: *Multi-dislocation dynamics and interactions*
T. N. Todorov: *Electrical conduction in nanoscale systems*
D. N. Theodorou, V. Mavrantzas: *Multiscale modelling of polymers*

Interatomic Forces in Condensed Matter

Mike Finnis
Queen's University Belfast

Great Clarendon Street, Oxford, OX2 6DP,
United Kingdom

Oxford University Press is a department of the University of Oxford.
It furthers the University's objective of excellence in research, scholarship,
and education by publishing worldwide. Oxford is a registered trade mark of
Oxford University Press in the UK and in certain other countries

© Oxford University Press, 2003

The moral rights of the author have been asserted

First published 2003
First published in paperback 2010

All rights reserved. No part of this publication may be reproduced, stored in
a retrieval system, or transmitted, in any form or by any means, without the
prior permission in writing of Oxford University Press, or as expressly permitted
by law, by licence or under terms agreed with the appropriate reprographics
rights organization. Enquiries concerning reproduction outside the scope of the
above should be sent to the Rights Department, Oxford University Press, at the
address above

You must not circulate this work in any other form
and you must impose this same condition on any acquirer

Published in the United States of America by Oxford University Press
198 Madison Avenue, New York, NY 10016, United States of America

British Library Cataloguing in Publication Data
Data available

Library of Congress Cataloging in Publication Data
Data available

ISBN 978–0–19–958812–1

PREFACE

In this book, I demonstrate how various more approximate schemes of total energy and force calculation can be derived. I hope it will be helpful to graduate students and others who would like to understand better the status of various models of interatomic forces which abound in the literature. It can be disconcerting to meet such models; one suspects they might be totally arbitrary in form, and pay no respect to the principles of quantum mechanics. Sometimes a closer examination confirms this first impression! But several models can be derived by a sequence of approximations from first principles. Such derivations are not always easy to find, and I was motivated to write the book by the feeling that it would be useful to take a unified approach to the nature and derivation of interatomic forces. Furthermore, as far as I know some of the material is not in the literature at all.

Most of us were introduced to the idea of interatomic forces in condensed matter by way of the Lennard–Jones potential:

$$V(r) = -\frac{A}{r^6} + \frac{B}{r^{12}}.$$

It describes the potential energy of interaction of two noble gas atoms. A noble gas atom has a stable, closed shell of electrons surrounding a core. In the case of helium, the core is just a pair of protons and neutrons forming a nucleus, whereas in the heavier noble gas atoms the word core refers to an ion: a nucleus surrounded by one or more closed shells of electrons. The outer closed shell of electrons on each atom is a fluctuating object which at any instant has a particular dipole moment. As fast as light can travel from this atom to the other, its instantaneous dipole moment induces and interacts with the dipole moment on the other atom; it is easy to show that the resulting attractive energy of interaction falls off with separation r as r^{-6}. On the other hand a repulsion beween ion cores makes itself felt at short range and it is represented by the r^{-12} term. In a solid noble gas, this pairwise potential model is not quite so good as it is for a pair of atoms, because the real interactions are not simple, linearly additive potentials. Nevertheless it is not too bad. However, for any other material than a noble gas, this is a very poor model indeed. Moreover, it cannot be fixed by choosing some other functional forms for the attractive and repulsive parts. One can prove, simply from the elastic properties of many materials, that they could not possibly be described accurately by *any* pairwise interaction between atoms (a key observation here is the violation of the Cauchy relation, discussed in Section 5.3). There are other properties of materials which cannot be described even qualitatively by a pair potential model, as became apparent once people began to determine experimentally the formation energies of defects, the positions of atoms at surfaces, and many other characteristics. On the other hand, there are situations where the pair potential model has been quite useful over the years for giving some qualitative insights. For example, some of the first computer simulations using pair potentials discovered the phenomenon of *focussed collision sequences* in solids under irradiation, whereby momentum is rapidly transferred down a chain of atoms in a crystalline solid to create a defect some distance away from the original impact of a neutron or electron. Furthermore in some cases, such as simple metals or highly ionic crystals, *for a restricted range of atomic positions which do not change the density*, one can justify a pair potential model. Pair potentials are the simplest among the variety of models of interatomic forces which this book describes.

When atoms approach each other to form a metal, their electrons generally become delocalized as the atomic orbitals overlap. How can one even speak of 'interatomic forces' in such a case? Should we not consign the whole concept of 'interatomic' to the dustbin in the light of the quantum theory of solids? Emphatically not. We have to focus our attention on the forces acting on the atomic nuclei. The force on a nucleus is just the electrostatic force from all the other nuclei and electrons in the system. The use of the term 'interatomic' to describe it is a matter of convenience, not necessarily implying that the contributions to the force can be attributed to distinct atoms. We might calculate the total energy of the material as a function of the positions of the nuclei, making the assumption that the electrons are in wavefunctions enslaved to these atomic positions (the adiabatic approximation), and the resulting force we could calculate on any nucleus would be a function of all the other nuclear positions. It cannot in general be represented by a sum of pairwise forces between nuclei, but the purpose of this book is to show that all is not lost. We can still derive functions to describe interatomic forces which are much simpler than you might expect, even if they are not pairwise, but depend on the coordinates of many of the atoms. In general they might be called many-body potentials.

There are at least two good reasons why we want models of interatomic forces of the general type I have described, the scientific reason and the practical reason. The scientific reason is that there is often more physics in a simple, approximate model of a force than there is in an accurate computer calculation of its value. This is an argument against reductionism; I am taking the view that 'understanding' involves a process of simplification in which the essential physics or chemistry is identified and other details which determine the precise numerical value of some quantity are thrown away. The practical reason is that one wants to study structures, mechanisms or processes in materials which involve so many atoms that an accurate total energy and force calculation would be impractical with present computers.

Over the broad range of ionic, covalent and metallic systems, the only practical method of total energy and force calculation powerful enough to work for each is based on the *Hohenberg–Kohn–Sham density functional theory* (DFT). It is an *ab initio*, or first-principles approach, in the sense that the only assumed property of the material is its composition (to be precise, the charge on each nucleus) and the positions of the nuclei (which I will refer to as the atomic positions). The DFT is in principle exact, but in practice some approximation has to be made for the so-called exchange and correlation potential (described below). Such approximations go by the name of local density approximation (LDA) and generalised gradient approximation (GGA). There are several standard computer codes, some of them free, for doing DFT calculations in molecules and solids, and they are used routinely by hundreds or perhaps thousands of academic and commercial research groups around the world. In most codes, the atomic positions can be varied automatically either to seek a configuration of lowest energy or to simulate the dynamics of the system.

Working with a model of interatomic forces can be rather like dealing with a carpet which has been imperfectly cut and fitted to your room. One has a model with a certain number of free parameters which have been fitted to some experimental data such as the atomic volume and elastic moduli; the carpet looks wonderful, entirely free from bumps. But you are not content just to admire the room as it is, you now want to arrange your furniture differently. How confident are you that you can move the sofa from one side of the room to the other without exposing a hidden bump? Perhaps you can cover the bump with a chair and admire the rest of the floor, but sooner or later you will need to move that chair or something else. Can one be any more confident in applying empirical interatomic potentials?

Now, having a model which has been derived from DFT does not remove all the bumps in the carpet but I believe they are likely to be smaller and spread more evenly. A model which contains parameters that are fitted to some data, and which successfully predicts data of some other kind, is said to be transferable. Transferability is the goal of all model makers. Surely the more physics can be included in the *form* of a model, the more likely it is to be transferable.

The schemes or models I will discuss are:

1. *Pair Potentials.* A part of the total energy of simple (s–p bonded) metals, namely a part describing the energy of different configurations of the atoms at constant volume, can be written as a sum of pairwise potentials, although they look quite different from Lennard–Jones potentials. Furthermore they are volume-dependent.
2. *Tight Binding.* Traditionally this kind of model has been applied to transition metals, semiconductors and their alloys. It has also found applications to more ionic systems. Tight-binding models are implemented in many ways and to various degrees of approximation, as discussed in Chapter 7. Because of its versatility the tight-binding approach deserves the largest chapter in the book.

3. *Hybrid Schemes.* Under this heading I have discussed two important schemes that in different ways combine the concepts of previous chapters. One is the Generalized Pseudopotential Theory, known in its simplified form as the Model Generalized Pseudopotential Theory, which loosely speaking can be thought of as a hybrid of tight binding with pair potentials. The second is effective medium theory (EMT). There have been very many versions and applications of models which are based on some kind of effective medium theory. Such a model characterizes the energy of an atom as a function of the electron density at its location in the solid prior to placing it there. A simplified and empirical form of the EMT is widely used and referred to as the embedded atom model (EAM).
4. *Ionic Models.* The original ionic model was due to Max Born. It has undergone many refinements over the years, including compressible and polarisable ions, and, most recently, variable ionic charges. The present treatment makes a close link between ionic models and the tight-binding model.

The book is intended to be self-contained, in the sense that it requires no knowledge on the part of the reader beyond elementary quantum mechanics. The quantum mechanical tools that are required are described in Part I, and the derivation of models is in Part II. You will not find detailed information here about the implementation of any models, or their applications, for which I refer to the literature. This book is rather an overview of the *principles* of different schemes.

Acknowledgements

Many friends and colleagues have contributed wittingly or not to this book. No one I know has contributed more to the subject generally than Volker Heine, who first inspired my interest in it over 30 years ago. David Pettifor, Adrian Sutton and Tony Paxton have also been seminal influences, collaborators, and critical discussion partners. Robert Rudd carefully read the entire first draft and helped me smooth its rough edges. These friends have made numerous useful suggestions and corrections to preliminary versions of the manuscript, but of course bear no responsibility for the errors it still contains. Among others from whom I have learned and who might recognise their work reflected here I would particularly like to mention Graeme Ackland, Stefano Fabris and Lisa Moore.

I am especially grateful to my dear friend Eleanor Woods who provided me with an ideal haven for writing in Wolvercote, where I completed most of the manuscript during a sabbatical year granted by Queen's University Belfast. I dedicate the book to my wife Susanne.

GLOSSARY OF SYMBOLS

α_{M}	Madelung constant
$\Delta E_{\mathrm{f}}^{\mathrm{i}}$	Interstitial formation energy
$\Delta E_{\mathrm{f}}^{\mathrm{v}}$	Vacancy formation energy
$\Delta H_{\mathrm{f}}^{\mathrm{v}}$	Vacancy formation enthalpy
$\Delta S_{\mathrm{f}}^{\mathrm{v}}$	Vacancy formation entropy
$\Delta V_{\mathrm{f}}^{\mathrm{v}}$	Vacancy formation volume
ϵ_{e}	Dielectric constant of electrons
ϵ_{F}	Fermi energy
ϵ^{jel}	Jellium energy per electron
ϵ_{H}	Hartree dielectric constant
ϵ_n	nth energy eigenvalue
ϵ_{xc}	Exchange and correlation energy per electron in jellium
μ	Label to distinguish local orbitals
ρ	Electron density
$\rho^{(1)}$	First-order electron density
$\rho^{(2)}$	Second-order electron density
ρ^{in}	Input electron density
ρ^{val}	Valence electron density
ρ^{ex}	Exact (self-consistent) electron density
τ_I	Type of atom at site I
χ_{e}	Exact response function of electrons
χ_{s}	Response function of non-interacting 'electrons'
BOP	Bond Order Potential
B^{jel}	Bulk modulus of jellium
B^{pair}	Pair potential contribution to bulk modulus
B^{vol}	Contribution to bulk modulus of volume dependent energy
$D(\epsilon)$	Total density of states
$D_{I_\mu}(\epsilon)$	Local density of states on site I, orbital μ
$\mathcal{D}_n(\epsilon)$	Polynomials arising in recursion method (see eq. 7.115)
$\mathbf{E}$	Electric field
E^{rep}	Repulsive energy
$E_{\mathrm{band}}^{\mathrm{rep}}$	Repulsive energy in band model
E_{rim}	Energy of rigid-ion model

E_{TF}	Thomas–Fermi energy
E_{TFD}	Thomas–Fermi–Dirac energy
E^{tot}	Total energy
E^{vol}	Volume dependent energy
E_{x}	Exchange energy
E_{xc}	Exchange and correlation energy
$E_{\mathrm{xc}}^{\mathrm{ex}}$	Exact (self-consistent) exchange and correlation energy
E_{xc}^{I}	Exchange and correlation energy of atom I
$E_{\mathrm{xc}}^{\mathrm{in}}$	Input exchange and correlation energy
E_{ZZ}	Ion–ion (or nucleus–nucleus) Coulomb repulsion energy
EAM	Embedded Atom Method
EMT	Effective Medium Theory
f_{GSP}	Goodwin–Skinner–Pettifor cutoff function
$\mathcal{G}$	Local field correction
GPT	Generalized Pseudopotential Theory
H^{in}	Hamiltonion built from input charge density
$\mathbf{k}_{\mathrm{F}}$	Fermi wavevector
MGPT	Model Generalized Pseudopotential Theory
N	Number of electrons in the system
N_{a}	Number of atoms in the system
N_{orb}	Number of local atomic orbitals in the system
$\mathbf{O}$	Matrix of off-diagonal elements of $\mathbf{S}$
P_{jel}	Pressure arising from E_{jel}
P^{pair}	Pressure arising from E^{pair}
R_L	Real regular spherical harmonic
q_{TF}	Thomas–Fermi wavenumber
$\boldsymbol{R}_I$	Position vector of atom at site I
$\boldsymbol{R}_{IJ}$	Vector $\boldsymbol{R}_I - \boldsymbol{R}_J$
R_{IJ}	Distance $\lvert\boldsymbol{R}_I - \boldsymbol{R}_J\rvert$
V_{IJ}	Pairwise potential between atoms at sites I and J
V_{xc}	Exchange and correlation contribution to effective potential
$\mathbf{S}$	Overlap matrix
SCTM	Self-consistent charge transfer model
TB	Tight Binding
TBBM	Tight-Binding Bond Model
v_{C}	Coulomb potential $1/\lvert\mathbf{r} - \mathbf{r}'\rvert$
v_{e}	Coulomb potential $1/r$

CONTENTS

PART I

THE FRAMEWORK

1

ESSENTIAL QUANTUM MECHANICS

In this chapter, I will review the general concepts in quantum mechanics that are essential for the purpose of deriving models of interatomic forces. It is not meant for complete beginners, and takes for granted the usual material of a first course in quantum mechanics. Nevertheless, the content is by no means advanced, and a lot will be familiar if you have already studied something of the quantum mechanics of more than one particle. In that case the chapter may serve you mainly as a guide to my notation.

1.1 The Time-independent Schrödinger Equation

We have to start somewhere, and it will be with non-relativistic electrons which are confined by some potential. The potential is provided by a number of atomic nuclei at positions $\mathbf{R}_I$, where if there are M atoms $1 \leq I \leq M$. The nuclear positions will be treated here as if they are classical particles, which is a valid assumption for most elements in the periodic table at high temperatures. Each nucleus generates a Coulomb potential $eZ_I/|\mathbf{r} - \mathbf{R}_I|$, where e is the electronic charge, $\mathbf{r}$ is the position of an electron and Z_I is the atomic number. If we were to start including relativity, which as Dirac showed is essential for the consistent picture of electrons with spin, the simplicity of most of the useful results would probably be lost. I will henceforth have very little to say about spin in this book, except at the most elementary level, and correspondingly will say nothing about magnetic fields and their effect on interatomic forces. Furthermore, the effect of electric currents on interatomic forces is beyond the scope of this book. There are important recent developments in the subject of forces modified by electric currents and these will be described by Tchavdar Todorov, in another book in this series. So, to be specific, the starting point for us is the time-independent Schrödinger equation. In the process of introducing it now we will establish some more notation.

The simplest form of time-independent Schrödinger equation in SI units is

$$\hat{\mathcal{H}}\Psi = E\Psi \tag{1.1}$$

which if we explicitly write out the Hamiltonian operator $\hat{\mathcal{H}}$ looks like

$$\left(\sum_i -\frac{\hbar^2}{2m}\nabla_i^2 + \frac{1}{2}\sum_i \sum_j{}' \frac{e^2}{4\pi\epsilon_0|\mathbf{r}_i - \mathbf{r}_j|} + \sum_i \sum_I \frac{-e^2 Z_I}{4\pi\epsilon_0|\mathbf{r}_i - \mathbf{R}_I|}\right)\Psi = E\Psi. \tag{1.2}$$

The subscripts i and j in (1.2) label the electrons. The first term is the operator for the total kinetic energy of the electrons, acting on the wave function, or state, $\Psi(\mathbf{x}_1, \ldots, \mathbf{x}_N)$. The position of electron i is

$$\mathbf{r}_i \equiv (x_i, y_i, z_i), \tag{1.3}$$

and the operator for the kinetic energy of electron i is

$$-\frac{\hbar^2}{2m}\nabla_i^2 = -\frac{\hbar^2}{2m}\left(\frac{\partial^2}{\partial x_i^2} + \frac{\partial^2}{\partial y_i^2} + \frac{\partial^2}{\partial z_i^2}\right). \tag{1.4}$$

The arguments $\mathbf{x}_i$ of the wave function stand for the pair of arguments $(\mathbf{r}_i, s_i)$, where s_i stands for the spin of the electron. The spin introduces into the wavefunction two components for each electron, which we denote $\alpha(s_i)$ and $\beta(s_i)$. We think of them as the components of 'up' and 'down' spin associated with electron i. For our purposes we might think of them as components of a two-dimensional vector $\mathbf{s}_i$, but spin is not really a vector because it does not transform like one under rotation of the axes. The second term in (1.2) is the operator for the electron–electron interaction. The factor $\frac{1}{2}$ in front of the second term corrects for the double counting of each i, j pair which the summation runs over twice. The prime on the summation signifies that the term $j = i$ is to be excluded. The final term is the operator for the potential felt by the electrons from the nuclei. The electronic charge e is by convention a positive number, so that the charge on the electron is $-e$ Coulombs. Hence the negative sign of the last term, which represents negative charges $-e$ interacting with positive charges eZ_I.

In this simple form of the Schrödinger equation, there are no spin dependent operators, so the spins do not effect the energy of the system directly or explicitly. The only role of the spin arguments is in relation to the symmetry of the wave function, but this has a major effect on the energy indirectly as we shall see. The point here is that the wave function $\Psi(\mathbf{x}_1, \ldots, \mathbf{x}_N)$ must be antisymmetric with respect to interchange of any electron pair. That is another way of saying electrons are fermions.

$$\Psi(\mathbf{x}_1, \ldots, \mathbf{x}_i, \ldots, \mathbf{x}_j, \ldots, \mathbf{x}_N) = -\Psi(\mathbf{x}_1, \ldots, \mathbf{x}_j, \ldots, \mathbf{x}_i, \ldots, \mathbf{x}_N). \tag{1.5}$$

The wavefunction which is a solution of eq. (1.2) can be scaled by any constant and it would still be a solution, but by convention the scaling is chosen so that the wavefunction is *normalized*, meaning that

$$\int |\Psi|^2 \, d\mathbf{x}_1, \ldots, d\mathbf{x}_N = 1. \tag{1.6}$$

The integral over a coordinate $\mathbf{x}_i$ is shorthand for the integral over the spacial coordinates $\mathbf{r}_i$ and the summation over the spin components $\alpha(s_i)$ and $\beta(s_i)$. This normalization leads to an expression for the density of electrons at a position $\mathbf{r}_1$ which is:

$$\rho(\mathbf{r}_1) = N \int |\Psi|^2 \, ds_1, d\mathbf{x}_2, \ldots, d\mathbf{x}_N. \tag{1.7}$$

1.1.1 The Born–Oppenheimer Approximation

Equation (1.2) is an eigenvalue equation for the energy which neglects the effect of the motion of the nuclei. This is a crucial approximation in the theory of interatomic forces. A second, related approximation is to ignore all but the lowest energy solution out of the infinite number of solutions which exist for a given configuration of the nuclei $\{\mathbf{R}_I\}$. This solution is called the *ground state* and it also plays a fundamental part in the development of the theory. We label the states by a suffix such that the ground state energy and wavefunction are E_0 and Ψ_0, and the excited states have energies E_1, E_2, etc. By ignoring excited states of the electrons, even when we want to consider the dynamics of the atoms, we are making the so-called *Born–Oppenheimer Approximation* (Ziman, 1972, p. 200). It is rather a good approximation, since it depends on the mass of the electrons being much smaller than the masses of the nuclei, and on the velocity of the nuclei being low compared to the velocity of the electrons. The first condition is always satisfied and the latter condition is very often satisfied in practice. I shall not describe the derivation here, since it is very well covered in Todorov's volume. Its consequence is that the theory of interatomic forces will be restricted to situations in which we can assume that the electrons follow the motion of the nuclei adiabatically, and excluding any situations where electronic excitations are important. It is indeed possible to extend the theory to include excited states, but to do so in a rigorous manner would take us beyond the scope of this book.

Within the Born–Oppenheimer approximation, the total energy of the system is a unique function of the positions of the nuclei, independent of their velocities or history. Only with this assumption is it possible to derive interatomic potentials and interatomic forces which are functions only of the nuclear positions.

1.1.2 Atomic Units

At this point, we are going to bid farewell to SI units, because it makes life easier to work in *Hartree atomic units*, in which $e = m = \hbar = 4\pi\epsilon_0 = 1$. The unit of length is the Bohr radius $a_0 = 4\pi\epsilon_0\hbar^2/(me^2) = 0.5292$ Å. Instead of *Hartree* atomic units, some authors use *Rydberg* atomic units, and the two systems are easily confused. In Hartree units the unit of energy is the Hartree. One Hartree $= e^2/(4\pi\epsilon_0 a_0) = 27.2$ eV, which is equal to two Rydbergs. Rydberg units differ from Hartree units in that $m = \frac{1}{2}$, $e^2 = 2$, and the unit of energy is *one* Rydberg. In Hartree atomic units the Schrödinger equation takes the form

$$\left(\sum_i -\frac{1}{2}\nabla_i^2 + \frac{1}{2}\sum_i \sum_j \frac{1}{|\mathbf{r}_i - \mathbf{r}_j|} + \sum_i \sum_I \frac{-Z_I}{|\mathbf{r}_i - \mathbf{R}_I|} \right) \Psi = E\Psi. \tag{1.8}$$

The difference to Rydberg units would appear only in the first term, which is the kinetic energy operator. In Rydberg units it appears without the factor $\frac{1}{2}$. This extra simplicity of Rydberg units makes them very popular for deriving formulae and implementing them in computer codes. However, it is my personal preference to keep the $\frac{1}{2}$ in the kinetic energy operator, because I do not like the electron to have a charge $\sqrt{2}$ which Rydberg units require! So I will use Hartree atomic units throughout this book, and refer to them simply as atomic units.

1.1.3 Parts of the Total Energy

We will label the states in order of increasing energy, so E_0 is the energy of the ground state, which has a wavefunction Ψ_0, and E_n for $n > 0$ is the energy of the nth excited state Ψ_n.

Corresponding to the three terms in the Hamiltonion in eq. (1.8) the energy can be split into three parts: the kinetic energy, given the symbol T, the electron–electron interaction energy E_{ee} and the electron–nucleus energy E_{eZ}.

$$E = T + E_{ee} + E_{eZ}. \tag{1.9}$$

These can be expressed as expectation values:

$$T = \int \Psi^* \left(\sum_i -\frac{1}{2}\nabla_i^2 \right) \Psi \, d\mathbf{x}_1 \cdots d\mathbf{x}_N, \tag{1.10}$$

$$E_{ee} = \int \Psi^* \left(\frac{1}{2}\sum_i \sum_j \frac{1}{|\mathbf{r}_i - \mathbf{r}_j|} \right) \Psi \, d\mathbf{x}_1 \cdots d\mathbf{x}_N, \tag{1.11}$$

$$E_{eZ} = \int \Psi^* \left(\sum_i \sum_I \frac{-Z_I}{|\mathbf{r}_i - \mathbf{R}_I|} \right) \Psi \, d\mathbf{x}_1 \cdots d\mathbf{x}_N. \tag{1.12}$$

Any potential acting on the electrons, apart from that due to the electrons themselves, is referred to as the external potential, V_{ext}, and in this case it is just the potential of the nuclei:

$$V_{\text{ext}}(\mathbf{r}) = \sum_I \frac{-Z_I}{|\mathbf{r} - \mathbf{R}_I|}. \tag{1.13}$$

This leads to a simple expression for E_{eZ}. Inserting (1.13) into (1.12) gives

$$E_{eZ} = \int \Psi^* \left(\sum_i V_{\text{ext}}(\mathbf{r}_i) \right) \Psi \, d\mathbf{x}_1 \cdots d\mathbf{x}_N. \tag{1.14}$$

Combining this with the expression (1.7) for the electron density gives

$$E_{eZ} = \sum_i \int \frac{1}{N} \rho(\mathbf{r}_i) V_{\text{ext}}(\mathbf{r}_i) \, d\mathbf{r}_i = \int \rho(\mathbf{r}) V_{\text{ext}}(\mathbf{r}) \, d\mathbf{r}. \tag{1.15}$$

This result is the classical expression for the interaction energy.

1.2 Wave-mechanics of Non-interacting Fermions

1.2.1 Mean Field Theory

It is surprisingly useful to consider the case of non-interacting fermions, that is to examine the solution of the Schrödinger equation when the interaction between the electrons is switched off. Actually, we can easily do a lot better than to ignore the electron–electron interaction completely. Rather than simply setting the second term in eq. (1.8), the electron–electron interaction, to zero, it is replaced by a term which represents the interaction of the electrons with an external source of potential energy, representing the mean potential created by the electrons themselves. This source of potential is added to the potential generated by the nuclei, the third term of (1.8), to give a total *effective potential*, $V_{\text{eff}}(\mathbf{r})$, seen by the non-interacting 'electrons'. Within this *mean field approximation* eq. (1.8) becomes the simpler equation

$$\left(\sum_i -\frac{1}{2} \nabla_i^2 + \sum_i V_{\text{eff}}(\mathbf{r}_i) \right) \Psi = E\Psi. \tag{1.16}$$

This is an old idea, but it took on a new life in the nineteen sixties when it was realised by Hohenberg and Kohn that it can be the basis of accurate total energy calculations. The proper justification for it is the business of density functional theory, and we will be returning to that later.

In eq. (1.16) the many-body Hamiltonian $\mathcal{H}$ has been simplified to a sum of single-particle Hamiltonians H_i. We now have to solve

$$\left(\sum_i \hat{H}_i\right)\Psi = E\Psi, \tag{1.17}$$

in which

$$\hat{H}_i \equiv -\tfrac{1}{2}\nabla_i^2 + V_{\text{eff}}(\mathbf{r}_i). \tag{1.18}$$

This is a key property of the mean field approximation, and we shall now investigate it further.

1.2.2 Spacial and Spin Parts in Single-particle Wavefunctions

Suppose that we have solutions $\psi_{n\sigma}(\mathbf{x})$ which satisfy the equation

$$\hat{H}_i\psi_{n\sigma}(\mathbf{x}_i) = \epsilon_n\psi_{n\sigma}(\mathbf{x}_i). \tag{1.19}$$

These single-particle wavefunctions can be chosen to form an orthonormal set, meaning that scalar products are given by:

$$\int \psi_{n\sigma}^*(\mathbf{x})\psi_{m\mu}(\mathbf{x})\,\mathrm{d}\mathbf{x} = \delta_{mn}\delta_{\sigma\mu}. \tag{1.20}$$

The index n labels the eigenvalues and eigenfunctions of the single-particle Hamiltonian H_i such that ϵ_0 and $\psi_{0\sigma}$ are its ground state energy and wave function. For a single particle, the wave function factorizes into its spacial component and its spin component, that is to say we can write

$$\psi_{n\sigma}(\mathbf{x}_i) = \phi_n(\mathbf{r}_i)\sigma(s_i) \tag{1.21}$$

where σ is one of the 'up' or 'down' components α or β respectively. There are therefore two solutions to eq. (1.19) for each state labelled n, they have the same energy and spacial part of the wavefunction but opposite spins. The spacial and spin parts have their separate orthonormality rules:

$$\begin{aligned}\int \phi_n^*(\mathbf{r})\phi_m(\mathbf{r})\,\mathrm{d}\mathbf{r} &= \delta_{mn}\\ |\alpha(s_i)\alpha(s_i)| &= |\beta(s_i)\beta(s_i)| = 1\\ |\alpha(s_i)\beta(s_i)| &= 0.\end{aligned} \tag{1.22}$$

The summation over the spin components in the scalar product of two wavefunctions, denoted for electron 1 by the symbolic $\mathrm{d}s_1$ in eq. (1.7), is interpreted with the help of these orthonormality rules. Thus, the scalar product of a spin state $\sigma_1 = a_1\alpha(s_i) + b_1\beta(s_i)$ with another spin state $\sigma_2 = a_2\alpha(s_i) + b_2\beta(s_i)$ is given by

$$\int \sigma_1(s_i)^*\sigma_2(s_i)\mathrm{d}s_i = a_1^*a_2 + b_1^*b_2. \tag{1.23}$$

1.2.3 Determinant Wavefunctions

If we did not have the antisymmetry of the wavefunction Ψ to worry about we would be able to write the ground state wavefunction Ψ_0 as the product $\Pi_{i=1}^{N}\psi_{0\sigma}(\mathbf{x}_i)$ and the values of the spins would be irrelevant! All the electrons would be in the same spacial wave function, and the energy would be $N\epsilon_0$. Similarly any other state could be constructed as a product of single-particle states, with an energy equal to the sum of the energies of those states. But this violates (1.5), and something a bit cleverer has to be done. The correct solution is to construct a *determinant* rather than a single product, because a determinant will guarantee the antisymmetry automatically. A determinant made up of single-particle states is called a *Slater determinant*. This would be a good point to review the properties of determinants, if you happen to be hazy about them. The important thing to note is that if two rows or columns of a determinant are interchanged, the effect is to change the sign. Let us consider the case of just two electrons, occupying the single-particle states ψ_{A} and ψ_{B}, where A and B are combined labels for spacial and spin parts of the single-particle wavefunction. The determinant wavefunction for the two electrons is

$$\begin{aligned}\Psi(\mathbf{x}_1,\mathbf{x}_2) &= \sqrt{\tfrac{1}{2}}\,\det(\psi_{\mathrm{A}}(\mathbf{x}_1),\psi_{\mathrm{B}}(\mathbf{x}_2))\\ &\equiv \sqrt{\tfrac{1}{2}}\begin{vmatrix}\psi_{\mathrm{A}}(\mathbf{x}_1) & \psi_{\mathrm{B}}(\mathbf{x}_1)\\ \psi_{\mathrm{A}}(\mathbf{x}_2) & \psi_{\mathrm{B}}(\mathbf{x}_2)\end{vmatrix}\\ &\equiv \sqrt{\tfrac{1}{2}}\,(\psi_{\mathrm{A}}(\mathbf{x}_1)\psi_{\mathrm{B}}(\mathbf{x}_2)-\psi_{\mathrm{B}}(\mathbf{x}_1)\psi_{\mathrm{A}}(\mathbf{x}_2))\,. \end{aligned} \tag{1.24}$$

The first line in this formula introduces a shorthand notation for the determinant, which is then written out in full. The ground state for this case would be

$$\begin{aligned}\Psi(\mathbf{x}_1,\mathbf{x}_2) &= \sqrt{\tfrac{1}{2}}\,\det(\phi_0(\mathbf{r}_1)\alpha(s_1),\phi_0(\mathbf{r}_2)\beta(s_2))\\ &\equiv \sqrt{\tfrac{1}{2}}\begin{vmatrix}\phi_0(\mathbf{r}_1)\alpha(s_1) & \phi_0(\mathbf{r}_1)\beta(s_1)\\ \phi_0(\mathbf{r}_2)\alpha(s_2) & \phi_0(\mathbf{r}_2)\beta(s_2)\end{vmatrix}\\ &\equiv \sqrt{\tfrac{1}{2}}\,\phi_0(\mathbf{r}_1)\phi_0(\mathbf{r}_2)\,(\alpha(s_1)\beta(s_2)-\alpha(s_2)\beta(s_1))\,. \end{aligned} \tag{1.25}$$

The wavefunction is symmetric in the spacial coordinates but antisymmetric in the spin coordinates, which makes it overall antisymmetric.

Another common situation for two electrons is that they have the same spin, and therefore different spacial states. The wavefunction looks like

$$\begin{aligned}\Psi(\mathbf{x}_1,\mathbf{x}_2) &= \sqrt{\tfrac{1}{2}}\det(\phi_A(\mathbf{r}_1)\alpha(s_1),\phi_B(\mathbf{r}_2)\alpha(s_2)) \\ &\equiv \sqrt{\tfrac{1}{2}}\begin{vmatrix}\phi_A(\mathbf{r}_1)\alpha(s_1) & \phi_B(\mathbf{r}_1)\alpha(s_1) \\ \phi_A(\mathbf{r}_2)\alpha(s_2) & \phi_B(\mathbf{r}_2)\alpha(s_2)\end{vmatrix} \\ &\equiv \sqrt{\tfrac{1}{2}}\,(\phi_A(\mathbf{r}_1)\phi_B(\mathbf{r}_2)-\phi_B(\mathbf{r}_1)\phi_A(\mathbf{r}_2))\,\alpha(s_1)\alpha(s_2). \end{aligned} \tag{1.26}$$

In this case the wavefunction is antisymmetric in the spacial coordinates.

The two electron wavefunctions illustrate how the electron density is constructed from eq. (1.7) as follows. Consider first the spacially symmetric wavefunction (1.25). Take the squared modulus of (1.25) and apply (1.7) and the orthogonality rules (1.22):

$$\begin{aligned}\rho(\mathbf{r}_1) &= 2\int |\Psi|^2\,\mathrm{d}s_1\,\mathrm{d}\mathbf{x}_2 \\ &= \int |\phi_0(\mathbf{r}_1)\phi_0(\mathbf{r}_2)|^2\,\mathrm{d}\mathbf{r}_2 \times \left(|\alpha(s_1)|^2|\beta(s_2)|^2 + |\alpha(s_2)|^2|\beta(s_1)|^2\right) \\ &= 2|\phi_0(\mathbf{r}_1)|^2. \end{aligned} \tag{1.27}$$

We can interpret (1.27) as the density created by two independent electrons of opposite spin occupying the same single-particle wavefunction $\phi_0(\mathbf{r})$, so that the resulting density is exactly twice that associated with a single electron.

Now consider two electrons in wavefunction (1.26). Following the same procedure as above we find

$$\begin{aligned}\rho(\mathbf{r}_1) &= \int |\phi_A(\mathbf{r}_1)\phi_B(\mathbf{r}_2)-\phi_B(\mathbf{r}_1)\phi_A(\mathbf{r}_2)|^2\mathrm{d}\mathbf{r}_2 \times |\alpha(s_1)|^2|\alpha(s_2)|^2 \\ &= |\phi_A(\mathbf{r}_1)|^2 + |\phi_B(\mathbf{r}_1)|^2. \end{aligned} \tag{1.28}$$

For more than two electrons the determinant is constructed in an analogous way by identifying each column of the determinant with a different single-particle state and each row with a different electron. An important case is the ground state of N non-interacting electrons where N is even. The $N/2$ single-particle spacial states of lowest energy are each doubly occupied with electrons of up and down spin, and the Slater determinant has the form:

$$\Psi(\mathbf{x}_1,\ldots,\mathbf{x}_N) = \sqrt{\frac{1}{N!}}\begin{vmatrix}\phi_0(\mathbf{r}_1)\alpha(s_1) & \phi_0(\mathbf{r}_1)\beta(s_1) & \cdots & \phi_{N/2-1}(\mathbf{r}_1)\alpha(s_1) & \phi_{N/2-1}(\mathbf{r}_1)\beta(s_1) \\ \vdots & \vdots & & \vdots & \vdots \\ \phi_0(\mathbf{r}_N)\alpha(s_N) & \phi_0(\mathbf{r}_N)\beta(s_N) & \cdots & \phi_{N/2-1}(\mathbf{r}_N)\alpha(s_N) & \phi_{N/2-1}(\mathbf{r}_N)\beta(s_N)\end{vmatrix} \tag{1.29}$$

The energy of this ground state is the sum of the single-particle energies:

$$E_{s0} = 2 \sum_{n=0}^{N/2-1} \epsilon_n, \tag{1.30}$$

where the factor of two counts the up and down spins. The charge density associated with it is easily shown to be

$$\rho(\mathbf{r}) = 2 \sum_{n=0}^{N/2-1} |\phi_n(\mathbf{r})|^2, \tag{1.31}$$

where as before the factor 2 arises from the spins. The normalization of the single-particle wavefunctions ensures that the density integrated over space gives the correct total number of electrons.

$$\int \rho(\mathbf{r})\mathrm{d}\mathbf{r} = N. \tag{1.32}$$

1.3 Basis Vectors and Representations

In this section, we are going to develop some useful ideas and notation to do with single-particle wavefunctions, which are essential ingredients of all theories of electronic structure and total energy.

The idea of scalar products of vectors is very important in quantum mechanics, as are other less elementary bits of linear algebra. Equations (1.22) already have the flavour of the scalar product of orthonormal vectors. The single-particle wavefunctions ϕ_n are individually analogous to vectors with an infinite number of dimensions; the components of the vector ϕ_n are $\phi_n(\mathbf{r})$, and the integral in (1.22) is multiplying together equivalent components and adding the products. The complex conjugate in the product ensures that the result is real. Any vector in the vector space can be represented by its components in a chosen *basis*, where as usual a basis is a set of vectors which has the property of completeness. $\phi(\mathbf{r})$ can be thought of as a component of the vector ϕ in a basis each member of which is like a delta function in $\mathbf{r}$-space. The solutions to the single-particle Schrödinger equation form another important complete set of vectors which can be used as a basis. These rather vague notions can be made more precise and powerful by introducing some more notation. This notation will also make life easier, in case you are wondering how to tell at a glance whether some quantity is a vector or a scalar or the component of a vector.

1.3.1 Bras and Kets

Quantum mechanics requires the concept of a vector space to be generalized to include infinitely many basis elements (eigenstates). *Hilbert space* is the name of the abstract vector space which has the required mathematical properties for the machinery of quantum mechanics to work. Wavefunctions, the states of electrons, live in it. To think of electron states as vectors in a Hilbert space it is useful to adopt Dirac's 'bra' and 'ket' notation. This section introduces the relevant notation and properties that will be useful later. For readers interested in the mathematical subtleties, Dirac's book would be a good place to look. In this way of thinking, states χ are not explicitly functions of position $\mathbf{r}$. The Hamiltonion operator is also a more abstract object than the one we are used to in elementary quantum mechanics, and it is also not necessarily an explicit function of $\mathbf{r}$. Although the ideas and notation are introduced here for single particle states, the same concepts are powerful when dealing with many-particle states; the dimensions of the Hilbert space must be increased accordingly.

The Dirac notation for a state ϕ_n which is a solution of some particular single-particle Schrödinger equation is $|\phi_n\rangle$ or simply $|n\rangle$. These are called the *eigenstates* of the Hamiltonian. If $\hat{H}$ denotes the particular single-particle Hamiltonian operator, the Schrödinger equation looks like

$$\hat{H}|n\rangle = \epsilon_n|n\rangle. \tag{1.33}$$

The Hamiltonian can be split into its kinetic energy part ($\hat{T}$) and its potential energy part:

$$\hat{H} = \hat{T} + V_{\text{eff}}, \tag{1.34}$$

where in this case the one electron is moving in an effective potential. Other states ϕ and χ which may be a solution of some other unspecified single-particle Schrödinger equation are denoted by $|\phi\rangle$ and $|\chi\rangle$. All such objects with right angle brackets $|\ldots\rangle$ are the 'kets'. In this abstract space there is also an infinite number of orthonormal vectors $|\mathbf{r}\rangle$. These can be regarded as the eigenstates of *position*.The scalar product of one of them with $|n\rangle$ is written $\langle\mathbf{r}|n\rangle$, and this is equal to the scalar quantity $\phi_n(\mathbf{r})$. By analogy with finite vector spaces, we speak of $\phi(\mathbf{r})$ as being a state 'in the $\mathbf{r}$-representation', and we have the picture in mind that ϕ is being represented by its components in the basis set of kets $|\mathbf{r}\rangle$.

Quantities denoted by a left angle bracket such as $\langle\mathbf{r}|$ or $\langle\phi|$ are also defined. They inhabit what is called a dual space; these are the 'bras'. They are in one-to-one correspondence with the 'kets'. In Dirac notation we have for any single-particle state ϕ an infinite number of *scalar products* which specify it in the $\mathbf{r}$-representation:

$$\langle\mathbf{r}|\phi\rangle \equiv \phi(\mathbf{r}), \tag{1.35}$$

just as in everyday three-dimensional space the three scalar products of some vector $\mathbf{v}$ with unit vectors along the Cartesian axes completely specify the vector $\mathbf{v}$.

For the eigenstates of a single-particle Schrödinger equation the scalar product can be written as:

$$\langle \mathbf{r}|n\rangle \equiv \langle \mathbf{r}|\phi_n\rangle \equiv \phi_n(\mathbf{r}). \tag{1.36}$$

These scalar products may seem abstract, and they are, but we tend to have a very real picture of them based on our three-dimensional experience. Since $|\phi\rangle$ and $|\mathbf{r}\rangle$ live in the same space, and since the set of kets $|\mathbf{r}\rangle$ is *complete*, no aspect or property of the state is lost by representing it in terms of its components $\phi(\mathbf{r})$. In the same way a three-dimensional vector can be represented in spherical coordinates or Cartesian coordinates without loss of information.

The scalar product of ϕ_n with ϕ is denoted $\langle\phi_n|\phi\rangle$ or $\langle n|\phi\rangle$, and the set of $\langle n|\phi\rangle$ for all n specifies ϕ 'in the ϕ_n-representation'. Complex numbers enter rather naturally this Dirac formalism. They appear in the definition of the scalar product; namely, if the vector arguments in any scalar product are interchanged, the result is complex conjugation, e.g.:

$$\langle u|v\rangle = \langle v|u\rangle^*. \tag{1.37}$$

This gives us a consistent notation for expressing complex conjugates, e.g. $\langle\phi|\mathbf{r}\rangle = \phi(\mathbf{r})^*$. Another essential property of eigenstates of the single-particle Schrödinger equation is that the $|n\rangle$ like the $|\mathbf{r}\rangle$ form a *complete* and *orthonormal* set. The orthonormality can be expressed as

$$\langle n|m\rangle = \delta_{nm} \tag{1.38}$$

and

$$\langle \mathbf{r}|\mathbf{r}'\rangle = \delta(\mathbf{r}-\mathbf{r}') \tag{1.39}$$

and the completeness as

$$\sum_{n=0}^{\infty} |n\rangle\langle n| = \hat{\mathbf{1}} \tag{1.40}$$

and

$$\int_V |\mathbf{r}\rangle\langle\mathbf{r}|\,\mathrm{d}\mathbf{r} = \hat{\mathbf{1}}. \tag{1.41}$$

The range of the integration in (1.41) and in other integrals over space is defined by the volume V we choose to contain the subsystem of interest. Very often we are dealing in theory with an infinite system, but one which is periodically repeated in space. The repeating unit we choose as our subsystem, not necessarily the smallest repeating unit, is called a *supercell*, and it defines the volume V. On the other hand, if the subsystem of interest is bounded by vacuum, as in molecular systems, V could be the whole of space.

The quantities appearing in (1.40–1.41), like the Hamiltonian, are very particular examples of *operators*. In general operators act on a vector in the space and turn it into some other vector; in this case they are the *identity* operator, which leaves unchanged the vector on which it operates. Operators can always be written as a sum of terms of the form a ket followed by a coefficient, which in (1.40–1.41) is unity, followed by a bra. You have to think of operators differently to ordinary algabraic quantities, because their meaning is what they *do* rather than what they stand for. A symbol with a hat, such as $\hat{\mathbf{1}}$ will now denote an operator in its basis-independent form, rather than as in the previous use of $\hat{H}$, which expressed it in terms of the position cordinates $\mathbf{r}_i$. Without referring to position coordinates at all there is a simple way to write a Hamiltonian operator in terms of its eigenstates:

$$\hat{H} \equiv \sum_{n=0}^{\infty} |n\rangle \epsilon_n \langle n|. \tag{1.42}$$

A single ket followed by the equivalent bra, such as $|\mathbf{r}\rangle\langle\mathbf{r}|$, is an example of a *projection operator*. We can think of this operator as projecting a wavefunction, or picking out its component at $\mathbf{r}$.

To show an explicit case, here is the effect of the representation (1.40) of the identity operator acting on the ket $|\mathbf{r}'\rangle$:

$$\int |\mathbf{r}\rangle\langle\mathbf{r}|\, d\mathbf{r} |\mathbf{r}'\rangle = \int |\mathbf{r}\rangle \delta(\mathbf{r} - \mathbf{r}')\, d\mathbf{r} = |\mathbf{r}'\rangle. \tag{1.43}$$

More about operators later.

The scalar product of any single-particle wavefunction with itself is an important quantity which defines the normalisation of the wavefunction. A wavefunction $|\chi\rangle$ is normalised when

$$\langle\chi|\chi\rangle = 1. \tag{1.44}$$

The real space representation of this condition is obtained by inserting the identity operator (1.41) in the bra-ket:

$$\langle\chi|\chi\rangle = \langle\chi| \int |\mathbf{r}\rangle\langle\mathbf{r}|\, d\mathbf{r} |\chi\rangle = \int \chi(\mathbf{r})^* \chi(\mathbf{r})\, d\mathbf{r} = 1. \tag{1.45}$$

Notice that a solution of the Schrödinger equation is still a solution if it is scaled by an arbitrary factor. Therefore, strictly speaking the orthonormality of an eigenstate $|n\rangle$ is not automatic, it has to be imposed. But any $|\chi\rangle$ that is not normalised can always be made so by multiplying it by a factor $\langle\chi|\chi\rangle^{-1/2}$. Besides requiring normalisation, the solutions to the Schrödinger equation may be *degenerate*, that is to say, there is more than one solution with the same energy. In such a case, which is more common than not, the $|n\rangle$ may well not turn out to be orthogonal. However, they can always be *orthogonalized*. That is to say, linear combinations of the $|n\rangle$ can be found which are still solutions to the Schrödinger equation with the same ϵ_n but which are now orthogonal to each other.

1.3.2 Expansion Coefficients

Any vector in the Hilbert space can be expressed in a chosen basis, if that basis is complete, as a sum of *expansion coefficients* times basis functions. For example, any single-particle wavefunction $|\chi\rangle$ can be expanded in terms of the orthonormal eigenfunctions of some single-particle Hamiltonian:

$$|\chi\rangle = \sum_{n=0}^{\infty} C_n^{\chi} |n\rangle. \tag{1.46}$$

By this means the problem of solving the Schrödinger equation for the wavefunctions becomes a tractable problem of solving linear equations for the vectors C_n^{χ}, as we shall see. In this case formal expressions for the expansion coefficients are obtained by multiplying $\langle n|$ on each side and using the orthonormality property:

$$C_n^{\chi} = \langle n|\chi\rangle. \tag{1.47}$$

The resulting expansion

$$|\chi\rangle = \sum_{n=0}^{\infty} \langle n|\chi\rangle|n\rangle \equiv \sum_{n=0}^{\infty} |n\rangle\langle n|\chi\rangle \tag{1.48}$$

is also obtained directly by applying $\hat{\mathbf{1}}$ in the form (1.40) to $|\chi\rangle$. In practice, that is to say in numerical calculations when integrals have to be done, the expression (1.47) is evaluated in the **r**-representation. One inserts the identity operator to generate the required expression:

$$C_n^{\chi} = \langle n|\chi\rangle \equiv \int \langle n|\mathbf{r}\rangle\langle\mathbf{r}|\chi\rangle\, d\mathbf{r} \equiv \int \psi_n(\mathbf{r})^* \chi(\mathbf{r})\, d\mathbf{r}. \tag{1.49}$$

It can also be easily verified that

$$\langle\chi| = \sum_{n=0}^{\infty} C_n^{\chi *} \langle n|. \tag{1.50}$$

where $C_n^{\chi *}$ is the complex conjugate of C_n^{χ}.

With this way of looking at things, we can think of $\psi(\mathbf{r})$ as an expansion coefficient of $|\psi\rangle$ in the **r**-representation, where the expansion looks like:

$$|\psi\rangle = \int |\mathbf{r}\rangle\langle\mathbf{r}|\psi\rangle\, d\mathbf{r}.$$

1.3.3 Eigenstates of Momentum and Matrix Elements of Operators

Another complete set of basis vectors is formed by the eigenstates $|\mathbf{p}\rangle$ of the momentum operator in free space $\hat{\mathbf{p}}$. These are the solutions of the very abstract looking eigenvalue equation

$$\hat{\mathbf{p}}|\mathbf{p}\rangle = \mathbf{p}|\mathbf{p}\rangle. \tag{1.51}$$

In the $\mathbf{r}$-representation (1.51) becomes

$$\int \langle \mathbf{r}|\hat{\mathbf{p}}|\mathbf{r}'\rangle \langle \mathbf{r}'|\mathbf{p}\rangle \mathrm{d}\mathbf{r}' = \mathbf{p}\langle \mathbf{r}|\mathbf{p}\rangle. \tag{1.52}$$

The latter form has been obtained by pre-multiplying (1.51) by $\langle \mathbf{r}|$ and inserting the identity operator on the left-hand side. An object with the form 'bra' times operator times 'ket' is called a *matrix element* of the operator. In this case, the matrix elements of $\hat{\mathbf{p}}$ in the $\mathbf{r}$-representation are well known and are

$$\langle \mathbf{r}|\hat{\mathbf{p}}|\mathbf{r}'\rangle = -\mathrm{i}\hbar\nabla\delta(\mathbf{r}-\mathbf{r}'), \tag{1.53}$$

where ∇ is the gradient operator with respect to the coordinate $\mathbf{r}$. Although $\hbar = 1$ in atomic units, it is sometimes quite helpful to keep it in the formulae in order to highlight the essentially quantum mechanical parameters. The integral over $\mathbf{r}'$ in (1.52) can be performed, giving the familiar differential equation

$$-\mathrm{i}\hbar\nabla\langle \mathbf{r}|\mathbf{p}\rangle = \mathbf{p}\langle \mathbf{r}|\mathbf{p}\rangle \tag{1.54}$$

with the solutions

$$\langle \mathbf{r}|\mathbf{p}\rangle = \frac{1}{\sqrt{V}}\exp(\mathrm{i}\mathbf{k}\cdot\mathbf{r}), \tag{1.55}$$

where we have introduced the $\mathbf{k}$-vector,

$$\mathbf{k} = \frac{\mathbf{p}}{\hbar}. \tag{1.56}$$

There is a one-to-one correspondence between $\mathbf{k}$ and $\mathbf{p}$, and it will usually be better to label the same momentum eigenstate with its $\mathbf{k}$-vector rather than $\mathbf{p}$. I shall therefore use the notation $|\mathbf{k}\rangle$ or $|\mathbf{p}\rangle$ interchangeably to mean exactly the same state.

The function (1.55) has the form of a *plane wave*, with a propagation direction along the $\mathbf{k}$-vector and with a wavelength

$$\lambda = \frac{2\pi}{|\mathbf{k}|}. \tag{1.57}$$

The factor $1/\sqrt{V}$ is included to make the function normalisable; the integral over $\mathbf{r}$ has to be over a finite volume V:

$$\langle \mathbf{k}|\mathbf{k}\rangle = \int \langle \mathbf{k}|\mathbf{r}\rangle\langle \mathbf{r}|\mathbf{k}\rangle\,\mathrm{d}\mathbf{r} = \int |\langle \mathbf{r}|\mathbf{k}\rangle|^2\,\mathrm{d}\mathbf{r} = 1. \tag{1.58}$$

Now consider the orthogonality of these states. We can write:

$$\langle \mathbf{k}|\mathbf{k}'\rangle = \int \langle \mathbf{k}|\mathbf{r}\rangle\langle \mathbf{r}|\mathbf{k}'\rangle \mathrm{d}\mathbf{r} = \frac{1}{V}\int \exp\left(\mathrm{i}(\mathbf{k}' - \mathbf{k})\cdot \mathbf{r}\right)\,\mathrm{d}\mathbf{r}. \tag{1.59}$$

In order to proceed we need a textbook formula for the delta function in three dimensions, which is

$$\delta(\mathbf{r} - \mathbf{r}') = \frac{1}{(2\pi)^3}\int \exp\left(\mathrm{i}\mathbf{k}\cdot(\mathbf{r} - \mathbf{r}')\right)\,\mathrm{d}\mathbf{k}. \tag{1.60}$$

We change the arguments of this formula to fit the right-hand side of (1.59), which then becomes

$$\langle \mathbf{k}|\mathbf{k}'\rangle = \frac{(2\pi)^3}{V}\delta(\mathbf{k} - \mathbf{k}'). \tag{1.61}$$

Discussion of the orthogonality and completeness of the momentum basis functions introduces one of the basic concepts of solid state physics, which is **k**-space. We shall explore this concept further by considering the completeness relation, but first let us consider an important example of matrix elements of an operator.

1.3.4 Matrix Elements of the Coulomb Potential

The Coulomb potential of a unit negative charge at the origin as seen by an electron at **r** has the familiar form

$$v_{\mathrm{e}}(\mathbf{r}) = \frac{1}{r}. \tag{1.62}$$

We can think of it as an operator with matrix elements

$$\left\langle \mathbf{r}|\hat{v}_{\mathrm{e}}|\mathbf{r}'\right\rangle = \frac{1}{r}\delta(\mathbf{r} - \mathbf{r}'), \tag{1.63}$$

that is, it is diagonal in the **r**-representation. The Coulomb operator can also be expressed in the basis of momentum states, which requires us to make a Fourier transformation of the Coulomb potential. The matrix elements of $\hat{v}_{\mathrm{e}}$ in **k**-space are given by

$$\left\langle \mathbf{k}|\hat{v}_{\mathrm{e}}|\mathbf{k}'\right\rangle = \int \langle \mathbf{k}|\mathbf{r}\rangle \left\langle \mathbf{r}|\hat{v}_{\mathrm{e}}|\mathbf{r}'\right\rangle \langle \mathbf{r}'|\mathbf{k}'\rangle \mathrm{d}\mathbf{r}\mathrm{d}\mathbf{r}', \tag{1.64}$$

where we have inserted the identity operator on each side of $\hat{v}_{\mathrm{e}}$. Inserting (1.63) and the expressions for $\langle \mathbf{k}|\mathbf{r}\rangle$ and $\langle \mathbf{r}'|\mathbf{k}'\rangle$ from (1.55) and (1.37):

$$\left\langle \mathbf{k}|\hat{v}_{\mathrm{e}}|\mathbf{k}'\right\rangle = \frac{1}{V}\int \frac{1}{r}\exp\left(\mathrm{i}(\mathbf{k} - \mathbf{k}')\cdot \mathbf{r}\right)\mathrm{d}\mathbf{r}. \tag{1.65}$$

The Coulomb potential is clearly not diagonal in the **k**-representation, although it does have the simplifying feature that it only depends on the distance beween the ends of the vectors $\mathbf{k}$ and $\mathbf{k}'$ and not on their separate lengths or directions. The Fourier transform in (1.65) has a rather nasty singularity at $\mathbf{k} = \mathbf{k}'$ which we can circumvent by a trick. Instead of dealing with the bare Coulomb potential, let us consider a screened Coulomb potential (the Yukawa potential) given by

$$v_{e\lambda}(r) = \frac{\exp(-\lambda r)}{r}, \tag{1.66}$$

where $1/\lambda$ defines a *screening length*. In the present context we do not attribute any physical significance to this potential. If we now replace the Coulomb potential by the Yukawa potential the Fourier transform (1.65) can be carried out straightforwardly in spherical coordinates. We find

$$\left\langle \mathbf{k} \middle| \hat{v}_{e\lambda} \middle| \mathbf{k}' \right\rangle = \frac{1}{V} \frac{4\pi}{|\mathbf{k} - \mathbf{k}'|^2} \frac{1}{\{1 + \lambda^2/|\mathbf{k} - \mathbf{k}'|^2\}}. \tag{1.67}$$

Finally, as long as we realise that the result is not defined at $\mathbf{k} = \mathbf{k}'$ we can let λ go to zero and obtain:

$$\left\langle \mathbf{k} \middle| \hat{v}_{e} \middle| \mathbf{k}' \right\rangle = \frac{1}{V} \frac{4\pi}{|\mathbf{k} - \mathbf{k}'|^2}. \tag{1.68}$$

Sometimes you will see the above formula for the Fourier transform of the Coulomb potential without the factor $1/V$. Recall that this factor is there simply because the plane waves are normalised over a volume V, and it can be regarded as arbitrary. Since it does not correspond to any particular physical volume, it will never appear in any formulae for measurable quantities.

It is worth warning about a certain subtlety of the language in this subject when vectors are talked about. It has to do with the difference between the kets such as $|\mathbf{k}\rangle$ and the more familiar vectors such as $\mathbf{k}$. The objects $|\mathbf{k}\rangle$ and $|\mathbf{r}\rangle$ inhabit the same space, which is called *Hilbert space*. The sense of this is that any other objects in their space can be expressed as linear combinations of either $|\mathbf{k}\rangle$s or $|\mathbf{r}\rangle$s. The vectors $\mathbf{k}$ and $\mathbf{r}$ on the other hand inhabit their own spaces, $\mathbf{k}$- and $\mathbf{r}$-space, respectively. There is a one-to-one correspondence between $\mathbf{k}$ and $|\mathbf{k}\rangle$ but they live in entirely different spaces, as do $\mathbf{r}$ and $|\mathbf{r}\rangle$. $\mathbf{k}$ and $\mathbf{r}$ also live in completely different spaces; there is no linear combination of $\mathbf{r}$s which is equal to $\mathbf{k}$ or vice versa. However, because of the existence of a one-to-one correspondence between $\mathbf{k}$ and $|\mathbf{k}\rangle$, one sometimes loosely refers to $|\mathbf{k}\rangle$ as if it were also a vector in $\mathbf{k}$-space rather than Hilbert space.

1.3.5 Momentum Space and k-space

By analogy with eq. (1.41), there is a completeness relation for the momentum states $|\mathbf{p}\rangle$, which can be written as

$$\sum |\mathbf{p}\rangle\langle\mathbf{p}| = \hat{\mathbf{1}}. \tag{1.69}$$

In an infinite volume the allowed values of momentum are infinitely close together, so we should be able to pass to the limit

$$\sum |\mathbf{p}\rangle\langle\mathbf{p}| \to \int |\mathbf{p}\rangle\langle\mathbf{p}|\, d\mathbf{p},$$

where the integral is over all of *momentum space*. In the corresponding $\mathbf{r}$-space formula we simply defined the normalization so that (1.41) holds. Now we need to define a density of momentum states in momentum space (N_p) such that

$$N_p \int |\mathbf{p}\rangle\langle\mathbf{p}|\, d\mathbf{p} = \hat{\mathbf{1}}. \tag{1.70}$$

The value of N_p follows from the previous definitions as follows. To get it we again need to use the formula (1.60) for the delta function in three dimensions.

Now we sandwich (1.70) in $\langle\mathbf{r}|\cdots|\mathbf{r}'\rangle$, and insert (1.55) and (1.39) to give

$$\begin{aligned} N_p \int \langle\mathbf{r}|\mathbf{p}\rangle\langle\mathbf{p}|\mathbf{r}'\rangle\, d\mathbf{p} &= (N_p/V) \int \hbar^3 \exp\left(i\mathbf{k}\cdot(\mathbf{r}-\mathbf{r}')\right)\, d\mathbf{k} \\ &= \langle\mathbf{r}|\mathbf{r}'\rangle = \delta(\mathbf{r}-\mathbf{r}'), \end{aligned} \tag{1.71}$$

which by comparison with (1.60) provides us with the required value

$$N_p = \frac{V}{(2\pi\hbar)^3}. \tag{1.72}$$

We have used the correspondence between an element of volume $d\mathbf{p}$ in momentum space and the element $\hbar^3 d\mathbf{k}$ in $\mathbf{k}$-space, which follows from $\mathbf{p} = \hbar\mathbf{k}$. We can similarly define a density of states in $\mathbf{k}$-space, N_k such that

$$N_k d\mathbf{k} = N_p d\mathbf{p}$$

so that

$$N_k = \frac{V}{(2\pi)^3}. \tag{1.73}$$

The completeness of the momentum eigenstates is therefore expressed by:

$$\sum |\mathbf{k}\rangle\langle\mathbf{k}| \equiv \frac{V}{(2\pi)^3} \int |\mathbf{k}\rangle\langle\mathbf{k}|\, d\mathbf{k} = \hat{\mathbf{1}}. \tag{1.74}$$

This result for N_k is usually derived in textbooks on solid state physics by a different route, (Ashcroft and Mermin, 1976, p. 35), in which one considers the fitting of plane waves into a cubic box of side L, a condition which restricts the allowed values of $\mathbf{k}$. To do this it is necessary to introduce the idea of *periodic boundary conditions*. Since this idea turns out to be of very wide use, I will introduce it now formally.

1.4 Periodic Boundary Conditions

1.4.1 Fourier Transformation of a Wavefunction

First, let us pursue the idea of **k**-space and the density of states in it. The space within the cubic box of side L is supposed to be periodically repeated such that in the present case the wavefunctions $\langle \mathbf{r}|\mathbf{k}\rangle = (1/\sqrt{V})\exp(i\mathbf{k}\cdot\mathbf{r})$ have a periodicity L in x, y and z directions. Mathematically, a function $\psi(x, y, z)$ has a periodicity L in x, y and z directions if

$$\psi(x, y, z) = \psi(x + n_x L, y + n_y L, z + n_z L) \tag{1.75}$$

for any integers n_x, n_y, n_z. For $\langle \mathbf{r}|\mathbf{k}\rangle$ this imposes a condition on **k** which restricts it to a lattice of discrete points in **k**-space such that

$$k_x = \frac{2\pi n_x}{L}, \quad k_y = \frac{2\pi n_y}{L}, \quad k_z = \frac{2\pi n_z}{L}. \tag{1.76}$$

A slightly different way to see this is as follows. The sides of the cube must correspond to an integral number of wavelengths,

$$\begin{aligned} L &= n_x\lambda_x = n_x 2\pi/k_x \\ L &= n_y\lambda_y = n_y 2\pi/k_y \\ L &= n_z\lambda_z = n_z 2\pi/k_z. \end{aligned} \tag{1.77}$$

The spacing of these **k**-points is such that there are N_k per unit volume in **k**-space, as given by formula (1.73).

A state χ can be expanded in momentum eigenstates, a procedure which is the same as Fourier transformation. We write

$$|\chi\rangle = \sum_k C_k^\chi |\mathbf{k}\rangle \equiv \frac{V}{(2\pi)^3}\int C_k^\chi |\mathbf{k}\rangle\, d\mathbf{k}. \tag{1.78}$$

Notice that we are using the continuous limit of the distribution of **k**-points, so that C_k^χ is actually a continuous fuction of **k**. Pre-multiplying by $\langle \mathbf{k}'|$ (in other words, taking the scalar product with $|\mathbf{k}'\rangle$) and using (1.59) gives

$$C_k^\chi = \langle \mathbf{k}|\chi\rangle. \tag{1.79}$$

The identity operator in the form (1.41) can now be inserted between the bra and ket to give an expression for C_k^χ which is just the Fourier transform of the wavefunction:

$$C_k^\chi = \frac{1}{\sqrt{V}}\int \exp(-i\mathbf{k}\cdot\mathbf{r})\langle \mathbf{r}|\chi\rangle\, d\mathbf{r}. \tag{1.80}$$

The inverse Fourier transformation is obtained by taking the scalar product of (1.78) with $|\mathbf{r}\rangle$, giving

$$\langle \mathbf{r}|\chi\rangle = \frac{\sqrt{V}}{(2\pi)^3}\int \exp(\mathrm{i}\mathbf{k}\cdot\mathbf{r})C_k^\chi \, \mathrm{d}\mathbf{k}. \tag{1.81}$$

Notice the way the normalization appears in these formulae. Even if the wavefunction $\langle \mathbf{r}|\chi\rangle$ is infinite in extent, but has been normalized over the volume V, that is

$$\langle \chi|\chi\rangle = \int_V \langle \chi|\mathbf{r}\rangle\langle \mathbf{r}|\chi\rangle \, \mathrm{d}\mathbf{r} = 1, \tag{1.82}$$

then C_k^χ in (1.80) remains finite even if we let V tend to infinity. This works out because the wavefunction carries a factor $1/\sqrt{V}$ which together with the similar factor in front of the integral cancels the volume factor produced by the integration. If the wavefunction is bounded in space, we see from (1.80) that C_k^χ will scale as $1/\sqrt{V}$; in this case the product $C_k^\chi\sqrt{V}$ is independent of V provided V is at least large enough to entirely enclose $\langle \mathbf{r}|\chi\rangle$.

In the case of an electron in a constant potential, $V_{\text{eff}}(\mathbf{r}) = \text{constant}$, the eigenfunctions of momentum $|\mathbf{k}\rangle$ are also the eigenfunctions of the Hamiltonian with energy $\epsilon_k = \text{constant} + k^2/2$. This is only true because the operators $\hat{H}$ and $\hat{\mathbf{p}}$ commute when the potential is constant, and when operators commute, they have the same eigenfunctions. Strictly speaking, this is not automatically the case if there are degenerate eigenfunctions, but then the eigenfunctions can always be suitably chosen. In the constant potential case, corresponding to any value of $\mathbf{k}$ there are degenerate eigenfunctions of the Hamiltonian, two of which for example in the $\mathbf{r}$-representation are $\sin k_x x$ and $\cos k_x x$. Neither of these is an eigenfunction of the momentum operator, but the sums and differences of them are, namely $\exp(\mathrm{i}k_x x)$ and $\exp(-\mathrm{i}k_x x)$.

1.4.2 Cells and Supercells

A different and important aspect of periodic boundary conditions arises in the case when V_{eff} is not constant but periodic. This is the case of an infinite perfect crystal, with a repeating unit cell. It is important because the easiest way to calculate the properties of bulk crystals is to suppose that they are infinite in extent, and to calculate the properties per unit cell. The unit cell we shall refer to is is not necessarily the smallest, or primitive unit cell for the lattice in question. Let us describe it by the lattice vectors $(\mathbf{a}_1, \mathbf{a}_2, \mathbf{a}_3)$. For example in a face-centred cubic (FCC) crystal, the primitive unit cell would be described by the vectors

$$\begin{aligned}
\mathbf{a}_1 &= a\left(0, \tfrac{1}{2}, \tfrac{1}{2}\right), \\
\mathbf{a}_2 &= a\left(\tfrac{1}{2}, 0, \tfrac{1}{2}\right), \\
\mathbf{a}_3 &= a\left(\tfrac{1}{2}, \tfrac{1}{2}, 0\right),
\end{aligned} \tag{1.83}$$

where a is the side of the cubic cell, and the Cartesian components are given in brackets. On the other hand, we might find it more convenient to use a larger cell such as the cubic one

$$\begin{aligned}
\mathbf{a}_1 &= a\,(1, 0, 0), \\
\mathbf{a}_2 &= a\,(0, 1, 0), \\
\mathbf{a}_3 &= a\,(0, 0, 1).
\end{aligned} \tag{1.84}$$

For the purposes of simulating the crystal, when we might want to include defects or atomic displacements within the material, our unit cell will normally be much bigger than these, so as to contain many independent atoms. In such cases it is called a supercell. A supercell is so-called because its sides are lattice vectors which are integer multiples of some smaller lattice vectors, for example it might be described by

$$\begin{aligned}
\mathbf{a}_1 &= n_1 a\,(1, 0, 0), \\
\mathbf{a}_2 &= n_2 a\,(0, 1, 0), \\
\mathbf{a}_3 &= n_3 a\,(0, 0, 1).
\end{aligned} \tag{1.85}$$

The distinction between primitive and non-primitive cells is important if one is considering the symmetry of wave functions or the efficiency of a mathematical description in perfect crystals. However these are not such important matters in the simulation of condensed matter generally, so for our purposes the mathematical description will be the same whether we are dealing with primitive cells or supercells. Accordingly we shall refer to $(\mathbf{a}_1, \mathbf{a}_2, \mathbf{a}_3)$ as the *primitive vectors* of the cell, whether it be a primitive cell or a supercell. We shall order the vectors so that the basis is right-handed, meaning that the vector $\mathbf{a}_1 \times \mathbf{a}_2$ would be in the direction of $+\mathbf{a}_3$ in Cartesian axis, and in general its scalar product with $\mathbf{a}_3$ is positive.

The volume of the cell is given in terms of its primitive vectors by

$$V_c = \mathbf{a}_1 \cdot (\mathbf{a}_2 \times \mathbf{a}_3). \tag{1.86}$$

1.4.3 The Reciprocal Lattice

The reciprocal lattice is a particular set of points in **k**-space, which is incidentally also sometimes called reciprocal space. By analogy with a real lattice, or direct lattice, the reciprocal lattice points are generated by the set of reciprocal lattice vectors, which in turn are generated by three primitive vectors. The *primitive vectors* of the reciprocal lattice are given by

$$\mathbf{b}_1 = 2\pi \frac{\mathbf{a}_2 \times \mathbf{a}_3}{V_c},$$
$$\mathbf{b}_2 = 2\pi \frac{\mathbf{a}_3 \times \mathbf{a}_1}{V_c}, \qquad (1.87)$$
$$\mathbf{b}_3 = 2\pi \frac{\mathbf{a}_1 \times \mathbf{a}_2}{V_c}.$$

There is an orthogonality relation between the $\mathbf{b}_i$ and $\mathbf{a}_i$, namely

$$\mathbf{a}_i \cdot \mathbf{b}_j = 2\pi \delta_{ij}. \qquad (1.88)$$

Let us denote a vector of the direct lattice by **R** and a vector of the reciprocal lattice by **g**. Such general lattice vectors are multiples of the primitive lattice vectors:

$$\mathbf{R} = n_1\mathbf{a}_1 + n_2\mathbf{a}_2 + n_3\mathbf{a}_3 \qquad (1.89)$$

and

$$\mathbf{g} = g_1\mathbf{b}_1 + g_2\mathbf{b}_2 + g_3\mathbf{b}_3, \qquad (1.90)$$

where the n_i and g_i can be any integers.

The most important property of the reciprocal lattice concerns the scalar product of **g** with **R**; by applying the orthogonality relation (1.88):

$$\mathbf{g} \cdot \mathbf{R} = 2\pi(g_1 n_1 + g_2 n_2 + g_3 n_3), \qquad (1.91)$$

which is an integer multiple of 2π. This has the consequence that for any reciprocal and direct lattice vectors:

$$\exp(i\mathbf{g} \cdot \mathbf{R}) = 1. \qquad (1.92)$$

The statement that the atoms form a periodic strtucture with lattice vectors **R** is equivalent to saying that the Hamiltonian has this periodicity, in other words that it is invariant to translations **R**. Therefore the effective potential has the property that for any vector **R** of the form (1.89):

$$V_{\mathrm{eff}}(\mathbf{r} + \mathbf{R}) = V_{\mathrm{eff}}(\mathbf{r}). \qquad (1.93)$$

Lattice periodicity is also a propertiy of the electron density, but not a property of the wavefunctions as we shall see.

1.4.4 The First Brillouin Zone

The first Brilloun zone is the region of **k**-space which is nearer to the origin than to any other reciprocal lattice point. It can be generated as follows. Construct planes which are normal to and bisect each reciprocal lattice vector **g** from the origin. The smallest region including the origin which is enclosed by such planes is the first Brillouin zone.

It is the analogue in the reciprocal lattice of the Wigner–Seitz cell in the real lattice. One can construct a Wigner–Seitz cell or Brillouin zone taking *any* lattice, or reciprocal lattice, point as the origin. If one makes the construction about every reciprocal lattice point, the resulting Brillouin zones will be identical and they will fill the space. Each will therefore have the volume of the unit cell in reciprocal space, which is (Ashcroft and Mermin, 1976, p. 93)

$$V_{BZ} = \mathbf{b}_1 \cdot (\mathbf{b}_2 \times \mathbf{b}_3) = \frac{(2\pi)^3}{V_c}. \tag{1.94}$$

1.4.5 Bloch's Theorem

The solutions of the single-particle Schrödinger equation in a periodic potential (1.93) have a very particular form which is known as *Bloch's theorem*. The wavefunctions are by no means themselves necessarily periodic, but they can be described by a periodic function $u_k(\mathbf{r})$ modulated by a plane wave:

$$\langle \mathbf{r} | \psi_k \rangle = \frac{1}{\sqrt{V}} \exp(i\mathbf{k} \cdot \mathbf{r}) u_k(\mathbf{r}). \tag{1.95}$$

In this sense the wave functions ψ_k occupy points in **k**-space, just like the plane waves themselves. Furthermore the periodicity of $u_k(\mathbf{r})$ means that it can be represented by a Fourier series. This works as follows.

First, consider any **k**-vector $\mathbf{k}'$ that lies outside the first Brillouin zone. Find the reciprocal lattice vector **g** that lies closest to $\mathbf{k}'$. It follows that the vector **k**, defined as the difference $\mathbf{k}' - \mathbf{g}$, must lie in the first Brillouin zone. With this in mind, notice that if $\mathbf{k}'$ is a general **k**-point we can write the Bloch form of the wave function as:

$$\exp(i\mathbf{k}' \cdot \mathbf{r}) u_{k'}(\mathbf{r}) = \exp(i\mathbf{k} \cdot \mathbf{r}) u_{kn}(\mathbf{r}), \tag{1.96}$$

where, since $\exp(i\mathbf{g} \cdot \mathbf{r})$ is periodic, so is $u_{kn}(\mathbf{r}) = \exp(i\mathbf{g} \cdot \mathbf{r}) u_k(\mathbf{r})$. This result enables us to exchange $\mathbf{k}'$, the original **k**-vector which labels ψ_k, and which could be anywhere in **k**-space, for a **k**-vector within the first Brillouin zone, together with the index n which distinguishes all the wavefunctions which are now labelled with the same **k**-vector in the first Brillouin zone. The relabeled wave function is

$$\langle \mathbf{r} | \psi_{kn} \rangle = \frac{1}{\sqrt{V}} \exp(i\mathbf{k} \cdot \mathbf{r}) u_{kn}(\mathbf{r}). \tag{1.97}$$

The index n is called the *band* index, and it labels the wavefunctions at each **k**-point in order of increasing energy ϵ_{kn}. These single particle energies form a series of functions of **k** within the first Brillouin zone and these are known as the *energy bands*. The origin in **k**-space is known as the gamma point, and wave functions of all bands at the gamma point, and only there, are periodic with the same period as the potential.

An alternative statement of Bloch's theorem is in terms of the relative phase of the wavefunction at points in space separated by a lattice vector **R**, namely

$$\psi_{kn}(\mathbf{r}+\mathbf{R}) = \exp(\mathrm{i}\mathbf{k}\cdot\mathbf{R})\psi_{nk}(\mathbf{r}). \tag{1.98}$$

Exercise
Show that (1.97) and (1.98) are completely equivalent.

1.4.6 Expansion in Plane Waves

If we now express the wavefunction in **k**-space instead of **r**-space, something interesting happens. I will describe this in detail, because illustrates how we can move back and forth between **r**- and **k**-space. We write the wavefunction in the form (1.78), but now label the momemtum eigenstates by **q** to distinguish them from the **k** we are already using:

$$\langle \mathbf{r}|\psi_{kn}\rangle = \sum C_q^{kn}\langle \mathbf{r}|\mathbf{q}\rangle. \tag{1.99}$$

The expansion coefficients are given by (1.80):

$$C_q^{kn} = \sum \langle \mathbf{q}|\mathbf{r}\rangle\langle \mathbf{r}|\psi_{kn}\rangle = \frac{1}{V}\int_V \exp\left(\mathrm{i}(\mathbf{k}-\mathbf{q})\cdot\mathbf{r}\right) u_{kn}(\mathbf{r})\,\mathrm{d}\mathbf{r}. \tag{1.100}$$

Now we can exploit the periodicity of u_{kn} by writing the integral over V as a sum of integrals over each cell of volume V_c with its origin at the lattice vector **R**:

$$C_q^{kn} = \frac{1}{V}\sum_{\mathbf{R}} \exp\left(\mathrm{i}(\mathbf{k}-\mathbf{q})\cdot\mathbf{R}\right)\int_{V_c} \exp\left(\mathrm{i}(\mathbf{k}-\mathbf{q})\cdot\mathbf{r}\right) u_{kn}(\mathbf{r})\,\mathrm{d}\mathbf{r}. \tag{1.101}$$

Consider the sum over lattice vectors **R**. When $\mathbf{k}-\mathbf{q}$ is a reciprocal lattice vector, each term of the sum is unity, by virtue of (1.92). When $\mathbf{k}-\mathbf{q}$ is not a reciprocal lattice vector, the sum over **R** vanishes. To prove this, we have to picture the discrete, closely-spaced mesh of allowed points in **k**-space on which **k** and **q** can end rather than the continuous distribution of **k**-points which is convenient for other purposes. Suppose that the large repeating volume V comprises $N_1 \times N_2 \times N_3$ cells (or supercells). First notice that the primitive vectors of the reciprocal lattice of V are just those of the supercell V_c scaled by $1/N_1$, $1/N_2$ and $1/N_3$. Just check that these points have the right density in **k**-space, $V/(2\pi)^3$, and that the corresponding plane waves have the period of V. The allowed **k**-points therefore lie on a lattice in **k**-space defined by the vectors $m_1\mathbf{b}_1/N_1$, $m_2\mathbf{b}_2/N_2$ and $m_3\mathbf{b}_3/N_3$, where m_i are integers. Reciprocal lattice points **g** occur when m_i is an integer multiple of N_i for $i = 1, 2, 3$, including zero. The sum over **R** is

$$\sum_{\mathbf{R}} \exp\left(\mathrm{i}(\mathbf{k}-\mathbf{q})\cdot\mathbf{R}\right) = \sum_{n_1=0}^{N_1-1}\sum_{n_2=0}^{N_2-1}\sum_{n_3=0}^{N_3-1} \exp\left(\mathrm{i}2\pi\left(\frac{n_1m_1}{N_1}+\frac{n_2m_2}{N_2}+\frac{n_3m_3}{N_3}\right)\right), \tag{1.102}$$

where

$$\mathbf{k}-\mathbf{q} = m_1\mathbf{b}_1/N_1 + m_2\mathbf{b}_2/N_2 + m_3\mathbf{b}_3/N_3.$$

The right-hand side of (1.102) is the product of three factors, each of which is a geometric progression. Applying the standard formula to sum these geometric progressions we find:

$$\sum_{\mathbf{R}} \exp\left(\mathrm{i}(\mathbf{k}-\mathbf{q})\cdot\mathbf{R}\right)$$
$$= \frac{\left(\exp\left(\mathrm{i}2\pi m_1\right)-1\right)\left(\exp\left(\mathrm{i}2\pi m_1\right)-1\right)\left(\exp\left(\mathrm{i}2\pi m_1\right)-1\right)}{\exp\left(\mathrm{i}2\pi m_1/N_1-1\right)\exp\left(\mathrm{i}2\pi m_1/N_1-1\right)\exp\left(\mathrm{i}2\pi m_1/N_1-1\right)}. \tag{1.103}$$

When $\mathbf{k}-\mathbf{q}$ is a reciprocal lattice vector, there is a zero in both numerator and denominator in (1.103), and we have to step back to (1.102) to see that the sum over **R** is $N_1N_2N_3 = V/V_\mathrm{c}$. On the other hand when $\mathbf{k}-\mathbf{q}$ is not a reciprocal lattice vector, only the numerator in (1.103) vanishes and the sum is zero. The only **q** for which the sum over **R** does not vanish are therefore those for which $\mathbf{q} = \mathbf{k}+\mathbf{g}$ where **g** is some reciprocal lattice vector. Hence (1.99) becomes:

$$\langle\mathbf{r}|\psi_{kn}\rangle = \frac{1}{\sqrt{V}}\sum_{\mathbf{g}} C^n_{\mathbf{k}+\mathbf{g}} \exp\left(\mathrm{i}\left(\mathbf{k}+\mathbf{g}\right)\cdot\mathbf{r}\right), \tag{1.104}$$

and (1.101) becomes

$$C^n_{\mathbf{k}+\mathbf{g}} = \frac{1}{V_\mathrm{c}}\int_{V_\mathrm{c}} \exp\left(-\mathrm{i}\mathbf{g}\cdot\mathbf{r}\right) u_{kn}(\mathbf{r})\,\mathrm{d}\mathbf{r}. \tag{1.105}$$

Equation (1.104) is the expansion of a wavefunction in a basis of plane waves, which has proved very effective for the calculation of total energies and forces on atoms.

1.5 Local Orbitals and Spherical Harmonics

An important class of basis sets for practical calculations consists of so-called *local orbitals*. These are functions which are centred on the atomic nuclei, which have angular dependence in general, and which fall off with distance from the nuclei so that they are either zero or near enough zero after a nanometer or two. The orbitals of free atoms have these properties, and we often refer to local orbitals as atomic orbitals. The main value of local orbitals, compared to plane waves, is that since they already have some natural resemblance to the wavefunctions in the immediate vicinity of an atom, one expects that fewer of them will be required in a basis for expanding a wave function. However, free atom orbitals are not the best choice of local basis for a number of reasons, and there are a variety of other ways to specify a set of local orbitals. In all cases a compromise has to be struck between incompatible requirements. On the one hand, the shorter the range of the orbitals the better in terms of computational efficiency. In the matrix represention of the Schrödinger equation there are fewer matrix elements of the potential to be calculated. On the other hand, if they are too short ranged more local orbitals might be required to represent wavefunctions to a specified accuracy. We speak loosely in this sense about a basis being 'less complete' if it is worse at representing wave functions to a required accuracy.

The basis sets $\{|n\rangle\}$, $\{|\mathbf{k}\rangle\}$ and $\{|\mathbf{r}\rangle\}$ discussed so far each have the attractive property of being orthonormal. Local orbitals on the other hand, while they can always be normalised simply by scaling them, are not automatically orthogonal unless they are on the same atom. Atomic orbitals belonging to different atoms overlap, which is a basis for describing chemical bonding, but, especially if they are short-ranged, this means that they can only be made orthogonal by quite complicated linear transformations. Their non-orthogonality is a source of some difficulties which we will have to address in the calculation of total energies and forces.

1.5.1 Spherical Harmonics

Most commonly used local orbitals are eigenstates of the angular momentum operator. This means they have the form of the product of a radial function and a spherical harmonic, expressed in the usual polar coordinates and referred to an atom located at the origin:

$$\phi_{nL}(\mathbf{r}) = f_{nL}(r)Y_L(\theta, \varphi). \tag{1.106}$$

The index L is a composite index which stands for the orbital and magnetic indices l and m. l is a positive integer which counts the quanta of orbital angular momentum and m is the integer which quantizes their z component. m lies in the range $-l \leq m \leq +l$.

The index n is spare at this point, and simply distinguishes different members of a family of orbitals that have the same angular dependence L. The index n could for example distinguish the eigenstates $\phi_{nL}(\mathbf{r})$ of the single particle Schrödinger equation in the same spherically symmetric potential. This is not necessarily the Coulomb potential, but may be a specially designed potential for the purposes of generating basis functions. In this case the index n is familiar from atomic theory, and it is called the *principal quantum number*. It labels the 'shell' of states as in the elementary picture of the structure of an atom.

The lowest order spherical harmonics describe the 's' and the three 'p' states which have $l = 0$ and $l = 1$, respectively:

$$\begin{aligned} Y_{00}(\theta, \varphi) &= \frac{1}{\sqrt{4\pi}} \\ Y_{10}(\theta, \varphi) &= \sqrt{\frac{3}{4\pi}} \cos\theta \\ Y_{11}(\theta, \varphi) &= -\sqrt{\frac{3}{8\pi}} \sin\theta \, \mathrm{e}^{\mathrm{i}\varphi} \\ Y_{1,-1}(\theta, \varphi) &= \sqrt{\frac{3}{8\pi}} \sin\theta \, \mathrm{e}^{-\mathrm{i}\varphi}. \end{aligned} \tag{1.107}$$

The cumbersome multiples of $1/\sqrt{4\pi}$ which appear are there to enforce the normalization to unity when integrated over the surface of a sphere:

$$\int Y_{lm}^* Y_{l'm'} \sin\theta \, \mathrm{d}\theta \, \mathrm{d}\varphi = \delta_{ll'} \delta_{mm'}. \tag{1.108}$$

These factors are sometimes disposed of by working with the renormalised spherical harmonics of Racah:

$$C_{lm}(\theta, \varphi) = \sqrt{\frac{4\pi}{2l+1}} Y_{lm}(\theta, \varphi). \tag{1.109}$$

In these and the following definitions and notation I have followed Stone (1996). If we multiply the spherical harmonics by r^l or $r^{-(l+1)}$ we get the complete set of solutions of Laplace's equation in spherical polar coordinates. These useful functions of position are called the *regular* and *irregular* spherical harmonic polynomials, defined by

$$\begin{aligned} R_{lm}(\mathbf{r}) &= r^l C_{lm}(\theta, \varphi) \\ I_{lm}(\mathbf{r}) &= r^{-(l+1)} C_{lm}(\theta, \varphi). \end{aligned} \tag{1.110}$$

It is often more convenient to work with *real* spherical harmonics. Real orbitals are easier to visualize as well as to compute with than complex ones, and the tight-binding theory described later works with a basis set of real orbitals. An equivalent set of real spherical harmonics can easily be constructed from the complex ones. First note that the $+m$ and $-m$ functions for a given l are proportional to $(-1)^m \mathrm{e}^{\mathrm{i}\varphi}$ and $\mathrm{e}^{-\mathrm{i}\varphi}$ respectively. The $m = 0$ spherical harmonics are already real. For $m \neq 0$ we make functions which are proportional to $\cos\varphi$ and $\sin\varphi$, denoted by suffix c and s respectively, by the following linear combinations:

$$\begin{aligned} C_{lmc} &= \sqrt{\frac{1}{2}}\left[(-1)^m C_{lm} + C_{l,-m}\right], \\ C_{lms} &= \frac{1}{\mathrm{i}}\sqrt{\frac{1}{2}}\left[(-1)^m C_{lm} - C_{l,-m}\right], \end{aligned} \tag{1.111}$$

and

$$\begin{aligned} R_{lmc} &= \sqrt{\frac{1}{2}}\left[(-1)^m R_{lm} + R_{l,-m}\right], \\ R_{lms} &= \frac{1}{\mathrm{i}}\sqrt{\frac{1}{2}}\left[(-1)^m R_{lm} - R_{l,-m}\right]. \end{aligned} \tag{1.112}$$

Similar combinations apply to make the real irregular functions. In the real spherical harmonics, $m \geq 0$, but there are still $2l+1$ independent functions for each value of l. The real spherical harmonics are entirely equivalent to the original complex ones. They are not individually eigenstates of angular momentum, because m and $-m$ eigenstates have been mixed. But either the real or the complex set is a complete basis for expanding any well-behaved functions on the surface of a sphere. And the real or complex regular and irregular functions are a complete basis for expanding all functions of $\mathbf{r}$ which are solutions of Laplace's equation.

When it is clear from the context that we are dealing with real spherical harmonics let us allow the composite index L to stand for $l\kappa$, where κ in turn stands for mc or ms. With this shorthand notation you can easily show that they satisfy the orthogonality relations:

$$\int C_L C_{L'} \sin\theta \,\mathrm{d}\theta \,\mathrm{d}\varphi = \frac{4\pi}{2l+1}\delta_{LL'}, \tag{1.113}$$

where in a natural extension of the notation:

$$\delta_{LL'} \equiv \delta_{ll'}\delta_{\kappa\kappa'}.$$

Stone gives in Appendix E of his book a table of the R_L up to $l = 4$ expressed in terms of x, y and z. For future reference the s, p and d entries are:

$$
\begin{aligned}
s:&\quad R_{00} = 1 \\
p:&\quad \begin{cases} R_{10} = z \\ R_{11c} = x \\ R_{11s} = y \end{cases} \\
d:&\quad \begin{cases} R_{20} = \frac{1}{2}(3z^2 - r^2) \\ R_{21c} = \sqrt{3}xz \\ R_{21s} = \sqrt{3}yz \\ R_{22c} = \frac{1}{2}\sqrt{3}(x^2 - y^2) \\ R_{22s} = \sqrt{3}xy \; . \end{cases}
\end{aligned}
\tag{1.114}
$$

Exercise

Verify that the R_{L} and I_{L}, complex or real, are solutions of Laplace's equation, i.e. $\nabla^2 R_{\mathrm{L}} = 0$ and $\nabla^2 I_{\mathrm{L}} = 0$, where the Laplacian in spherical polar coordinates is

$$\nabla^2 = \frac{1}{r^2}\frac{\partial}{\partial r}r^2\frac{\partial}{\partial r} + \frac{1}{r^2}\frac{1}{\sin\theta}\frac{\partial}{\partial\theta}\sin\theta\frac{\partial}{\partial\theta} + \frac{1}{r^2}\frac{1}{\sin^2\theta}\frac{\partial^2}{\partial\varphi^2}. \tag{1.115}$$

1.5.2 Local Atomic-like Orbitals

Now we are going to discuss local orbitals on different sites, we shall have to add the label I for the site, which brings to four the total number of indices needed to label an orbital, namely the set $\{I, n, l, m\}$ or $\{I, n, l, \kappa\}$. I find it hard to read more than two indices at a time, so again I will introduce a compound index which stands for the set $\{n, l, \kappa\}$. Now we can distinguish a local orbital by two indices $|I\mu\rangle$. The I labels the atomic nucleus, or site, and the μ or another Greek letter labels the particular orbital on that site. These orbitals have been normalised, so that

$$\langle I\mu|I\nu\rangle = \delta_{\mu\nu}.$$

In condensed matter, the electronic states in the core of the atom, that is the filled shells, are very similar to the corresponding states in the free atoms, whereas the highest occupied electronic states bear little resemblence to those of free atoms. It nevertheless makes sense to use as a basis for all states in the material some atom-like states which have good quantum numbers l and m so that the basis functions are solutions of the Schrödinger equation in some spherically symmetric potential. To make an exact representation of the extended state of an electron in a solid we would in principle need an infinite number of such basis states *on each atom*! However, the utility of schemes based on local orbitals is that we can get away with a rather small number of local orbitals as basis states on each atom (in the most drastically simple model only one!) and still achieve a useful level of accuracy.

Although they are orthonormal on a given atomic site, as noted previously orbitals on different sites will not in general be orthogonal to each other, and this has important implications which we discuss here. A completely orthogonal basis of orbitals is mathematically much simpler to handle, but unfortunately not as accurate when the orbitals are short ranged as a basis in which overlapping orbitals are not orthogonal. When we use a non-orthogonal basis of local orbitals an important part will be played by the matrix formed by the overlap of these orbitals, **S**, which is defined by

$$S_{I\mu J\nu} = \langle I\mu | J\nu \rangle. \tag{1.116}$$

S is a square matrix of dimensions $N_{\text{orb}} \times N_{\text{orb}}$, where N_{orb} is the total number of basis orbitals in our system. A calculation of **S** involves integrating the products of overlapping orbitals. By inserting the identity operator in the form (1.41) into the bra-ket $\langle I\mu | J\nu \rangle$ the overlap matrix can be written explicitly in the **r**-representation as

$$S_{I\mu J\nu} = \langle I\mu | J\nu \rangle = \langle I\mu | \int |\mathbf{r}\rangle\langle\mathbf{r}|\, d\mathbf{r} | J\nu \rangle = \int \langle I\mu|\mathbf{r}\rangle\langle\mathbf{r}|J\nu\rangle d\mathbf{r} \tag{1.117}$$

Normally the chosen basis of orbitals $\langle \mathbf{r} | I\mu \rangle$ would be real functions of $\mathbf{r} \equiv (r, \theta, \phi)$, in which case

$$\langle I\mu | \mathbf{r} \rangle = \langle \mathbf{r} | I\mu \rangle.$$

An important case is when the local orbitals on different sites are orthogonal to each other, because it is used in the development of the simplest tight-binding models. Although quite difficult to achieve in practice, such an orthonormal basis may nevertheless be a useful approximation. The overlap matrix **S** in an orthonormal basis would be the identity matrix or unit matrix **1**, where

$$\mathbf{1}_{I\mu J\nu} = \delta_{IJ}\delta_{\mu\nu}.$$

The remainder of **S** in a non-orthonormal basis is therefore sometimes singled out and will be referred to as the **O**-*matrix*, where:

$$\mathbf{S} = \mathbf{1} + \mathbf{O}. \tag{1.118}$$

This can be a useful separation if we want to treat the non-orthogonality as small, so that **O** can be treated as a perturbation.

1.5.3 Expansion of a Wavefunction in a Non-orthogonal Basis

We can still expand any single-particle wavefunction in ket form as

$$|\chi\rangle = \sum_{I\mu} C^{\chi}_{I\mu} |I\mu\rangle, \tag{1.119}$$

or its bra (see eq. (1.50)) as

$$\langle\chi| = \sum_{I\mu} C^{\chi *}_{I\mu} \langle I\mu|, \tag{1.120}$$

but with non-orthogonal orbitals the expressions for the expansion coefficients involve the overlap matrix. If we multiply both sides of (1.119) by $\langle J\nu|$ we have

$$\langle J\nu|\chi\rangle = \sum_{I\mu} C^{\chi}_{I\mu} S_{J\nu I\mu}. \tag{1.121}$$

Now we only have to introduce the inverse of the matrix S to invert (1.121):

$$C^{\chi}_{I\mu} = \sum_{J\nu} S^{-1}_{I\mu J\nu} \langle J\nu|\chi\rangle. \tag{1.122}$$

The normalisation of the wavefunction can be expressed as

$$\langle\chi|\chi\rangle = 1 = \sum_{I\mu J\nu} C^{\chi *}_{I\mu} C^{\chi}_{J\nu} S_{I\mu J\nu}. \tag{1.123}$$

1.5.4 Expansion of an Operator

Any operator $\hat{A}$ in our Hilbert space can be written in the bra-ket form as

$$\hat{A} = \sum_{I\mu J\nu} |I\mu\rangle A^{I\mu J\nu} \langle J\nu|, \tag{1.124}$$

where the operator is completely specified by these coefficients $A^{I\mu J\nu}$. They can be expressed in terms of the matrix elements of the operator in the basis and the S matrix by pre- and post-multiplying (1.124) by a bra and ket from the basis:

$$\langle I'\mu'|\hat{A}|J'\nu'\rangle = \sum_{I\mu J\nu} S_{I'\mu' I\mu} A^{I\mu J\nu} S_{J\nu J'\nu'}, \tag{1.125}$$

and so

$$A^{I\mu J\nu} = \sum_{I'\mu' J'\nu'} S^{-1}_{I\mu I'\mu'} \langle I'\mu'|\hat{A}|J'\nu'\rangle S^{-1}_{J'\nu' J\nu}. \tag{1.126}$$

We can, if we like, think of the coefficients $A^{I\mu J\nu}$ as matrix elements of a matrix $\mathbf{A}$, but they are not the same as $\langle I\mu|\hat{A}|J\nu\rangle$, the matrix elements of the operator $\hat{A}$, unless the basis is orthonormal. Using the above results we can write the identity operator as:

$$\hat{\mathbf{1}} = \sum_{I\mu J\nu} |I\mu\rangle S^{-1}_{I\mu J\nu} \langle J\nu|, \qquad (1.127)$$

It will be convenient to use suffices in order to distinguish the matrix elements of an operator from its expansion coefficients, so we will write

$$A_{I\mu J\nu} \equiv \langle I\mu|\hat{A}|J\nu\rangle.$$

Such raised and lowered indices may be familiar to some readers in the context of covariant and contravariant vector spaces, where they have a similar function.

1.5.5 Spherical Harmonics and Multipoles

So far we just seem to have introduced more and more functions without any payback in terms of physical insight. In this section, let us discuss a bit more of the physics of spherical harmonics and the special properties which make them beautifully suited as the angular parts of basis functions for all kinds of applications. In particular we now show how they enable us to calculate the potential outside some localized distribution of charge, and the matrix elements of this potential in local orbitals. These are essential ingredients of any scheme of total energy and force calculation which makes use of local orbitals, and they will be required later in deriving tight-binding approximations.

First we define the concept of *multipoles*. Recall that we have defined regular and irregular spherical harmonics, R_L and I_L which are solutions of Laplace's equation. To be more precise, since the R_L are non-singular at the origin, they represent solutions of Laplace's equation *within* any sphere about the origin. The I_L on the other hand diverge at the origin and go to zero at infinity. They therefore represent solutions of Laplace's equation *outside* any sphere about the origin. In particular, if there is a multipole of charge centred at the origin, the electrostatic potential outside an arbitrary radius R_S is proportional to I_L, where the l-value tells us what kind of multipole we have. $l = 0$ represents a monopole, and its potential varies with r as the Coulomb potential r^{-1}, which is the behaviour of I_{00}. $l = 1$ represents a dipole, the potential of which falls off as r^{-2}, modulated by angular variation, and so on. Remember in the real functions we shall be using, L stands for the pair (l, κ), and therefore has $2l + 1$ components: 1 for a monopole, 3 for a dipole, 5 for a quadrupole, etc. The potential of an arbitrary charge distribution within R_S can be represented outside R_S as a series expansion or linear combination of as many I_L as required. Mathematically these ideas are expressed as follows. For each L the multipole associated with a charge density $\rho(\mathbf{r})$ within a region Ω around the point **R** is defined by:

$$Q_L(\mathbf{R}) = \int_\Omega R_L(\mathbf{r} - \mathbf{R})\rho(\mathbf{r})\,\mathrm{d}\mathbf{r}. \tag{1.128}$$

What is the potential outside a sphere enclosing Ω due to this multipole, indeed due to all the multipoles which taken together characterise the distribution $\rho(\mathbf{r})$? This question is answered with the help of the following formula, proved in many textbooks, which is an expansion of the Coulomb potential in spherical harmonics. See for example Jackson (1975, equation 3.70). In terms of real spherical harmonics (1.111) the Coulomb potential takes the form:

$$\frac{1}{|\mathbf{r} - \mathbf{r}'|} = \sum_L \frac{r_<^l}{r_>^{l+1}} C_L(\theta, \varphi) C_L(\theta', \varphi'). \tag{1.129}$$

The symbols $r_<$ and $r_>$ are conventionally used to denote the smaller and larger of the distances r and r'. The vectors $\mathbf{r}$ and $\mathbf{r}'$ have polar coordinates (r, θ, φ) and (r', θ', φ'), respectively. For our purpose we will take $\mathbf{r}'$ to be inside Ω and $\mathbf{r}$ outside, where the sphere of radius R_S centred on **R** encloses the volume Ω, so (1.129) becomes

$$\frac{1}{|\mathbf{r} - \mathbf{r}'|} = \sum_L I_L(\mathbf{r} - \mathbf{R}) R_L(\mathbf{r}' - \mathbf{R}). \tag{1.130}$$

Figure 1.1 may help visualize what is going on.

This formula tells us the potential at $\mathbf{r}$ due to a unit charge at $\mathbf{r}'$, so from it we can evaluate the potential at $\mathbf{r}$ due to the distribution of charge inside Ω. Making use of the definition of the multipoles we obtain

$$V_{\mathrm{H}}(\mathbf{r}) = \int_\Omega \frac{1}{|\mathbf{r} - \mathbf{r}'|}\rho(\mathbf{r}')\,\mathrm{d}\mathbf{r}' = \sum_L Q_L(\mathbf{R}) I_L(\mathbf{r} - \mathbf{R}). \tag{1.131}$$

To summarize, so far we have defined a multipole Q_L from an arbitrary localized charge distribution, such that if it is centred at a site **R**, it creates a potential

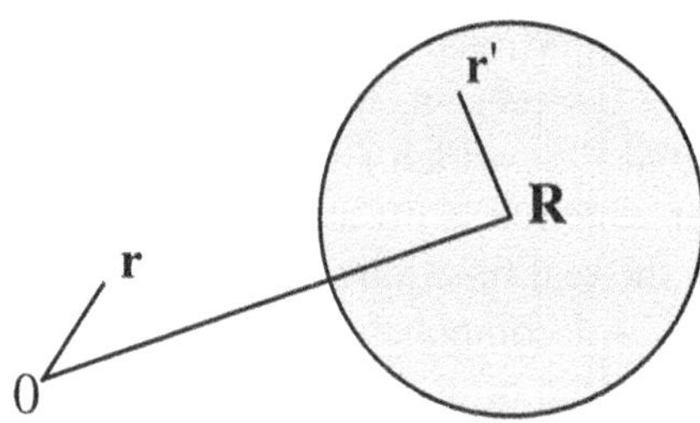

FIG. 1.1 A multipole is defined within the sphere centred on **R** and its potential acts at **r** according to eq. (1.131).

$Q_L I_L(\mathbf{r}-\mathbf{R})$, where $\mathbf{r}$ lies outside a sphere containing the charge distribution. Having done this, the next useful step is to represent the potential in the neighbourhood of the origin, due to a multipole at $\mathbf{R}$, as an expansion in $R_L(\mathbf{r})$. We will then have the tools to calculate the L-component of a potential at site $\mathbf{R}_I$ due to a localised distribution of charge centred at $\mathbf{R}_J$. For this purpose we make use of the following expansion:

$$I_L(\mathbf{r}-\mathbf{R}) = \sum_{L'} R_{L'}(\mathbf{r}) B_{L'L}(\mathbf{R}), \tag{1.132}$$

in which the coefficients $B_{LL'}(\mathbf{R})$ are rather complicated but well defined functions of the indices in L and L' and of $\mathbf{R}$. An elegant derivation of the form of $B_{L'L}$ has been given by Methfessel (1986, pp. 18–19); an alternative approach is given by Stone (1996). The following formula is essentially Methfessel's, which in the present notation becomes

$$B_{L'L}(\mathbf{R}) = \sum_{L''} \frac{(2l''+1)!!}{(2l+1)!!(2l'+1)!!} \sqrt{\frac{4\pi(2l+1)(2l'+1)}{(2l''+1)}} \times (-1)^l \delta_{l+l',l''} C_{LL''L'} I_{L''}(\mathbf{R}). \tag{1.133}$$

Notice that the sum over L'' is really only a sum over κ'' because of the δ-function factor, which imposes the restriction $l'' = l + l'$.

The quantities $C_{LL'L''}$ are known as Gaunt integrals (Thompson, 1994, p. 294; Gaunt, 1929).

$$C_{LL''L'} = \int Y_L Y_{L''} Y_{L'} \sin\theta \, d\theta \, d\varphi. \tag{1.134}$$

In this case the spherical harmonics are of the real kind. The same formula expressed in the original complex spherical harmonics gives the quantities usually known as *Clebsch–Gordan* or *Wigner coefficients*. They are of major importance in the theory of crystal fields as well as atomic physics generally. One of the important symmetry properties of the Gaunt integrals, real or complex, is that l, l' and l'' must satisfy the *triangle inequality*, namely that none of the three should exceed the sum of the other two, otherwise the corresponding integral vanishes.

In general the potential at the origin can be expanded as

$$V_{\mathrm{H}}(\mathbf{r}) = \sum_L V_{HL} R_L(\mathbf{r}). \tag{1.135}$$

Inserting (1.132) into (1.131) and comparing with (1.135) we find

$$\sum_L V_{HL} R_L(\mathbf{r}) = \sum_{LL'} R_{L'}(\mathbf{r}) B_{L'L}(\mathbf{R}) Q_L(\mathbf{R}). \tag{1.136}$$

Hence the components of $V_\mathrm{H}(\mathbf{r})$ are given by

$$V_{HL} = \sum_{L'} B_{LL'}(\mathbf{R})Q_{L'}(\mathbf{R}). \tag{1.137}$$

This is a powerful result. To put it another way, if we have a multipole $Q_{L'}(\mathbf{R}_I)$ on site I, it creates components of potential on site $J \neq I$ given by

$$V_{HL}(\mathbf{R}_J) = B_{LL'}(\mathbf{R}_I - \mathbf{R}_J)Q_{L'}(\mathbf{R}_I). \tag{1.138}$$

Thus the electrostatic energy of interaction between two multipoles is

$$Q_L(\mathbf{R}_J)B_{LL'}(\mathbf{R}_I - \mathbf{R}_J)Q_{L'}(\mathbf{R}_I). \tag{1.139}$$

Referring to the 'components' of the potential at the site J is analogous to the way we talk about vectors, but V_H is very definitely a scalar quantity, so you might think this is a loose way of talking. In fact it has a very precise meaning which has to do with the way the potential behaves under a rotation of the axes. If the frame of reference is rotated, the $2l + 1$ quantities $V_{Hl\kappa}$, where $\kappa = 0, 1c, 1s, \ldots, lc, ls$, become linear combinations of the unrotated quantites. A general rotation can be specified by the three Euler angles (α, β, γ) (see e.g. Stone, 1996, p. 8). There is a matrix D depending on these angles which describes the new functions obtained when any of the original functions $V_{Hl\kappa}$ is rotated with respect to the axes, namely:

$$V_{Hl\kappa} \to \sum_{\kappa'} V_{Hl\kappa'} D^l_{\kappa'\kappa}(\alpha, \beta, \gamma). \tag{1.140}$$

The point is that *all* the real spherical harmonics, regular and irregular, multipoles and the L-components of the potential or of any other scalar field transform with exactly the same matrices, which simply mix the κ-components, but do not mix any functions with different l-values. A similar set of matrices defines how complex spherical harmonics behave under rotation, which mixes the m-values but not the l-values. These are known as *Wigner rotation matrices*. Any quantity with $2l + 1$ components which transform in the same way as a spherical harmonic is called a *spherical tensor* of rank l.

As an example consider the spherical tensor $V_{H1\kappa}$. Its contribution to the potential $V_\mathrm{H}(\mathbf{r})$ is $V_{H10}z + V_{H1c}x + V_{H1s}y$. These components of the potential therefore describe a uniform electric field which in Cartesian coordinates would be written $\mathbf{E} = -\nabla V_\mathrm{H} = -(V_{H1c}, V_{H1s}, V_{H10})$. Likewise $V_{H2\kappa}$ are the five components of a quadrupole field, which represents the curvature of the potential as a function of displacement from the origin. As such it is equivalent to the 3×3 matrix of second derivatives $\partial^2 V_\mathrm{H}/\partial x_\alpha \partial x_\beta$. The latter matrix has just five independent elements, the same as the number of components of the spherical tensor of rank 2. Where does the number five come from? First, because it is symmetric, the number of its independent elements is reduced from nine to six. Second, Laplaces equation $\nabla^2 V_{\rm H} = 0$ means that the sum of its diagonal elements vanishes, further reducing the number of independent elements from six to five.

Now consider the electrostatic energy of interaction between $V_{\rm H}(\mathbf{r})$ and a charge density $\rho(\mathbf{r})$ which is localized within a region Ω about the origin. By expanding the potential in regular spherical harmonics we have

$$\int_{\Omega} \rho(\mathbf{r}) V_{\rm H}(\mathbf{r})\, \mathrm{d}\mathbf{r} = \sum_{L} Q_L V_{HL}. \tag{1.141}$$

This simple equation shows that components of the potential only couple to multipoles of the charge with exactly the same symmetry, that is with equivalent components of the multipole. It is an example of a *contraction* or scalar product of two spherical tensors, which produces a rotationally invariant quantity, or scalar, in this case the interaction energy. In general, the scalar product of two spherical tensors A_L and B_L in the real representation is

$$\sum_{\kappa} A_{l\kappa} B_{l\kappa}. \tag{1.142}$$

In the complex representation, by direct substitution this is easily shown to be

$$\sum_{\kappa} A_{l\kappa} B_{l\kappa} = \sum_{m} (-1)^m A_{lm} B_{l,-m} \equiv \sum_{m} A_{lm} B^*_{lm}. \tag{1.143}$$

1.6 The Variational Principle and the Schrödinger Equation

1.6.1 The Single-particle Schrödinger Equation

The variational principle is of great use in quantum mechanics generally, and is central to density functional theory. In this section, we will see how by solving the single-particle Schrödinger equation you are also solving the following variational problem. Find a state $|\chi\rangle$ which minimises the quantity $\langle\chi|\hat{H}|\chi\rangle$ subject to the normalization condition $\langle\chi|\chi\rangle = 1$. The solution is $|0\rangle$, where $|0\rangle$ or ϕ_0 is the eigenfunction corresponding to the lowest eigenvalue ϵ_0. This is the ground state of the particle. This variational approach only works because the Hamiltonian is a Hermitian operator. The proof is simple if we expand the wavefunction in terms of the eigenstates of $\hat{H}$, eqs. (1.47) and (1.50), and exploit their orthonormality:

$$\langle\chi|\hat{H}|\chi\rangle = \sum_{m}\sum_{n} C^{\chi *}_m \langle m|\epsilon_n C^{\chi}_n|n\rangle = \sum_{n} |C^{\chi}_n|^2 \epsilon_n. \tag{1.144}$$

Here, as elsewhere in the book, we shall proceed on the assumption that the eigenfunctions are not degenerate. Allowing for the possibility of degenerate states would complicate the analysis for little gain. Even though real systems very often display degeneracies, due to some symmetry of the arrangement of atoms, we can always imagine this symmetry to be broken by a small perturbation, removing the degeneracy, and then suppose this perturbation to become vanishingly small. With (1.144) the problem has become: find the set of $\{C_n^\chi\}$ which minimizes (1.144) subject to the normalization requirement:

$$\sum_n |C_n^\chi|^2 = 1. \tag{1.145}$$

The solution is

$$\begin{aligned} C_0^\chi &= e^{i\phi} \\ C_n^\chi &= 0, \quad \{n > 0\}. \end{aligned} \tag{1.146}$$

$e^{i\phi}$ is an arbitrary phase factor. To see that this is so, consider that any change in this solution means adding some weight times ϵ_m where $m > 1$ to (1.144) while subtracting the same weight times ϵ_0 in order to preserve the normalization. Since $\epsilon_m > \epsilon_0$ by definition, any such change can only increase the value of $\langle\chi|\hat{H}|\chi\rangle$. We can also approach the problem more formally with the machinery of the calculus of variations. In this case, the condition for an extremum is given by the following set of equations, in which the Lagrange multiplier λ is introduced corresponding to the normalization constraint:

$$\frac{\partial\langle\chi|\hat{H}|\chi\rangle}{\partial|C_m^\chi|} - \lambda\frac{\partial\sum_n |C_n^\chi|^2}{\partial|C_m^\chi|} = 0, \tag{1.147}$$

hence

$$|C_m^\chi|\epsilon_m = \lambda|C_m^\chi|. \tag{1.148}$$

This has to be true for each state m. But if it is true for a particular one with a non-vanishing coefficient, say n with $C_n^\chi \neq 0$, then $\lambda = \epsilon_n$, and the only way it can also be true for all other m for which $\epsilon_m \neq \epsilon_n$ is if the coefficients C_m^χ vanish. What we have just proved is that the quantity $\langle\chi|\hat{H}|\chi\rangle$ is stationary with respect to variations in $|\chi\rangle$ if and only if $|\chi\rangle$ is any one of its eigenstates. It follows that the minimum value of $\langle\chi|\hat{H}|\chi\rangle$ is ϵ_0, corresponding to $|\chi\rangle = |0\rangle$, which is the ground state.

The same result could be obtained without introducing the Lagrange multiplier by minimizing

$$\epsilon = \frac{\langle\chi|\hat{H}|\chi\rangle}{\langle\chi|\chi\rangle} \tag{1.149}$$

with respect to $|\chi\rangle$. The denominator now takes care automatically of the normalization. The left-hand side of (1.149) is an energy and it is a function of the state. In the $\mathbf{r}$-representation the state itself is a function of $\mathbf{r}$, namely $\chi(\mathbf{r})$. A function of a function is called a *functional*, a concept which plays a large part in this book.

Finally note that the idea that the ground state energy is obtained by minimizing a functional of the form (1.149) is not restricted to the energy of a single particle in an external potential. The same applies to the energy of a full many-body wavefunction describing interacting electrons, in which case the energy functional to be minimized could be written:

$$E = \frac{\langle\Psi|\hat{\mathcal{H}}|\Psi\rangle}{\langle\Psi|\Psi\rangle}, \tag{1.150}$$

where by $\hat{\mathcal{H}}$ we now mean the full Hamiltonian as given in the $\mathbf{r}$-representation in eq. (1.8), and $|\Psi\rangle$ is an N-electron state.

1.6.2 The Hartree–Fock Approximation

The Hartree–Fock (HF) approximation relies on the above variational principle. If we approximate Ψ by a determinant wave function, then there is an orthonormal set of single-particle wave functions which minimizes (1.150). If we place no other restrictions on this set of wave functions this is called the *unrestricted HF approximation*. A simpler version of it to assume that the single particle states are each doubly occupied which is equivalent to choosing a determinant of the form (1.29), and this is called the *restricted HF approximation*. We have seen that the problem of minimising a functional of the form (1.150) is in general equivalent to solving the Schrödinger equation. In the HF approximation it transforms to the problem of solving a certain set of Schrödinger-like differential equation for the single-particle wave functions. These are the HF equations. Let us for simplicity stay within the framework of the restricted HF approximation. I omit any details of the derivations here, since they are in standard text books such as Parr and Yang (1989), and simply summarize the results. This introduces a little more useful terminology. The simplest operator form of a HF equation for the single-particle states looks like:

$$(\hat{T} + \hat{V}_{\text{ext}} + \hat{G})|n\rangle = \epsilon_n|n\rangle. \tag{1.151}$$

The potential energy operator has now been split into two parts. The external potential $V_{\text{ext}}(\mathbf{r})$ appears as an operator which acts on a state $|n\rangle$ simply by scaling it in the real space representation; that is in real space $\hat{V}_{\text{ext}}$ is diagonal:

$$\langle\mathbf{r}|\hat{V}_{\text{ext}}|\mathbf{r}'\rangle = V_{\text{ext}}(\mathbf{r})\delta(\mathbf{r} - \mathbf{r}'). \tag{1.152}$$

The effect of this operator on any state, say $|\chi\rangle$ in the **r**-representation is to multiply it by $V_{\text{ext}}(\mathbf{r})$:

$$\langle\mathbf{r}|\hat{V}_{\text{ext}}|\chi\rangle = \langle\mathbf{r}|\hat{V}_{\text{ext}}\int|\mathbf{r}'\rangle\langle\mathbf{r}'|\mathrm{d}\mathbf{r}'|\chi\rangle = V_{\text{ext}}(\mathbf{r})\langle\mathbf{r}|\chi\rangle. \tag{1.153}$$

A potential energy operator which is diagonal in real space is also referred to as a *local* operator. Another example is the effective potential V_{eff} introduced before. Unfortunately the operator $\hat{G}$, which describes all the electron–electron interactions, does not have this simple property. It can be written as:

$$\hat{G} = \hat{J} - \hat{K}, \tag{1.154}$$

where $\hat{J}$ and $\hat{K}$ are known as the Coulomb and exchange operators respectively and have the effect

$$\langle\mathbf{r}|\hat{J}|n\rangle = 2\sum_{m=0}^{N/2-1}\int\frac{|\langle\mathbf{r}'|m\rangle|^2\langle\mathbf{r}|n\rangle}{|\mathbf{r}-\mathbf{r}'|}\,\mathrm{d}\mathbf{r}' \tag{1.155}$$

and

$$\langle\mathbf{r}|\hat{K}|n\rangle = \sum_{m=0}^{N/2-1}\int\frac{\langle\mathbf{r}|m\rangle\langle m|\mathbf{r}'\rangle\langle\mathbf{r}'|n\rangle}{|\mathbf{r}-\mathbf{r}'|}\,\mathrm{d}\mathbf{r}'. \tag{1.156}$$

The Coulomb operator is local; it is just the Coulomb potential of the charge density (1.31), called the Hartree potential V_{H}:

$$V_{\text{H}}(\mathbf{r}) = \int\frac{\rho(\mathbf{r}')}{|\mathbf{r}-\mathbf{r}'|}\,\mathrm{d}\mathbf{r}'. \tag{1.157}$$

Without the exchange operator this would be the *Hartree approximation*, in which each electron sees the mean field of the other electrons as if it were due to the density $\rho(\mathbf{r})$; individual electrons are uncorrelated. Within the Hartree approximation we would have an effective potential for a single-particle Schrödinger equation given by:

$$\hat{V}_{\text{eff}}(\mathbf{r}) = \hat{V}_{\text{ext}}(\mathbf{r}) + V_{\text{H}}(\mathbf{r}). \tag{1.158}$$

However, the exchange operator gives us a non-local exchange potential. The set of $|n\rangle$ that are the solutions to (1.151) are called HF orbitals, and there is no Schrödinger equation with a local effective potential for which these HF orbitals are solutions. Another complication, shared even by the Hartree approximation, is that these operators with which we want to solve for the wavefunctions $\langle\mathbf{r}|n\rangle$ themselves depend on the wavefunctions! This means that solving HF equations has to be done by an iterative procedure which ends up with a self-consistent solution. To implement such a procedure the wavefunctions are expanded in *basis functions* as discussed in Section 1.10.

Once the solutions, the wavefunctions $\langle \mathbf{r}|n\rangle$ and energies ϵ_n have been found, the total energy can be evaluated by inserting the Slater determinant into eq. (1.150). The result has the form:

$$E_{\mathrm{HF}} = T_s + E_{eZ} + E_{\mathrm{H}} + E_x. \tag{1.159}$$

The first term in (1.159) is the kinetic energy of the occupied single-particle states:

$$T_s = 2\sum_{n=0}^{N/2-1} \langle n| - \tfrac{1}{2}\nabla^2|n\rangle. \tag{1.160}$$

The second term is the energy from the expectation value of the external potential; in other words it is the electron–nucleus interaction. It is given in eq. (1.15). The third term, known as the Hartree energy, is just the Coulomb self energy of the distribution $\rho(\mathbf{r})$, which can be written as

$$E_{\mathrm{H}} = \frac{1}{2}\int \frac{\rho(\mathbf{r})\rho(\mathbf{r}')}{|\mathbf{r}-\mathbf{r}'|}\,\mathrm{d}\mathbf{r}\,\mathrm{d}\mathbf{r}' = \frac{1}{2}\int \rho(\mathbf{r})V_{\mathrm{H}}(\mathbf{r})\,\mathrm{d}\mathbf{r}. \tag{1.161}$$

Finally, E_{X} is the part of the total energy involving the exchange operator and is called the *exchange energy*, given by

$$E_{\mathrm{X}} = -\sum_{n=0}^{N/2-1}\sum_{m=0}^{N/2-1}\int \frac{\langle n|\mathbf{r}\rangle\langle \mathbf{r}|m\rangle\langle m|\mathbf{r}'\rangle\langle \mathbf{r}'|n\rangle}{|\mathbf{r}-\mathbf{r}'|}\,\mathrm{d}\mathbf{r}\,\mathrm{d}\mathbf{r}'. \tag{1.162}$$

Notice that exchange lowers the total energy from what it would otherwise be. This is because the exchange term arises from the antisymmetry of the N-electron wavefunction. The antisymmetry keeps particles of like spin automatically out of each other's way, so their Coulomb repulsion energy is lowered. By comparing the form of E_{X} with E_{H} we can also see that the effect of E_{X} includes subtracting from E_{H} the following unphysical Coulomb self-energy for each electron in state n and spin α:

$$E_{n\alpha} = \tfrac{1}{2}\int \frac{\langle n|\mathbf{r}\rangle\langle \mathbf{r}|n\rangle\langle n|\mathbf{r}'\rangle\langle \mathbf{r}'|n\rangle}{|\mathbf{r}-\mathbf{r}'|}\,\mathrm{d}\mathbf{r}\,\mathrm{d}\mathbf{r}'. \tag{1.163}$$

which was included in E_{H} automatically, although spuriously.

The Coulomb repulsion itself of course also helps to keep individual electrons out of each other's way, whatever their spin. There are thus *two* effects which lead to correlation in the positions of electrons, however only the first, the exchange, is included in the HF model. The rest of the energy which would be accounted for in a more exact theory than HF is called the *correlation energy*. To be precise the exact correlation energy E_c is defined in terms of the exact energy E_{exact} and the HF energy E_{HF} by

$$E_c = E_{\mathrm{exact}} - E_{\mathrm{HF}}. \tag{1.164}$$

An important feature of this and other variational methods is that the resulting value of E is an *upper bound* to the exact ground state energy which can often be a good enough estimate of it because the error in E is only second order in the error in Ψ and not first order.

1.7 The Density Matrix and the Charge Density

This section introduces a useful new concept, the *density matrix*. A full discussion of density matrices would take us into interesting realms of quantum statistical mechanics, but this is not where we are going. We only need the density matrix which is associated with a determinant wave function of the type given in eq. (1.29). We define a density operator:

$$\hat{\rho} = \sum_{n=0}^{\infty} f_n |n\rangle\langle n| \tag{1.165}$$

which is a sum over those single particle states which are occupied by electrons, in which the occupancy of each state is given by the factor f_n. In the complete theory of density matrices for fermions, what we consider here would be called the *first-order* density matrix. This means that it is acting in the single particle Hilbert space, although it carries information about the many particles which may be in the system, the nature of which determines the states and the occupancies.

Note that the density matrix contains no more or less information than the wavefunction, which is the determinant built from the occupied states. However, if fractional occupations are allowed, it actually carries information about systems which are open, and cannot fully be described by their own wavefunction. We shall not be concerned with this generality except to note the case of finite temperature, when it becomes important in practice as we shall see.

1.7.1 The Fermi Distribution

For a system in its ground state we would be summing over the $N/2$ states of lowest energy (assuming an even number of electrons, which is only a restriction if the system is very small) and each occupancy would be 2:

$$f_n = 2, \quad n < N/2. \tag{1.166}$$

However, we can generalize the density operator to finite temperature as follows. Within the model of independent particles, the f_n follow a *Fermi distribution*:

$$f_n = 2 f_{\mathrm{F}}(\epsilon_n) \tag{1.167}$$

and the *Fermi distribution function* is given by

$$f_{\mathrm{F}}(\epsilon) = \frac{1}{1 + \exp\left((\epsilon - \epsilon_{\mathrm{F}})/k_{\mathrm{B}}T\right)}. \tag{1.168}$$

The quantity ϵ_{F} is the *Fermi energy*. It is defined so as to ensure the correct total number of electrons, that is to say it is defined implicitly by

$$N = \sum_{n=0}^{\infty} 2 f_{\mathrm{F}}(\epsilon_n). \tag{1.169}$$

1.7.2 Matrix Representations of the Density Operator

The density matrix, which is the name given to the matrix representation of the density operator, is obtained as for all operators by premultipling by a bra and postmultiplying by a ket. The result is a matrix element. In the **r**-representation the matrix element is a function of two positions, and has the form:

$$\rho(\mathbf{r}, \mathbf{r}') = \langle \mathbf{r}|\hat{\rho}|\mathbf{r}'\rangle = \sum_{n=0}^{\infty} f_n \langle \mathbf{r}|n\rangle\langle n|\mathbf{r}'\rangle \equiv \sum_{n=0}^{\infty} f_n \psi_n(\mathbf{r})^* \psi_n(\mathbf{r}'). \tag{1.170}$$

The electron density at **r** is just a diagonal element of this matrix:

$$\rho(\mathbf{r}) = \rho(\mathbf{r}, \mathbf{r}). \tag{1.171}$$

At this point it is convenient to introduce the idea of the *trace* (denoted Tr) of an operator or its matrix representation. In an orthonormal basis, the trace is the sum of the diagonal elements. The trace of the density operator gives the total number of electrons N. In the ψ_n representation, the density matrix is already diagonal; directly from (1.165) we see that its elements are:

$$\langle n|\hat{\rho}|n\rangle = f_n \tag{1.172}$$

so

$$\mathrm{Tr}\,\hat{\rho} = \sum_{n=0}^{\infty} f_n = N. \tag{1.173}$$

In the **r**-representation, in which the sum becomes an integral:

$$\mathrm{Tr}\,\hat{\rho} = \int \rho(\mathbf{r}, \mathbf{r})\,\mathrm{d}\mathbf{r} = N. \tag{1.174}$$

The trace of the density matrix (or indeed of any operator) does not depend on whether it is in the **r**-representation or the ψ_n-representation. This is an example of the general result that the trace of an operator is the same in any complete orthonormal basis.

Another important trace we will come across later is the trace of the product of the operators $\hat{\rho}$ and $\hat{H}$:

$$\begin{aligned}\mathrm{Tr}(\hat{\rho}\hat{H}) &= \sum_n \langle n| \left\{ \sum_m |m\rangle f_m \langle m| \sum_{m'} |m'\rangle \epsilon_{m'} \langle m'| \right\} |n\rangle \\ &= \sum_n f_n \epsilon_n . \end{aligned} \tag{1.175}$$

This is just the sum of the energies of the electrons within the model of independent particles. It is also called the *band energy*, and will be denoted E^{band}.

Other representations of the density operator can be useful, such as a basis of local orbitals. It is straightforward to get an expression as follows. The quantities that are actually obtained by calculation are the coefficients $C^n_{I\mu}$ of the local orbitals in the expansion of each eigenfunction $|n\rangle$. From (1.121), given that

$$|n\rangle = \sum_{I\mu} C^n_{I\mu} |I\mu\rangle, \tag{1.176}$$

we can express the density operator as

$$\hat{\rho} = \sum_{I\mu J\nu} |I\mu\rangle \rho^{I\mu J\nu} \langle J\nu|. \tag{1.177}$$

where

$$\rho^{I\mu J\nu} = \sum_{n=0}^{\infty} f_n C^n_{I\mu} C^{n*}_{J\nu} \tag{1.178}$$

The charge density is therefore

$$\rho(\mathbf{r}) = \sum_{I\mu J\nu} \langle \mathbf{r}|I\mu\rangle \rho^{I\mu J\nu} \langle J\nu|\mathbf{r}\rangle. \tag{1.179}$$

It is a well-known result of linear algebra that the trace of an operator is invariant to the basis, however when the basis is not orthonormal the trace is not simply the sum of the diagonal matrix elements. To get the trace in our basis of local orbitals, first write it in terms of the orthonormal basis $|n\rangle$:

$$\mathrm{Tr}\,\hat{\rho} = \sum_n \langle n| \left\{ \sum_{I\mu J\nu} |I\mu\rangle \rho^{I\mu J\nu} \langle J\nu| \right\} |n\rangle. \tag{1.180}$$

Now move the factor $\langle n|I\mu\rangle$ to the end and notice that $\sum |n\rangle\langle n|$ is just the identity operator, see eq. (1.40). We find

$$\mathrm{Tr}\,\hat{\rho} = \sum_{I\mu J\nu} \rho^{I\mu J\nu} S_{J\nu I\mu}, \tag{1.181}$$

a result we could also obtain by integrating the charge (1.179) over $\mathbf{r}$. A similar expression is valid if we replace $\hat{\rho}$ by any other one-electron operator. Making use of (1.126) we can also express the trace of an operator in terms of its matrix elements, e.g. (1.181) becomes

$$\mathrm{Tr}\,\hat{\rho} = \sum_{I\mu J\nu} \rho_{I\mu J\nu} S^{-1}_{J\nu I\mu}. \tag{1.182}$$

We shall later be using the expression for the band energy $\mathrm{Tr}\,\hat{\rho}\hat{H}$ in this basis. The trace of $\hat{\rho}$ times any other operator has a particularly simple form. By the same route as we used to obtain (1.181) we find

$$\mathrm{Tr}\hat{\rho}\hat{H} = \sum_{I\mu J\nu} \rho^{I\mu J\nu} H_{J\nu I\mu}. \tag{1.183}$$

Notice that the status of the representations of $\hat{\rho}$ and of $\hat{H}$ entering this equation are different. $\rho^{I\mu J\nu}$ is an *expansion coefficient* of the density operator, while $H_{J\nu I\mu}$ is a *matrix element* of the Hamiltonian. Remember, these two terms *expansion coefficient* and *matrix element* only mean the same thing when the basis in which they are expressed is orthonormal.

1.7.3 Mulliken Charges, Bond Charges and Bond Orders

If we suppose we have an orthonormal basis of local orbitals, the total charge obtained by integrating (1.179) is given precisely by the sum of atom-centred charges ρ_I, where:

$$\rho_I = \sum_{\mu} \rho^{I\mu I\mu} = \sum_{n=0}^{\infty} \sum_{\mu} f_n |C^n_{I\mu}|^2. \tag{1.184}$$

These are a special case of *Mulliken charges*, after the Nobel prize winning chemist Robert Mulliken who helped us understand chemical bonding with these concepts. A general approach to them is given by Mayer (1983). If the basis is not orthogonal, the total number of electrons (the total electronic charge) can still be written as a sum of these atom-centred charges with the addition of 'bond' charges:

$$N = \sum_{n=0}^{\infty} \sum_{I\mu J\nu} f_n C^n_{I\mu} C^{n*}_{J\nu} S_{J\nu I\mu} = \sum_{I\mu J\nu} \rho^{I\mu J\nu} S_{J\nu I\mu} = \sum_I \rho_I + \frac{1}{2} \sum_{I \neq J} \rho_{IJ}. \tag{1.185}$$

It follows from (1.185) that the bond charges are given by

$$\rho_{IJ} = \sum_{\mu\nu} \left(\rho^{I\mu J\nu} O_{J\nu I\mu} + \rho^{J\nu I\mu} O_{I\mu J\nu} \right). \tag{1.186}$$

The atom-centred charges and bond charges are one way of splitting up the total charge. Often, however, we want to allocate all the charge to charges q_I on individual ions. The charge q_I is a characteristic of the valency of the ion, but it is more or less modified by the kind and number of its nearest neighours. We want to allocate *all* the charge, including the bond charges (but excluding core charge if we are only describing valence electrons explicitly, for example with pseudopotentials), even when the orbitals are not orthogonal ($O \neq 0$). From the above formulae a suitable definition of q_I is

$$q_I = \rho_I + \frac{1}{2} \sum_J \rho_{IJ}. \tag{1.187}$$

These are the Mulliken charges. All quantities describing numbers of electrons are guaranteed to be real, whether or not complex basis functions are used. However, the numbers must be interpreted with caution, because they clearly depend on the choice of basis functions, which is not unique. At best they can be used as an indication of the trends in charge transfer between atoms. For example if we compare the charge on a titanium atom in stoichiometric TiO_2 with its value in Ti_2O_3 we expect from the nominal valency we would have Ti^{4+} and Ti^{3+} ions respectively, with divalent O^{2-}. Accordingly, we might expect to find $q_{Ti} - Z_{Ti} = -4$ and $q_{Ti} - Z_{Ti} = -3$, with $q_O - Z_O = 2$. However, if we choose any reasonable basis such large charge transfers are not predicted by real density functional theory calculations. In order to account for all the electrons, the Mulliken prescription allocates the bond charge to particular atoms by dividing it equally between the participants in a bond, which is certainly not a unique prescription. Reassuringly, Mayer (1983) has shown that the Mulliken prescription does have a special fundamental significance in terms of the expectation value of the number operator on a site. Nevertheless, however the bond charge is divided, within reasonable limits, real DFT calculations always give numbers that are less than the nominal valencies; a typical value for the oxygen ion for example would take its charge into the range 1–1.8, where these numbers depend on the basis set. We should not feel uneasy about this. The point is that in condensed matter the charge on an individual atom is not really a physical, measurable quantity. It is useful to remember this point when you come across spectroscopies with high resolution such as electron energy loss spectroscopy (EELS) which purport to give information about the local charge on an atom. What they are actually measuring is another integrated quantity, about which there is no ambiguity to do with basis sets, and which can be calculated directly. It approximates a partial density of states (see Section 1.8) weighted by some matrix element; see for example Paxton *et al.* (2000) for a detailed discussion.

The off-diagonal expansion coefficients of $\hat{\rho}$ are called *bond orders*, a term that was first introduced by Coulson (1939) in his description of chemical bonding. Incidentally, different authors have also used the term 'bond order' to mean slightly different things, but we shall not need any other definition than this:

$$\rho^{I\mu J\nu} = \sum_{n=0}^{\infty} f_n C^n_{I\mu} C^{n*}_{J\nu} \quad \{I \neq J\}. \tag{1.188}$$

As we shall see later, bond orders are an important indicator of the strength of bonding, so it is important to grasp their physical significance. Each term in (1.188) is non-vanishing if the corresponding state labelled n satisfies two conditions. First, it must be occupied, $f_n > 0$, secondly it must engage the orbitals on *both* atoms I and J, that is both the coefficients must be non-zero. However, there can still be a lot of cancellation between individual states in the sum.

The example usually given is of a diatomic molecule, hydrogen being the archetypical and simplest case. With one s-orbital on each atom, ϕ_{A} and ϕ_{B}. There are bonding and antibonding states:

$$\begin{aligned} \psi_b &= \frac{1}{\sqrt{2}} \left(\phi_{\mathrm{A}} + \phi_{\mathrm{B}} \right), \\ \psi_a &= \frac{1}{\sqrt{2}} \left(\phi_{\mathrm{A}} - \phi_{\mathrm{B}} \right), \end{aligned} \tag{1.189}$$

such that in the ground state the bonding state ψ_b is occupied by two electrons, $f_0 = 2$, and the antibonding state ψ_a is unoccupied, $f_1 = 0$. The factors C^0_{A} and C^0_{B} are both $1/\sqrt{2}$, the factor $C^1_{\mathrm{A}} = 1/\sqrt{2}$ and $C^1_{\mathrm{B}} = -1/\sqrt{2}$. Putting these factors together we see that the bond order is unity, which is characteristic of a *saturated* bond. On the other hand let us now change the occupancies, putting an electron into the antibonding state, so that $f_0 = 1$ and $f_1 = 1$. The bond order is now zero, which accords with there being no net bonding of the atoms. In this case, when the two coefficients have opposite sign they *reduce the bond order*. With a basis of s-orbitals, when the coefficients of orbitals on neighbouring atoms have the opposite sign, this is generally the indication of an antibonding orbital.

When orbitals of lower symmetry are involved this sign difference is not a direct indicator of whether a state is bonding or antibonding. For instance consider two real p-orbitals on atoms A and B pointing the same way along the axis of a diatomic molecule. The bonding state in this case is

$$|b\rangle = \frac{1}{\sqrt{2}}\left(|p_{\mathrm{A}}\rangle - |p_{\mathrm{B}}\rangle\right) \tag{1.190}$$

in which the coefficients of the orbitals have *opposite* sign, so the lobes with the same sign are pointing towards each other.

1.8 The Density of States

1.8.1 Definition

Supposing we have all the eigenstates $|n\rangle$ and energies ϵ_n of a single electron Hamiltonian, it is useful to define the *total density of states* of the system $D(\epsilon)$ by the relation

$$D(\epsilon) = \sum_n \delta(\epsilon - \epsilon_n). \tag{1.191}$$

This should be understood in the operational sense that if we integrate $D(\epsilon)$ over an energy range ϵ_1 to ϵ_2 we obtain the total number of states within that energy range. The total number of electrons is given by

$$N = \sum_n f_n = 2\int_{-\infty}^{\infty} f_{\mathrm{F}}(\epsilon) D(\epsilon)\, \mathrm{d}\epsilon. \tag{1.192}$$

We now introduce the idea of a *local* density of states. This must be defined in such a way that when it is integrated over all space it gives the total density of states. It should also measure in some way the relative weight with which each state at energy ϵ is represented at a given point in space. In the **r**-representation the function that fills the bill is:

$$D_r(\mathbf{r}, \epsilon) = \sum_n |\langle \mathbf{r}|n\rangle|^2 \delta(\epsilon - \epsilon_n). \tag{1.193}$$

You can verify that

$$\int D_r(\mathbf{r}, \epsilon)\, \mathrm{d}\mathbf{r} = D(\epsilon). \tag{1.194}$$

We can think of the local density of states as the diagonal elements in the **r**-representation of the operator

$$\hat{D}(\epsilon) = \sum_n |n\rangle \delta(\epsilon - \epsilon_n) \langle n|. \tag{1.195}$$

The total density of states is the *trace* of this operator

$$D(\epsilon) = \mathrm{Tr}\, \hat{D}(\epsilon). \tag{1.196}$$

Another useful representation of the density of states is obtained by using the following identity to represent the delta function:

$$\delta(\epsilon - \epsilon_n) \equiv -\frac{1}{\pi} \lim_{\eta \to 0^+} \mathrm{Im} \frac{1}{\epsilon + i\eta - \epsilon_n}. \tag{1.197}$$

In this expression η is a small *positive* quantity which is taken to zero from above (hence the notation 0^+). In the following expressions containing η, it will be understood implicitly that this limit is to be taken. Thus we write

$$\hat{D}(\epsilon) = -\frac{1}{\pi} \mathrm{Im} \sum_n |n\rangle \frac{1}{\epsilon + i\eta - \epsilon_n} \langle n|. \tag{1.198}$$

The density of states gives us another useful tool for summing over single-particle states. For example the band energy in (1.175) could be written

$$E^{\mathrm{band}} = \mathrm{Tr}(\hat{\rho}\hat{H}) = 2 \int_{-\infty}^{\infty} f_{\mathrm{F}}(\epsilon) \epsilon D(\epsilon) \, \mathrm{d}\epsilon. \tag{1.199}$$

At zero temperature the integral over the occupied density of states takes the simpler form

$$\mathrm{Tr}(\hat{\rho}\hat{H}) = 2 \int_{-\infty}^{\epsilon_{\mathrm{F}}} \epsilon D(\epsilon) \, \mathrm{d}\epsilon. \tag{1.200}$$

It is also sometimes convenient to think of the energy of the occupied states as the integral of a locally varying energy, by making use of the local density of states, whereby (1.199) becomes:

$$E^{\mathrm{band}} = \mathrm{Tr}(\hat{\rho}\hat{H}) = \int_V \left\{ 2 \int_{-\infty}^{\infty} f_{\mathrm{F}}(\epsilon) \epsilon D_r(\mathbf{r}, \epsilon) \, \mathrm{d}\epsilon \right\} \mathrm{d}\mathbf{r}. \tag{1.201}$$

The term in braces is a local projection at $\mathbf{r}$ of the band energy. An important physical concept is expressed here, which is particularly evident in metallic systems. As an example think of a large metallic crystal containing a single defect. Far from a defect the local density of states will be indistinguishable from that in a perfect crystal; the perturbation of a defect is more or less local, because it is screened by the conduction electrons. The same cannot be said of individual, extended wave functions. In theory, the nth wavefunction, which is extended over the entire crystal, may look entirely different at every point $\mathbf{r}$, however far from the defect, as compared to the corresponding wavefunction in the perfect crystal. Density matrices and densities of states therefore seem to be more physical quantities for describing extended systems than individual wavefunctions.

1.8.2 The Green Function

The density of states can also be expressed in terms of the Hamiltonian matrix, as follows. Consider the operator $\hat{G}$ defined by

$$\hat{G}(\epsilon) = \sum_n |n\rangle \frac{1}{\epsilon - \epsilon_n} \langle n|. \tag{1.202}$$

This is the *inverse* of the operator $(\epsilon\hat{\mathbf{1}} - \hat{H})$, and it is called the *Green function* corresponding to the single particle Hamiltonian. I have not defined the inverse of an operator, but it is as intuitive as the product of two operators. It means that if we multiply $(\epsilon\hat{\mathbf{1}} - \hat{H})$ into $\hat{G}$ we get the identity operator:

$$(\epsilon\hat{\mathbf{1}} - \hat{H}) \cdot \hat{G} = \hat{\mathbf{1}}. \tag{1.203}$$

To see this, write $(\epsilon\hat{\mathbf{1}} - \hat{H})$ in the eigenstate representation, introduced in (1.42):

$$\epsilon\hat{\mathbf{1}} - \hat{H} = \sum_m |m\rangle(\epsilon - \epsilon_m)\langle m|. \tag{1.204}$$

If you multiply this by $\hat{G}$ in (1.202) and use the orthonormality of the eigenstates the result, when $\eta \to 0^+$, is the identity operator in the eigenstate representation, (1.40). The order of multiplication does not matter here, as you can easily verify; an operator commutes with its inverse. From the expression (1.197) for the delta function we can also write

$$-\frac{1}{\pi}\mathrm{Im}\,\hat{G}(\epsilon + \mathrm{i}\eta) = \sum_n |n\rangle\delta(\epsilon - \epsilon_n)\langle n|. \tag{1.205}$$

We have now reached the important relation between the density of states and $\hat{G}$, namely

$$\hat{D}(\epsilon) = -\frac{1}{\pi} \lim_{\eta\to 0^+} \mathrm{Im}\,\hat{G}(\epsilon + \mathrm{i}\eta). \tag{1.206}$$

We can also now obtain an expression for the density operator in terms of $\hat{G}$. We just have to multiply (1.205) by $2f_{\mathrm{F}}(\epsilon)$ and integrate over all energies, which gives:

$$-\frac{2}{\pi}\int_{-\infty}^{+\infty} f_{\mathrm{F}}(\epsilon)\mathrm{Im}\,\hat{G}(\epsilon + \mathrm{i}\eta)\mathrm{d}\epsilon = \sum_n |n\rangle f_n \langle n| = \hat{\rho}. \tag{1.207}$$

$\hat{G}$ can also be expanded in a local basis:

$$\hat{G}(\epsilon) = \sum_{I\mu J\nu} |I\mu\rangle G^{I\mu J\nu}(\epsilon)\langle J\nu| \tag{1.208}$$

and it follows that

$$-\frac{1}{\pi}\mathrm{Im}\, G^{I\mu J\nu}(\epsilon + \mathrm{i}\eta) = \sum_{n} C^{n}_{I\mu} C^{n*}_{J\nu} \delta(\epsilon - \epsilon_n). \tag{1.209}$$

From the definition of the expansion coefficients of the density operator (1.178), eq. (1.209) provides an important relation between the density matrix or bond order and the Green function:

$$\begin{aligned} \rho^{I\mu J\nu} &= -\frac{2}{\pi}\int_{-\infty}^{\infty} f_{\mathrm{F}}(\epsilon)\mathrm{Im}\, G^{I\mu J\nu}(\epsilon + \mathrm{i}\eta)\,\mathrm{d}\epsilon \\ &= -\frac{2}{\pi}\int_{-\infty}^{\epsilon_{\mathrm{F}}} \mathrm{Im}\, G^{I\mu J\nu}(\epsilon + \mathrm{i}\eta)\mathrm{d}\epsilon \quad \text{at } T = 0\,\mathrm{K}. \end{aligned} \tag{1.210}$$

As an aside, looking back at (1.202) and (1.203) you might be tempted also to write

$$\hat{G} = \frac{1}{(\epsilon\hat{\mathbf{1}} - \hat{H})}. \tag{1.211}$$

This is often done, but it is a bit dangerous since operators certainly do not obey all the rules of ordinary algebra, and I would advise against it.

We have now established that the density of states is given by

$$D(\epsilon) = -\frac{1}{\pi}\mathrm{Im}\,\mathrm{Tr}\,\hat{G}(\epsilon + \mathrm{i}\eta). \tag{1.212}$$

Remember these are spacial states, we still need a factor of 2 for spins if we are counting electrons in the states. The local density of states is derived from (1.193) in a similar fashion if we do not take the trace but just one diagonal element in the **r**-representation:

$$D_r(\mathbf{r}, \epsilon) = -\frac{1}{\pi}\mathrm{Im}\langle \mathbf{r}|\hat{G}(\epsilon + \mathrm{i}\eta)|\mathbf{r}\rangle. \tag{1.213}$$

A local density of states can also be defined in terms of the local orbitals. We simply have to write the trace of $\hat{D}$ in the local orbital basis:

$$D(\epsilon) = \sum_{I\mu J\nu} D^{I\mu J\nu}(\epsilon) S_{J\nu I\mu} \tag{1.214}$$

from which the density of states on orbital $I\mu$ can be defined as

$$D_{I\mu}(\epsilon) = \sum_{J\nu} D^{I\mu J\nu}(\epsilon) S_{J\nu I\mu}. \tag{1.215}$$

In the special case of orthogonal orbitals we have

$$D_{I\mu}(\epsilon) = D^{I\mu I\mu}(\epsilon) = -\frac{1}{\pi}\mathrm{Im}\, G^{I\mu I\mu}(\epsilon + \mathrm{i}\eta) = -\frac{1}{\pi}\mathrm{Im}\, G_{I\mu I\mu}(\epsilon + \mathrm{i}\eta). \tag{1.216}$$

This formula has been very useful in the development of tight-binding models, as we shall see.

Finally we can express the charge density in terms of D and G as follows. From (1.193) and (1.213):

$$\rho(\mathbf{r}) = 2\int_{-\infty}^{+\infty} f_{\mathrm{F}}(\epsilon) D_r(\mathbf{r}, \epsilon)\mathrm{d}\epsilon = -\frac{2}{\pi}\mathrm{Im}\int_{-\infty}^{+\infty} f_{\mathrm{F}}(\epsilon)\langle \mathbf{r}|\hat{G}(\epsilon + \mathrm{i}\eta)|\mathbf{r}\rangle\, \mathrm{d}\epsilon. \tag{1.217}$$

In the limit of zero temperature, the Fermi factor would be 1 for $\epsilon < \epsilon_{\mathrm{F}}$ and 0 for $\epsilon > \epsilon_{\mathrm{F}}$.

1.9 Jellium

The jellium model has an important part to play in all the density functional theories. This is the model in which the charges on the nuclei are spread out into a uniform distribution of density ρ, referred to as the uniform positive background. The electrons then form a gas of uniform density, also ρ. This is an infinite homogeneous jellium. The textbooks, for example Ashcroft and Mermin (1976), describe this model in their early chapters as the Sommerfeld theory, or the free electron model. It is the simplest quantum mechanical model of a metal and superceded Drude's classical theory of free electrons.

When it is used as a description of metals, not all the electrons are included in the jellium model. It would be absurd to describe the lowest energy states, the bound electrons deep in the core of an atom, as free. They are usually lumped together with the nuclei to provide a reduced effective positive charge, and only the higher energy electrons (the *conduction* electrons) are treated within the jellium model. It is not a trivial matter to know where to draw the line between such 'core' electrons and conduction electrons, and this problem must be faced when *pseudopotentials* are used, a subject I shall return to. In the present context however the point is not to use jellium as an approximation to our material. We regard it rather as a theoretical substance in its own right, which we study in order to extract formulae for the kinetic and potential energy of electrons, and these formulae will be used in the density functional models to be described.

Bounded jelliums have also been studied. For instance to make a simple model of a metal surface we would define the posiitive background as exising only in the $z < 0$ half space. The electron density becomes inhomogeneous at the surface so defined, and varies between a constant value ρ within the jellium to zero outside the jellium. For present purposes we only have to deal with the uniform infinite jellium in which the electrons have the same constant average density as the positive background.

1.9.1 Cancellation of Electrostatic Energies

So far we have referred to the total energy E of the electrons in a system, the eigenvalue of the many-electron Schrödinger equation. However, in any calculation of total energies and forces a vital contribution is also made by the direct nucleus–nucleus interaction:

$$E_{ZZ} = \frac{1}{2}\sum_{IJ} \frac{Z_I Z_J}{|\mathbf{R}_I - \mathbf{R}_J|}. \tag{1.218}$$

In jellium, in which the discrete positive charges have been smeared out, it is identical to the Hartree energy (1.161):

$$E_{ZZ} \equiv \frac{1}{2}\int \frac{\rho(\mathbf{r})\rho(\mathbf{r}')}{|\mathbf{r}-\mathbf{r}'|}\,\mathrm{d}\mathbf{r}\,\mathrm{d}\mathbf{r}' = E_{\mathrm{H}}. \tag{1.219}$$

Both these terms diverge to $+\infty$ as the size of the system, proportionally to the square of the number of nuclei N_{a} (or electrons, N). However, we know that physically in electrically neutral systems of macroscopic size the total energy should only increase linearly with the size of the system, and in this limit the energy per electron or nucleus is invariant to system size. That is a cornerstone of thermodynamics; the internal energy of a system is an *extensive* quantity. There must also be a term in the total energy that scales with sytem size as $N^{2/3}$, due to the surfaces, which becomes negligible in comparison to the bulk term. As long as the system is uncharged the energy cannot include a term scaling like N^2.

This problem is taken care of by the third electrostatic term E_{eZ} (see eq. (1.15), which in jellium takes the simple form

$$E_{eZ} = -\int \frac{\rho(\mathbf{r})\rho(\mathbf{r}')}{|\mathbf{r}-\mathbf{r}'|}\,\mathrm{d}\mathbf{r}\,\mathrm{d}\mathbf{r}'. \tag{1.220}$$

Thus in jellium the three terms which scale as N^2 exactly cancel each other:

$$E_{ZZ} + E_{\mathrm{H}} + E_{eZ} = 0. \tag{1.221}$$

The total energy of jellium, ignoring surfaces, is just the sum of kinetic and exchange and correlation energies:

$$E^{\mathrm{jel}} = T + E_{\mathrm{x}} + E_{\mathrm{c}}. \tag{1.222}$$

1.9.2 The Kinetic Energy

At this point there is a shift in definitions. In the modern theory, it turns out to be very useful to divide the total energy slightly differently by defining the kinetic energy T_s of an identical number of hypothetical non-interacting electrons. These hypothetical electrons move in a single-particle effective potential $V_{\rm eff}(\mathbf{r})$, which for jellium is constant. T_s is a good approximation to the true kinetic energy T of the electrons in jellium. To make the total energy exactly right, the error in $T - T_s$ is included in a redefined exchange and correlation function $E_{\rm XC}$, so that the following equation can be regarded as a definition of the exchange and correlation energy:

$$E^{\rm jel} = T_s + E_{\rm XC}. \tag{1.223}$$

To avoid arbitrariness in the size of the system we can define $E^{\rm jel}$ to refer to unit volume. We also introduce the energies per electron:

$$\epsilon^{\rm jel} = t_s + \epsilon_{\rm XC}, \tag{1.224}$$

where $\epsilon^{\rm jel} = E^{\rm jel}/\rho$, $t_s = T_s/\rho$ and $\epsilon_{\rm XC} = E_{\rm XC}/\rho$. These are all functions of the jellium density ρ. The kinetic energy T_s can be derived in analytic form as follows. We want to add up the kinetic energies of all the electrons treated as independent particles obeying Fermi statistics. There are two electrons in each momentum eigenstate $|\mathbf{k}\rangle$, and $V/(2\pi)^3$ states per unit volume in $\mathbf{k}$-space. States in $\mathbf{k}$-space must be occupied in increasing order of kinetic energy $\frac{1}{2}k^2$ and all the occupied states are contained by a sphere of radius $k_{\rm F}$. This is called the *Fermi sphere*, and the highest energy of occupied states is called the *Fermi energy*, $\epsilon_{\rm F}$, where setting the constant potential to zero:

$$\epsilon_{\rm F} = \tfrac{1}{2}k_{\rm F}^2. \tag{1.225}$$

The total number of electrons must therefore satisfy

$$N = 2 \cdot \frac{V}{(2\pi)^3} \cdot \frac{4\pi k_{\rm F}^3}{3}. \tag{1.226}$$

The first factor of 2 counts the two electrons of each spin occupying each spacial state $|\mathbf{k}\rangle$. Equation (1.226) can be turned round to give us an expression for $k_{\rm F}$ in terms of the density $\rho = N/V$:

$$k_{\rm F} = (3\pi^2\rho)^{1/3}. \tag{1.227}$$

Now we can calculate the kinetic energy by an integration within the Fermi sphere. The number of states between k and $k + {\rm d}k$ is $4\pi k^2 {\rm d}k$ times the number of states per unit volume in $\mathbf{k}$-space. Hence

$$T_s = 2 \cdot \frac{V}{(2\pi)^3} \int_0^{k_{\rm F}} \tfrac{1}{2}k^2 4\pi k^2 {\rm d}k = \tfrac{3}{5}N\epsilon_{\rm F} = \tfrac{3}{10}N(3\pi^2\rho)^{2/3} \tag{1.228}$$

and

$$t_s = \tfrac{3}{5}\epsilon_{\rm F} = \tfrac{3}{10}(3\pi^2\rho)^{2/3}. \tag{1.229}$$

1.9.3 Exchange and Correlation Energy

No such exact and simple formula is available for the local exchange and correlation energy ϵ_{xc}. There is, however, an exact formula for the exchange part of it (Dirac, 1930; Parr and Yang, 1989, p. 108):

$$\epsilon_x = -\frac{3}{4}\left(\frac{3\rho}{\pi}\right)^{1/3}. \tag{1.230}$$

As for the remainder, the correlation energy per electron ϵ_c, the best estimates are from numerical calculations based on the quantum Monte–Carlo method, tabulated or fitted as a function of ρ (Ceperley and Alder, 1980; Perdew and Zunger, 1981). These results have been very widely used in the applications of density functional theory.

1.10 The Matrix Eigenvalue Problem

Usually, the first step in solving the single-particle Schrödinger equation is to convert it into a matrix eigenvalue problem. Let us see how this works with local orbitals. We consider a one-electron Hamiltonian operator $\hat{H}$ and the task is to solve the equation

$$\hat{H}|n\rangle = \epsilon_n|n\rangle \tag{1.231}$$

for the state n and its corresponding eigenvalue, the single particle energy ϵ_n.

In the $\mathbf{r}$-representation, this Schrödinger equation is the familiar single-particle Schrödinger equation we introduced with eq. (1.18):

$$\left(-\tfrac{1}{2}\nabla^2 + V_{\mathrm{eff}}(\mathbf{r})\right)\langle\mathbf{r}|n\rangle = \epsilon_n\langle\mathbf{r}|n\rangle. \tag{1.232}$$

If we expand the wavefunction $|n\rangle$ in local orbitals it looks like

$$\sum_{J\nu}\hat{H}C^n_{J\nu}|J\nu\rangle = \epsilon_n\sum_{J\nu}C^n_{J\nu}|J\nu\rangle. \tag{1.233}$$

If we now project onto $\langle I\mu|$ it becomes

$$\sum_{J\nu}\langle I\mu|\hat{H}|J\nu\rangle C^n_{J\nu} = \epsilon_n\sum_{J\nu}\langle I\mu|J\nu\rangle C^n_{J\nu}. \tag{1.234}$$

This is now a 'generalised' matrix eigenvalue equation of the form

$$\text{matrix} \times \text{vector} = \text{eigenvalue} \times \text{matrix} \times \text{vector}.$$

In the more usual matrix eigenvalue equations there is no matrix on the right hand side. We could write it in a notation more familiar in algebra as

$$\sum_{J\nu} H_{I\mu J\nu} C^n_{J\nu} = \epsilon_n \sum_{J\nu} S_{I\mu J\nu} C^n_{J\nu}. \tag{1.235}$$

H is the Hamiltonian matrix in a local orbital representation, and S is the overlap matrix we introduced previously in eq. (1.116).

Within this basis we can express a local and *partial* density of states exactly analogous to the one we defined in the basis of $|\mathbf{r}\rangle$, as one of the terms in the trace, eq. (1.212), using expression (1.181) for the trace in a non-orthogonal basis:

$$D_{I\mu}(\epsilon) = -\frac{1}{\pi} \mathrm{Im} \sum_{J\nu} G^{I\mu J\nu}(\epsilon) S_{J\nu I\mu}. \tag{1.236}$$

In this formula, we are decomposing the total density of states into contributions from each individual orbital rather than from each point in space. The non-orthogonality has the effect of extending the locality from a single site to the range of the S-matrix, but this can still be just one or two shells of neighbours. Using this expression we might try to express formally the sum of the electron energies, $\mathrm{Tr}\,\hat{\rho}\hat{H}$ in eq. (1.199), as a sum of local quantities:

$$\mathrm{Tr}\,\hat{\rho}\hat{H} = \sum_{I\mu} 2 \int_{-\infty}^{+\infty} f_{\mathrm{F}}(\epsilon)\epsilon D_{I\mu}(\epsilon)\,\mathrm{d}\epsilon. \tag{1.237}$$

The physical sense of this is that these local quantities are actually not sensitive to the positions of atoms far away, but depend most strongly on the positions of their nearest neighbours. A conseqence we shall be exploring later is that the total energy has a local interpretation, in other words the whole energy is the sum of its parts. This can only be true of course if the influence of any long-ranged electrostatic interactions is screened out, so it can be a useful concept in metallic systems or semiconductors, but not in ionic systems.

1.10.1 Local Orbitals or Plane Waves?

The matrix Schrödinger equation (1.235) has to be solved for its eigenvalues ϵ_n and the eigenvectors C^n. The form (1.235) is useful for practical calculations, because there are established algorithms for solving problems in linear algebra of this type. Although we set out to express the Schrödinger equation in a basis of local orbitals, there is nothing in the derivation of this matrix equation that says the orbitals have to be local. In fact the Schrödinger equation can be cast in the form of a matrix eigenvalue problem in terms of any complete basis, local, extended or a mixture of the two, and several kinds have turned out to be particularly useful in practice. One is the set of plane waves, the momentum eigenfunctions. There are a number of reasons why this is a very widely used basis. Not least is the fact that they are orthogonal, so S is the unit matrix. Secondly there is a useful connection to the ideas of Fourier transformation, which is essentially what we are doing when we obtain the coefficients of a wavefunction in a basis of plane-waves. In Fourier transformation, general functions are represented by plane-waves, and there are fast algorithms for making the transformation backwards and forwards. The techniques of fast Fourier transformation make the method technically attractive. A third attractive feature is that only wavelengths up to a certain cutoff are included in the basis, and this gives a single parameter, the cutoff wavevector k_c, such that plane waves with $k > k_c$ are not included. There is therefore a unique way to improve the basis set; simply increase k_c. It is a disadvantage with a basis of localised orbitals that there are choices to be made when one wants to make the basis set more complete: one could add radial functions for different n or l and the functional forms can be varied, but it is not obvious what to do. A fourth advantage is that plane waves are not biased to any particular atom; the whole of space is treated on an equal footing. Finally it is useful that the kinetic energy is diagonal in a plane wave basis, so the matrix elements of $\hat{T}_s$ are simply $\frac{1}{2}k^2$.

After reading these remarks you might ask why bother with local orbitals at all. There are two main reasons for using local orbitals. A major disadvantage of plane waves as a basis is that a large number of them are needed, compared to the number of atomic-like orbitals which could describe the wavefunction to the same level of accuracy. To find the eigenvalues and orthogonalised eigenvectors of a large matrix is computationally demanding because it takes a time which scales as the cube of the number of basis functions. The other disadvantage is that having solved the eigenvalue problem you have very many numbers, the coefficients of plane waves, which have no particular physical interpretation. With local orbitals on the other hand, these numbers mean something interesting. They express how much of an atomic s, p or d orbital is involved in a particular wavefunction. We can see the participation of atomic orbitals in chemical bonds. Wavefunctions are often predominantly constructed of orbitals of just one or two types, such as the d-bands of transition metals, or particular p states near the top of the valence band in a semiconductor, and physical properties, including the crystal structure of a material, can best be understood in terms of a few such orbitals. A deeper account of this matter of the interpretation of local electronic structure may be found in Heine (1980).

1.10.2 Approaches to Solving the Matrix Eigenvalue Problem

How to solve (1.235) efficiently is a matter of great interest in practice. It is not only of interest from the point of view of computational efficiency; the physical interpretation of a model in terms of bond energies, such as the structure of pairwise and many-body forces, is connected with the kind of approximations made in its solution, as we shall see. In this book we shall not be dealing with methods of solution in any detail, but some general remarks seem appropriate here. In this section we have in mind a basis of local orbitals of the kind used in the tight-binding models to be described later.

If we are dealing with a molecule or a large cluster of atoms in general, terminated by a free surface, then the Hamiltonian is a square matrix of rank N_{orb}, where N_{orb} is the number of basis orbitals in the system. Increasing the size of the system increases N_{orb} in proportion to the number of atoms or electrons (N). The eigenvalues can be obtained by standard numerical methods. The computation time required scales as the cube of N, in the standard jargon it is an $O(N^3)$ process. While this may be quite acceptable for small systems of a few tens of atoms, it becomes unwieldy to use traditional matrix diagonalisation for a few hundred atoms, at least with the present generation of computers. So it is desirable to try other methods. Several approaches to finding the eigenvalues exploit the physical fact that the hopping integrals fall off sharply with distance R_{IJ} and can be truncated, even after the first or second shell of neighbours. Mathematically this means that we are dealing with a sparse matrix, and special methods of solution such as the Lanczos algorithm may be applied (Wilkinson, 1965; Heine, 1980). A second fact we can exploit is that the physical quantities of interest are usually local. At least in metals, the force on an atom cannot depend much on where other atoms are if they are further than a few nanometres away. This is because electric fields, which by the Hellmann–Feynman theorem tell us what the force is, are efficiently screened out by conduction electrons. More abstractly, the local densities of states are not sensitive to what the atoms are doing when they are more than a few nanometers away. The energy Tr $\hat{\rho}\hat{H}$ according to (1.183) can therefore be expressed as a sum of spacially localized contributions, which when they are far apart are independent of each other. It accords with our macroscopic intuition that the energy of a large system, like all the extensive quantities we use in thermodynamics, is the sum of its parts. This has motivated people to develop algorithms which are $O(N)$. These also lead to an appealing real space description of interatomic potentials, although not pairwise ones.

One method for dealing with large systems is much older than the others, namely the use of periodic boundary conditions. This is the method of choice for a crystalline material, in which the Bloch condition (1.97) can be applied. In a perfect crystal, the supercell is usually rather small; it is just the unit cell of the crystal. If we are dealing with a large system, such as we need to represent a defective crystal, it may still be advantageous to impose periodic boundary conditions, so that all the quantities involved are defined within a supercell of volume V_c. It is instructive to consider how this works in more detail with a local basis. As described in Section 1.4.5, the eigenfunctions can be labelled by a **k**-vector in the first Brillouin zone of the supercell and have the form of a periodic function $u_{kn}(\mathbf{r})$ modulated by a plane wave $\exp(i\mathbf{k}\cdot\mathbf{r})$. This is unsuitable for a local orbital representation, since we will get into difficulties when we try to represent the plane wave in terms of local orbitals. We need to look to the alternative formulation of the Bloch condition (1.98) in terms of the phase relation between the wave function at points separated by a lattice vector $\mathbf{R}$. Now with the following expression we can specify the entire wave function $\psi_{kn}(\mathbf{r})$ in all supercells by a set of coefficients of the local orbitals, where the coefficients run over just one supercell:

$$\psi_{kn}(\mathbf{r}) = \frac{1}{\sqrt{V}} \sum_{\mathbf{L}} \sum_{I\mu} \exp(i\mathbf{k}\cdot\mathbf{L}) C^{kn}_{I\mu} \phi_{I\mu}(\mathbf{r}-\mathbf{L}). \tag{1.238}$$

The first summation runs over all supercells, $\mathbf{L}$ being a lattice vector, and the second summation runs over all the atoms and associated orbitals within the supercell. If we shift $\mathbf{r}$ by a lattice vector $\mathbf{R}$, and define $\mathbf{L}' = \mathbf{L} - \mathbf{R}$, we find

$$\begin{aligned}
\psi_{kn}(\mathbf{r}+\mathbf{R}) &= \frac{1}{\sqrt{V}} \sum_{\mathbf{L}} \sum_{I\mu} \exp(i\mathbf{k}\cdot\mathbf{L}) C^{kn}_{I\mu} \phi_{I\mu}(\mathbf{r}+\mathbf{R}-\mathbf{L}) \\
&= \frac{1}{\sqrt{V}} \sum_{\mathbf{L}'} \sum_{I\mu} \exp(i\mathbf{k}\cdot\mathbf{R}) \exp(i\mathbf{k}\cdot\mathbf{L}') C^{kn}_{I\mu} \phi_{I\mu}(\mathbf{r}-\mathbf{L}') \\
&= \exp(i\mathbf{k}\cdot\mathbf{R}) \psi_{kn}(\mathbf{r}),
\end{aligned} \tag{1.239}$$

showing that the wave function in this form satisfies the Bloch condition (1.98). We will not be making further use of the periodic function $u_{kn}(\mathbf{r})$, but note for completeness that it is specified by

$$u_{kn}(\mathbf{r}) = \sum_{\mathbf{L}} \sum_{I\mu} \exp\left(i\mathbf{k}\cdot(\mathbf{L}-\mathbf{r})\right) C^{kn}_{I\mu} \phi_{I\mu}(\mathbf{r}-\mathbf{L}). \tag{1.240}$$

The wavefunction (1.238) is sometimes introduced as a summation over the set of Bloch states $\phi^{k}_{I\mu}$:

$$\psi_{kn}(\mathbf{r}) = \sum_{I\mu} C^{kn}_{I\mu} \phi^{k}_{I\mu}(\mathbf{r}), \tag{1.241}$$

where the Bloch states are defined by

$$\phi^{k}_{I\mu}(\mathbf{r}) = \frac{1}{\sqrt{V}} \sum_{\mathbf{L}} \exp\left(i\mathbf{k}\cdot\mathbf{L}\right)) \phi_{I\mu}(\mathbf{r}-\mathbf{L}). \tag{1.242}$$

A matrix Schrödinger equation can now be derived in a similar way to the steps between (1.233) and (1.235), we just need to replace each local orbital by its corresponding Bloch state. The result is

$$\sum_{J\nu} H^k_{I\mu J\nu} C^{kn}_{J\nu} = \epsilon_{kn} \sum_{J\nu} S^k_{I\mu J\nu} C^{kn}_{J\nu}, \tag{1.243}$$

where

$$H^k_{I\mu J\nu} = \sum_{\mathbf{L}} \exp(i\mathbf{k} \cdot \mathbf{L}) \int \phi^*_{I\mu}(\mathbf{r}) \hat{H} \phi_{J\nu}(\mathbf{r} - \mathbf{L})\, d\mathbf{r} \tag{1.244}$$

and

$$S^k_{I\mu J\nu} = \sum_{\mathbf{L}} \exp(i\mathbf{k} \cdot \mathbf{L}) \int \phi^*_{I\mu}(\mathbf{r}) \phi_{J\nu}(\mathbf{r} - \mathbf{L})\, d\mathbf{r}. \tag{1.245}$$

In the periodic system these matrices superscripted with k now play the role that the unsuperscripted matrices played in the finite system. Notice that for $k \neq 0$ the matrix elements are in general complex, even when the orbitals are real. In a large supercell containing many atoms, large that is compared to the range of the orbitals, most of the important matrix elements will be from overlapping orbitals within the same supercell. The overlaps between orbitals in different supercells will be weak unless the atoms happen to be close to each other on either side of the border between supercells. That is to say, most of the matrix elements will be just the original $H_{I\mu J\nu}$ and $S_{I\mu J\nu}$ of a finite system, and the value of $\mathbf{k}$ won't make much difference. This is just another aspect of the fact that the Brillouin zone for a large supercell is correspondingly small, so all the independent $\mathbf{k}$'s are close to the Γ-point ($\mathbf{k} = 0$). It therefore becomes a good approximation for large supercells to neglect the dispersion altogether and accurate calculations of the total energy and forces can be made assuming $\mathbf{k} = 0$ for all the wavefunctions. There are no general rules about whether tens, or hundreds of atoms per supercell are required for this to be reasonable, it is a matter that has to be tested on each particular system. However, usually metallic systems require a more detailed sampling of the Brillouin zone than semiconductors or insulators, because details of the Fermi surface are important for their properties.

Brillouin zone sampling is an important subject in its own right. The point is that many of the properties we calculate require us to sum over all $\mathbf{k}$-points in the Brillouin zone. Since the summation is over points which are infinitely close together, this really means we should be integrating some function of $\mathbf{k}$ over the volume of the Brillouin zone. This is best done, as for any numerical integration, by a judicious choice of points and weights over the range of the integral. The functions such as the wavefunction coefficients and eigenvalues are therefore normally evaluated on a mesh of $\mathbf{k}$-points which are often called *special points* (Monkhorst and Pack, 1976). By evaluating matrices and wavefunctions only at the Γ-point for example we are choosing a single special point. In fact in this particular case it is noteworthy that a better sampling in a cubic Brillouin zone would be obtained by using $\mathbf{k} = (1/4, 1/4, 1/4)$ in units of the side of the BZ, rather than $\mathbf{k} = 0$. The subject quickly becomes rather technical and we will not be concerned further with it in this book.

Exercise
Derive (1.243) along the lines of the derivation of (1.235).

1.11 Pseudopotentials

1.11.1 Basic Ideas

For many properties of materials, the inner shell, or core electrons are not explicitly relevant to the properties of the material. Changes in chemical bonding, and in condensed matter changes in the density of states, provide the driving forces for atomic rearrangement, involving energy scales from millivolts to a few electron volts. These changes involve the outer electrons (valence or conduction electrons). In large atoms, it would make very little difference to the world of materials if the $1s$, $2s$ and more electrons were simply replaced by some spherically symmetric distribution of negative charge. This would also make it easier to solve the Kohn-Sham equations, since fewer electrons means fewer variables, which means that all stages of the solution would be quicker and easier. So why not simply take the density of core electrons ρ^{core} calculated from a single atom, calculate its Hartree potential $V_{\text{H}}^{\text{core}}$ and add this in with the original external potential from the nuclei? Well, that would be a very crude pseudopotential, and one can do a lot better. A good introduction to the ideas of pseudopotential theory is given by Ziman (1972). An electron in an outer orbital, or propagating through the conduction band of a solid, sees not only the Hartree potential of the core electrons, it exchanges with them. In other words its wavefunction wriggles in such a way as to maintain orthogonality.

Let us think about Al as a standard example to illustrate some of the main ideas. A $2s$ orbital in an Al atom $\phi_{2s}(r)$ has a node, a value of r at which its sign changes, which allows it to be orthogonal to the $1s$ electrons, whose wavefunction is nodeless. The outermost s-electron, labelled $3s$, has a wavefunction $\phi_{3s}(r)$ with two nodes and its orbital is orthogonal to ϕ_{1s} and ϕ_{2s}. The $2p$ electrons have an orbital

$$\phi_{2p}(\mathbf{r}) = f_{2p}(r)Y_{1m}(\theta, \phi). \tag{1.246}$$

p-orbitals are automatically orthogonal to s-orbitals by virtue of their angular symmetry. Remember p-orbitals consist of two lobes of opposite sign on either side of the origin, so f_{2p} does not need to have a node.

Now a pseudopotential replaces the true single-particle wavefunction within some core radius R_τ by a *pseudowavefunction* which does not have the orthogonalizing wiggles. This is accomplished by replacing the true effective potential V_{eff} by the sum of a bare pseudopotential $v_\tau(\mathbf{r}, \mathbf{r}')$. for each ion of type τ, together with the potential $V_{\text{H}} + V_{\text{XC}}$ generated by the other pseudowavefunctions. The determination of the bare pseudopotential, which is by no means a uniquely defined potential, is carried out by calculations on the free atom, and there are various recipes for doing this. Because the pseudowavefunction is smoother than the true wavefunction within the core region its electron has less kinetic energy there. Since its total energy is the same as that of the true wavefunction, its potential energy within the core is correspondingly higher. That is to say, the pseudopotential is shallower than the true potential, and it does away completely with the $1/r$ singularity at the nucleus.

The matrix elements of a pseudopotential in **k**-space can be defined in the usual way by the Fourier transformation of the **r**-space represention, in the manner of (1.64) and (1.65). It will be convenient to redefine the pseudopotential in **k**-space by scaling it with the number of atoms so as to make it a finite quantity in the limit $V \to \infty$. Thus the normalised pseudopotential in **k**-space becomes

$$V_\tau(\mathbf{k}, \mathbf{k} + \mathbf{q}) = \frac{1}{\Omega} \int \exp(i(\mathbf{k} \cdot (\mathbf{r} - \mathbf{r}') - \mathbf{q} \cdot \mathbf{r}'))v_\tau(\mathbf{r}, \mathbf{r}')\, d\mathbf{r}\, d\mathbf{r}'. \tag{1.247}$$

where we have introduced the volume per atom $\Omega = V/N_{\text{a}} = V_{\text{c}}/N_{\text{c}}$. For accurate work non-local pseudopotentials are necessary. They can be represented in the space of atomic orbitals, and can have a different core radius for each angular momentum component. For qualitative purposes it is sufficient to use a simple, local potential or model potential.

1.11.2 The Ashcroft Empty-core Pseudopotential

One of the simplest forms of local model potential was introduced by Ashcroft. It is defined by:

$$v_\tau(r) = -\frac{Z_\tau}{r}\Theta(r - R_\tau), \tag{1.248}$$

where Θ denotes the Heaviside function. Within a core radius R_τ the potential is zero, while outside R_τ it reverts to the Coulomb potential.

Let us first obtain the equation for the pseudopotential in **k**-space. From (1.247) we have

$$\begin{aligned} V_\tau(\mathbf{q}) &= \frac{1}{\Omega} \int\limits_{r,r'>R_\tau} \exp(i(\mathbf{k} \cdot (\mathbf{r} - \mathbf{r}') - \mathbf{q} \cdot \mathbf{r}')) \left(-\frac{Z_\tau}{r}\delta(\mathbf{r} - \mathbf{r}')\right) d\mathbf{r}\, d\mathbf{r}' \\ &= -\frac{2\pi Z_\tau}{\Omega} \int_{R_\tau}^{\infty} \int_0^{\pi} \frac{\exp(-iqr\cos\theta) r^2 \sin\theta}{r}\, dr\, d\theta. \end{aligned} \tag{1.249}$$

This is an integral we have dealt with before. The θ integral is straightforward, but it is necessary to 'damp' the Coulomb interaction by making it temporarily a Yukawa potential, as we did in Section 1.3.4. The result is

$$V_\tau(q) = -\frac{4\pi}{\Omega q^2} Z_\tau \cos(q R_\tau). \tag{1.250}$$

We see that this is just the Fourier transform of the Coulomb potential, modulated by the cosine factor. At low q it tends asymptotically to the bare Coulomb potential $-4\pi Z_\tau/(\Omega q^2)$, which is the direct result of the $1/r$ tail of the potential in $\mathbf{r}$-space. Different elements are only distinguished by the ionic charge and the empty core radius, and the latter has been used as a fitting parameter in many studies (Cohen *et al.*, 1970). We shall use this pseudopotential later to illustrate the ideas of pairwise potentials in simple metals.

2

ESSENTIAL DENSITY FUNCTIONAL THEORY

The problem of calculating the total energy in the ground state of matter, and the corresponding interatomic forces, is a 'very many body indeed' problem, one of daunting complexity. Astonishingly, density functional theory was able to map this problem onto the problem of calculating the wavefunction and energy of a single electron, with no approximations! Of course, in any practical scheme of calculation, approximations do have to be made, as we shall see. In view of the underlying importance of density functional theory to so many kinds of models, I offer in this chapter a fairly self-contained description of it.

2.1 What is a Functional?

I briefly introduced an example of a functional in connection with minimizing the single-particle energy as a functional of a wave function, eq. (1.149). In density functional theory, we will be dealing with several functionals of the electron density, and it is worth spending a little time to summarize some properties of functionals in order to understand the manipulations that need to be done. This treatment follows that of Parr and Yang (1989). A very readable introduction to the modern theory is the book by Koch and Holthauser (2001).

We can think of a *function* as a rule for going from a variable which may be a scalar or a vector, to a number, such as $f(x)$ or $f(\mathbf{r})$. For example, x^2 and the scalar product $\mathbf{r} \cdot \mathbf{r}$ are functions of a scalar and a vector, respectively. In this sense a function is a mapping. In a similar way a *functional* is a rule for going from a function to a number. The argument of a functional is enclosed in square brackets to distinguish it from a function; thus $F[f]$ is a functional of the function f. The functional in (1.149) could be written as $\epsilon[\chi]$.

The central idea of density functional theory is that for a system of N electrons a functional $E[\rho]$ exists, such that when E is minimized with respect to variations in the electron density ρ, its value is the ground state energy of the system. The minimization of E must be carried out subject to the constraint

$$\int \rho(\mathbf{r})\, \mathrm{d}\mathbf{r} = N. \tag{2.1}$$

The rest, as they say, is history. Much of the theory of Hohenberg and Kohn and subsequent density functional theory was concerned with the existence of the functional $E[\rho]$ to be minimized with respect to density; its existence was quite a problem in itself, before even the question of what that functional actually looked like. The difficulties and solutions which emerged are fascinating, but for the purposes of this book I shall give a rather simplified version of the proof, to the extent that this is instructive.

To understand the nature of this theory and how one can proceed from it to the nature of interatomic forces we need a little familiarity with the mathematics of functionals.

2.2 Functional Derivatives

Consider making a small change to a function $f(x)$. We can write this change as $\epsilon\phi(x)$, where ϵ is a small parameter which we are going to let tend to zero and ϕ is some other function, which is completely arbitrary, except that it should be bounded. We now ask what is the corresponding change in the functional $F[f]$? In general this is

$$\Delta F = F[f + \epsilon\phi] - F[f]. \tag{2.2}$$

If the functions are not pathological, part of ΔF is proportional to ϵ, and this part becomes closer and closer to 100% of ΔF as ϵ gets smaller and smaller. There is now an analogy to defining the differentiation of functions if the limit $\lim_{\epsilon\to 0}(\Delta F/\epsilon)$ can be taken. But in this case *each* value of x may contribute to ΔF with its own weight. If the limit exists, we can *define* the functional derivative of F by the relation:

$$\lim_{\epsilon\to 0}\frac{\Delta F}{\epsilon} = \int_{-\infty}^{+\infty}\frac{\delta F}{\delta f(x)}\phi(x)\,\mathrm{d}x. \tag{2.3}$$

The quantity $\delta F/\delta f(x)$ is the functional derivative of F at x. Unlike a simple derivative of a function, this quantity may depend on the value of f at points other than x. That is to say, $\delta F/\delta f(x)$ itself is in general also a functional. Exactly the same formulae apply if we consider three-dimensional integrals, we must only replace x and $\mathrm{d}x$ by $\mathbf{r}$ and $\mathrm{d}\mathbf{r}$.

The integral in (2.3) is reminiscent of a scalar product, or more particularly of the definition of the derivative of a function of many variables $F(f_1, f_2, \ldots)$. In this case, if we make changes to the fs by arbitrary amounts $f_1 \to f_1 + \epsilon\phi_1$, $f_2 \to f_2 + \epsilon\phi_2$, etc., we would have the usual formula for the total differential of a function:

$$\Delta F = \frac{\partial F}{\partial f_1}\epsilon\phi_1 + \frac{\partial F}{\partial f_2}\epsilon\phi_2 + \cdots \tag{2.4}$$

and so

$$\lim_{\epsilon\to 0}\frac{\Delta F}{\epsilon} = \frac{\partial F}{\partial f_1}\phi_1 + \frac{\partial F}{\partial f_2}\phi_2 + \cdots \tag{2.5}$$

The formula (2.3) is rather like taking the limit of eq. (2.5) when 1, 2, etc. label adjacent values of x, the separation of which tends to zero and the total number of which tends to infinity.

As an important example of how to do a functional derivative in practice, consider the Hartree energy (1.161). We consider a small change $\delta\rho(\mathbf{r}) = \epsilon\Delta\rho(\mathbf{r})$ where $\Delta\rho$ is an arbitrary bounded function, and see what effect it has on $E_{\mathrm{H}}[\rho]$. Later we can impose the constraint of charge conservation on such changes $\delta\rho$ by means of a Lagrange multiplier, but for now let this not be a restriction. By substituting $\rho + \delta\rho$ for ρ in (1.161) and ignoring terms of order ϵ^2 we find:

$$E_{\mathrm{H}}[\rho+\delta\rho] - E_{\mathrm{H}}[\rho] = \Delta E_{\mathrm{H}} = \frac{1}{2}\int\left\{\frac{\rho(\mathbf{r})\delta\rho(\mathbf{r}') + \delta\rho(\mathbf{r})\rho(\mathbf{r}')}{|\mathbf{r}-\mathbf{r}'|}\right\}\mathrm{d}\mathbf{r}\,\mathrm{d}\mathbf{r}'. \tag{2.6}$$

The two terms in the braces make identical contributions; to see this we only need to interchange the dummy arguments $\mathbf{r}$ and $\mathbf{r}'$ in either term. The result is therefore:

$$\Delta E_{\mathrm{H}} = \int\frac{\rho(\mathbf{r}')\delta\rho(\mathbf{r})}{|\mathbf{r}-\mathbf{r}'|}\,\mathrm{d}\mathbf{r}\,\mathrm{d}\mathbf{r}' = \int V_{\mathrm{H}}(\mathbf{r})\delta\rho(\mathbf{r})\mathrm{d}\mathbf{r}, \tag{2.7}$$

where we have used the definition (1.157) of the Hartree potential V_{H}. Dividing both sides by ϵ and taking the limit as $\epsilon \to 0$:

$$\lim_{\epsilon\to 0}\frac{\Delta E_{\mathrm{H}}}{\epsilon} = \int V_{\mathrm{H}}(\mathbf{r})\Delta\rho(\mathbf{r})\,\mathrm{d}\mathbf{r}. \tag{2.8}$$

Hence from the definition of a functional derivative (2.3)

$$\frac{\delta E_{\mathrm{H}}[\rho]}{\delta\rho(\mathbf{r})} = V_{\mathrm{H}}(\mathbf{r}). \tag{2.9}$$

It is straightforward to define second functional derivatives. To evaluate the second functional derivative of E_{H} for example we have to remember that its first functional derivative $V_{\mathrm{H}}(\mathbf{r})$ is also a functional of ρ, which is given by (1.157). Then we find

$$\frac{\delta^2 E_{\mathrm{H}}[\rho]}{\delta\rho(\mathbf{r})\delta\rho(\mathbf{r}')} = \frac{1}{|\mathbf{r}-\mathbf{r}'|}. \tag{2.10}$$

Given suitable conditions of smoothness, a functional can be expanded about some function by analogy to a Taylor expansion, e.g.:

$$F[\rho+\delta\rho] = F[\rho]+\int \frac{\delta F}{\delta\rho(\mathbf{r})}\delta\rho(\mathbf{r})\,\mathrm{d}\mathbf{r}+\frac{1}{2}\int \frac{\delta^2 F}{\delta\rho(\mathbf{r})\delta\rho(\mathbf{r}')}\delta\rho(\mathbf{r})\delta\rho(\mathbf{r}')\,\mathrm{d}\mathbf{r}\,\mathrm{d}\mathbf{r}'+\cdots \tag{2.11}$$

We say the above series has been truncated to *second order* in $\delta\rho$. A term which is of the form

$$\int\int \frac{\delta^2 F}{\delta\rho_1(\mathbf{r})\delta\rho_2(\mathbf{r}')}\delta\rho_1(\mathbf{r})\delta\rho_2(\mathbf{r}')\,\mathrm{d}\mathbf{r}\,\mathrm{d}\mathbf{r}' + \text{higher order terms}$$

we shall refer to as *of order* $\delta\rho_1\delta\rho_2$, abbreviated as $O[\delta\rho_1\delta\rho_2]$. This nomenclature is extended also to terms of higher order, e.g. the terms omitted in the expansion (2.11) are $O[\delta\rho^3]$.

Often in this field a functional to be differentiated has the form:

$$F[\rho] = \int f(\rho(\mathbf{r}))\,\mathrm{d}\mathbf{r}, \tag{2.12}$$

where f is an ordinary function of the electron density. In this case it is easy to show that

$$\frac{\delta F[\rho]}{\delta\rho(\mathbf{r})} = \frac{\partial f(\rho(\mathbf{r}))}{\partial\rho(\mathbf{r})}, \tag{2.13}$$

so the functional derivative becomes a simple partial derivative.

Finding the stationary point of a functional $E[\rho]$ subject to the constraint (2.1) is a typical problem in variational calculus. It can be formulated as a differential equation, by analogy with finding the minimum of a function, namely:

$$\frac{\delta}{\delta\rho(\mathbf{r})}\left[E[\rho]-\mu\left(\int\rho(\mathbf{r})\,\mathrm{d}\mathbf{r}-N\right)\right] = 0, \tag{2.14}$$

leading to the *Euler–Lagrange equation*

$$\frac{\delta E[\rho]}{\delta\rho(\mathbf{r})} = \mu. \tag{2.15}$$

The quantity μ is a *Lagrange multiplier* which as usual in such problems has a very physical interpretation. It is the energy increase of the system on adding one more electron, which when the system is in its ground state has to be a constant, independent of $\mathbf{r}$.

2.3 The Thomas–Fermi Model

2.3.1 Description of the Thomas–Fermi Functional

Nowadays when people talk about density functional theory, they have in mind the classic papers of Hohenberg and Kohn (1964) and Kohn and Sham (1965). However, the earliest density functional approach for systems of electrons in an external potential provided by nuclei is the Thomas–Fermi model, proposed by Thomas (1927) and elaborated by Fermi (1927). It is also the simplest density functional theory, and a good way to learn about several concepts in interatomic forces. The idea is as follows.

The energy of a real inhomogeneous electron gas is written as a functional of the density $\rho(\mathbf{r})$ such that for each element of volume $d\mathbf{r}$ the electrons in that volume are supposed to have the kinetic energy per electron of a jellium with a *homogeneous* density equal to the density at the point $\mathbf{r}$, namely $\rho(\mathbf{r})$. To this is added the electrostatic energy $E_{\mathrm{H}} + E_{eZ} + E_{ZZ}$. Thus the total energy of the system is

$$E_{\mathrm{TF}}[\rho] = \int \rho(\mathbf{r}) t_s(\rho(\mathbf{r}))\, d\mathbf{r} + E_{\mathrm{H}}[\rho] + E_{eZ}[\rho] + E_{ZZ}. \tag{2.16}$$

The first term is the kinetic energy functional, in which the kinetic energy per electron is given by the jellium formula (1.229). The electrostatic terms are the usual functionals (1.161), (1.15) and the pairwise summation (1.218). The ground state energy for given positions of the nuclei is obtained by minimizing the functional $E_{\mathrm{TF}}[\rho]$ with respect to variations in the density, subject to the constraint of charge conservation (2.1).

A more sophisticated energy functional is the so-called Thomas–Fermi–Dirac functional, which is obtained by including Dirac's exchange energy, eq. (1.230), in the same way as the kinetic energy:

$$E_{\mathrm{TFD}}[\rho] = \int \rho(\mathbf{r}) t_s(\rho(\mathbf{r}))\, d\mathbf{r} + \int \rho(\mathbf{r}) \epsilon_x(\rho(\mathbf{r}))\, d\mathbf{r} + E_{\mathrm{H}}[\rho] + E_{eZ}[\rho] + E_{ZZ}. \tag{2.17}$$

2.3.2 The Euler–Lagrange Equation

To minimize the Thomas–Fermi energy with respect to ρ we derive the corresponding Euler–Lagrange equation. It is just as easy to do this with the Dirac exchange term included, so let us start with the formula (2.17), including all the density dependence explicitly:

$$E_{\text{TFD}}[\rho] = \frac{3}{10}(3\pi^2)^{2/3} \int \rho(\mathbf{r})^{5/3}\,d\mathbf{r} - \frac{3}{4}\left(\frac{3}{\pi}\right)^{1/3} \int \rho(\mathbf{r})^{4/3}\,d\mathbf{r}$$
$$+ \frac{1}{2}\int \frac{\rho(\mathbf{r})\rho(\mathbf{r}')}{|\mathbf{r}-\mathbf{r}'|}\,d\mathbf{r}\,d\mathbf{r}' + \int \rho(\mathbf{r})V_{\text{ext}}(\mathbf{r})\,d\mathbf{r} + E_{ZZ}. \quad (2.18)$$

Differentiating with respect to ρ gives us the required equation:

$$\frac{\delta E_{\text{TFD}}[\rho]}{\delta\rho(\mathbf{r})} = \frac{1}{2}(3\pi^2)^{2/3}\rho(\mathbf{r})^{2/3} - (3/\pi)^{1/3}\rho(\mathbf{r})^{1/3} + \int \frac{\rho(\mathbf{r}')}{|\mathbf{r}-\mathbf{r}'|}d\mathbf{r}' + V_{\text{ext}}(\mathbf{r})$$
$$= \frac{1}{2}(3\pi^2)^{2/3}\rho(\mathbf{r})^{2/3} + V_{\text{X}}(\mathbf{r}) + V_{\text{H}}(\mathbf{r}) + V_{\text{ext}}(\mathbf{r}), \quad (2.19)$$

where we have introduced the *exchange potential* $V_{\text{X}}(\mathbf{r})$ defined here as the functional derivative of the exchange energy. In this case the exchange potential is obtained from the Dirac formula for the exchange energy density, which gives the explicit formula

$$V_{\text{X}}(\mathbf{r}) = -\left(\frac{3}{\pi}\right)^{1/3} \rho(\mathbf{r})^{1/3}. \quad (2.20)$$

The Euler–Lagrange equation (2.15) takes the form:

$$\tfrac{1}{2}(3\pi^2)^{2/3}\rho(\mathbf{r})^{2/3} + V_{\text{eff}}(\mathbf{r}) = \mu_{TFD}, \quad (2.21)$$

in which we have defined a new effective potential as the sum of the three potentials in (2.19). This equation has to be solved self-consistently, since V_{eff} is a functional of ρ, and the solution must be carried out by an iterative numerical method except in very simple cases such as a single atom.

The Thomas–Fermi model is interesting and instructive, and it has evolved further than simply the addition of Dirac exchange to include more exact functionals. However it has fundamental limitations which prevent it from being really useful as a quantitative method for the calculation of interatomic forces. Even with the addition of a correlation functional, a major failing of the Thomas–Fermi model is that it always predicts that atoms repel each other! This statement can be proved rigorously, but its plausibility will become obvious later when we start considering how to calculate interatomic forces.

2.3.3 The Local Density Approximation

This way of dealing with kinetic energy or exchange energy, namely by treating it as a function of the *local* electron density, and integrating over the system, is called a *local density approximation* (LDA) for the quantity. The most serious approximation of the Thomas–Fermi model is the LDA for the kinetic energy of the electrons. This is put right by the Hohenberg–Kohn–Sham (HKS) density functional theory, as we shall see. However, the more modern HKS theory, at least in its original form and one that is still very widely used, still makes use of the LDA for both the exchange and correlation contributions to the total energy.

2.4 The Kohn–Sham Equations

A major and inescapable error of the Thomas–Fermi approach and its refinements still resides in the LDA for the kinetic energy. This was partially corrected by including a term in the kinetic energy functional depending on the gradient of the electron density. The correction, called a *Weizsacker correction,* led to a family of functionals, described by Parr and Yang (1989), in which the kinetic energy density was augmented by a gradient term. None however has proved to match the accuracy of the Kohn–Sham functional, which started from the idea of building a good non-local kinetic energy functional in a completely novel way.

2.4.1 The Existence of a Density Functional

The following proposition is a more general version of Hohenberg and Kohn's original insight. These ideas are discussed in Parr and Yang's book. *For any reasonable density $\rho(\mathbf{r})$ there is an antisymmetric wave function $|\Psi\rangle$ describing N electrons whose density is $\rho(\mathbf{r})$.* There is no suggestion that there is only one such antisymmetric wave function. 'Reasonable' here means that the function $\rho(\mathbf{r})$ is non-negative, continuous and normalized (Parr and Yang, 1989). Such a density is referred to as 'N-representable'. In addition, this wave function may be the solution of the Schrödinger equation for these electrons if they move in an external potential $v(\mathbf{r})$. In the latter case the density is referred to as 'v-representable'. Let us take it for granted that this very plausible assumption is indeed true. The proposition is also true if the electrons are of the hypothetical variety that do not interact with each other, in which case $\langle \mathbf{x}_1, \ldots \mathbf{x}_N|\Psi\rangle$ is a Slater determinant.

Now recall the variational principle, (1.150). This tells us that for N electrons

$$\langle\Psi|\hat{T} + \frac{1}{2}\sum_i\sum_j \frac{1}{|\mathbf{r}_i - \mathbf{r}_j|} + \sum_i V_{\text{ext}}(\mathbf{r}_i)|\Psi\rangle \geq E_0, \tag{2.22}$$

where E_0 is the ground state energy. Since we have specified that the wavefunctions are to produce the same density ρ, from (1.15) we can write (2.22) as

$$\langle\Psi|\hat{T} + \frac{1}{2}\sum_i\sum_j \frac{1}{|\mathbf{r}_i - \mathbf{r}_j|}|\Psi\rangle + \int \rho(\mathbf{r})V_{\text{ext}}(\mathbf{r})d\mathbf{r} \geq E_0 \tag{2.23}$$

and the last term on the left-hand side does not depend explicitly on Ψ, just on ρ.

The equality in (2.22) and (2.23) only holds when the wavefunction is the exact ground state wavefunction, in which case ρ is the exact ground state density. The first term of (2.23) does depend on the particular Ψ which is chosen, supposing there exist more than one. We can make this term unique if we choose whatever wavefunction minimizes it, subject as usual to constant N but in this case also to constant ρ. This defines a functional

$$F[\rho] = \min_{\Psi \to \rho} \langle \Psi | \hat{T} + \frac{1}{2} \sum_i \sum_j \frac{1}{|\mathbf{r}_i - \mathbf{r}_j|} | \Psi \rangle = T[\rho] + E_{ee}[\rho], \tag{2.24}$$

where $\Psi \to \rho$ indicates that the minimization is with respect to all Ψ which generate the given density ρ. Having defined $F[\rho]$ we see that the functional we seek having the property that its minimum value is E_0 when ρ is the exact ground state density is:

$$E[\rho] = F[\rho] + E_{eZ}[\rho], \tag{2.25}$$

where E_{eZ} is the functional specified by (1.15). Thus in the ground state this functional has to satisfy

$$\frac{\delta E[\rho]}{\delta \rho(\mathbf{r})} = \mu = \frac{\delta F[\rho]}{\delta \rho(\mathbf{r})} + \frac{\delta E_{eZ}[\rho]}{\delta \rho(\mathbf{r})}. \tag{2.26}$$

Since the functional derivative of $E_{eZ}[\rho]$ is simply $V_{\text{ext}}(\mathbf{r})$ the ground state condition becomes

$$\frac{\delta F[\rho]}{\delta \rho(\mathbf{r})} + V_{\text{ext}}(\mathbf{r}) = \mu. \tag{2.27}$$

2.4.2 The Hohenberg–Kohn–Sham Functional

The next crucial step, taken by Kohn and Sham, is to make use of the properties of a reference system of non-interacting electrons with density ρ. In this reference system the functional F contains only the kinetic energy, which we write as $T_{\text{s}}[\rho]$. The HKS functional F^{HKS} includes T_{s} to represent the kinetic energy part and takes the form:

$$F^{\text{HKS}}[\rho] = T_{\text{s}}[\rho] + E_{\text{H}}[\rho] + E_{\text{xc}}[\rho]. \tag{2.28}$$

The term T_{s} was also included in the Thomas–Fermi functional, where it was treated within the LDA. This local approximation to the major part of the kinetic energy is avoided in the Kohn–Sham approach, as we shall see. If we think of the true kinetic energy functional $T[\rho]$ as $T_{\text{s}}[\rho] + \Delta T$, the difference ΔT must be included in the exchange and correlation functional $E_{\text{xc}}[\rho]$, which will have to be approximated. So far nothing seems to have been achieved except some more definitions, and these would be of no interest were it not for two related facts. The most important is that there is a beautiful route to the calculation of T_{s}. The other is that E_{xc}, which contains besides the actual exchange and correlation the rather small error in T_s, is amenable to rather good approximations, of which the most famous is the LDA, based on the energy of jellium.

The HKS functional is therefore

$$E^{\mathrm{HKS}}[\rho] = T_s[\rho] + E_{\mathrm{H}}[\rho] + E_{\mathrm{xc}}[\rho] + E_{eZ}[\rho] + E_{ZZ}. \tag{2.29}$$

If the number of atoms is infinite, this energy would be infinite just as in jellium. In that case we have to refer it to a system within a volume V or V_c which is periodically repeated. Even so, the electrostatic energy would diverge as V^2, as it did in the case of jellium, were it not for the cancellations brought about by including the nucleus–nucleus interactions E_{ZZ}. The inclusion of E_{ZZ} therefore not only completes the description of the total energy, it ensures that we are dealing with the electrostatic energy of a system which is electrically neutral overall.

The ground state is found by setting the functional derivative of the HKS functional to zero, including the constant number of electrons with the Lagrange multiplier μ:

$$\frac{\delta E^{\mathrm{HKS}}}{\delta\rho} = \frac{\delta T_s}{\delta\rho} + V_{\mathrm{H}}(\mathbf{r}) + V_{\mathrm{xc}}(\mathbf{r}) + V_{\mathrm{ext}}(\mathbf{r}) = \mu, \tag{2.30}$$

where V_{xc} is defined as the functional derivative of E_{xc}:

$$V_{\mathrm{xc}}(\mathbf{r}) = \frac{\delta E_{\mathrm{xc}}[\rho]}{\delta\rho(\mathbf{r})}. \tag{2.31}$$

2.4.3 The Kohn and Sham Trick

Before considering the problem of what E_{xc} and hence V_{xc} actually is, we still have the problem of how to calculate $T_s[\rho]$, without which we cannot possibly solve (2.30). But in fact we can proceed without ever knowing $T_s[\rho]$! Kohn and Sham noticed that the form of eq. (2.30) for the ground state density is identical to the equation for the ground state density of the reference system of non-interacting electrons in a potential V_{eff}, namely:

$$\frac{\delta T_s}{\delta\rho} + V_{\mathrm{eff}}(\mathbf{r}) = \mu, \tag{2.32}$$

and it will be exactly the same equation with the same solution for $\rho(\mathbf{r})$ if we define

$$V_{\mathrm{eff}}(\mathbf{r}) = V_{\mathrm{H}}(\mathbf{r}) + V_{\mathrm{xc}}(\mathbf{r}) + V_{\mathrm{ext}}(\mathbf{r}). \tag{2.33}$$

For the system of non-interacting electrons, V_{eff} plays the role of an external potential, they see no exchange or correlation potentials. We know how to solve eq. (2.32), at least in principle; there is a Schrödinger equation to be solved:

$$\left(-\tfrac{1}{2}\nabla^2 + V_{\text{eff}}(\mathbf{r})\right)\psi_n(\mathbf{r}) = \epsilon_n \psi_n(\mathbf{r}) \tag{2.34}$$

and from its solutions we can construct the kinetic energy as in HF theory

$$T_s = \sum_n f_n \langle n|\hat{T}|n\rangle = \sum_n f_n \int \psi_n^*(\mathbf{r})\left(-\frac{1}{2}\nabla^2\right)\psi_n(\mathbf{r})\,d\mathbf{r}, \tag{2.35}$$

and the charge density is given by

$$\rho(\mathbf{r}) = \sum_n f_n |\psi_n(\mathbf{r})|^2. \tag{2.36}$$

The single-particle Schrödinger equations in this context are known as the *Kohn–Sham equations*. No approximations have been made at all. So by solving a certain Schrödinger equation for a system of non-interacting electrons we will have solved the real problem of the system of interacting electrons! It seems like we are getting something for nothing, but the truth is that all the approximations which will have to be made in practice have now been swept into the functional $E_{\text{xc}}[\rho]$. Furthermore, the solution of the Kohn–Sham equations requires a process of iteration to self-consistency, as did the full Thomas–Fermi equation, because the effective potential V_{eff} itself depends on the charge density ρ which can only be obtained from the solutions ψ_n.

This trick leads to a useful way to express the HKS functional. First, consider the quantity $\int \rho(\mathbf{r})V_{\text{eff}}(\mathbf{r})\,d\mathbf{r}$ which we can write as

$$\int \rho(\mathbf{r})V_{\text{eff}}(\mathbf{r})\,d\mathbf{r} = \sum_n f_n \langle n|\hat{V}_{\text{eff}}|n\rangle. \tag{2.37}$$

As this formula shows, we can regard V_{eff} as an operator $\hat{V}_{\text{eff}}$ which is diagonal in the $\mathbf{r}$-representation. The same applies to the other potentials we have met so far; they are all *local* operators. Thus we can write the kinetic energy (2.35) as

$$\begin{aligned} T_s &= \sum_n f_n \langle n|\hat{T}|n\rangle \\ &= \sum_n f_n \langle n|\hat{T} + \hat{V}_{\text{eff}}|n\rangle - \int \rho(\mathbf{r})V_{\text{eff}}(\mathbf{r})\,d\mathbf{r}. \end{aligned} \tag{2.38}$$

By adding and subtracting (2.37) we have expressed T_s in terms of the total energy of the non-interacting electrons, less their interaction with the effective potential. This is true for any set of $|n\rangle$, that is when the $|n\rangle$ belong to any HF wavefunction, which is compatible with the charge density ρ according to (2.36). Thus, we have constructed a functional $T_s[\rho]$ which is not a functional purely of ρ but which also involves the states from which ρ is built. Since these states are in one-to-one correspondence with the determinental wavefunction, which in turn is determined by V_{eff} which in turn is determined by ρ, our T_s is still in effect a functional of ρ. At the minimum of $E^{\text{HKS}}[\rho]$, where the $|n\rangle$ satisfy the Kohn–Sham equations, it takes the value

$$T_s[\rho] = \sum_n f_n \epsilon_n - \int \rho(\mathbf{r}) V_{\text{eff}}(\mathbf{r})\, d\mathbf{r}. \tag{2.39}$$

There is a subtle difference between this version of the functional and a general kinetic energy functional. We are now, in theory, restricting ourselves to the class of ρ that are v-representable, because the associated wavefunction is determined by V_{eff}, rather than the larger class of ρ which are N-representable. However this theoretical limitation can be removed by a different derivation, which selects $T_s[\rho]$ as the smallest in magnitude among all non-interacting electron kinetic energies with the same density ρ (see chapter 7 of Parr and Yang if you are interested in these subtleties). We can now put together these ingredients into a modified form of the HKS functional, which after some trivial rearrangements, replacing V_{eff} with its definition (2.33), becomes

$$E^{\text{HKS}}[\rho] = \sum_n f_n \left\langle n|\hat{T} + \hat{V}_{\text{eff}}|n\right\rangle - \frac{1}{2}\int \rho V_{\text{H}} - \int \rho V_{\text{xc}} + E_{\text{xc}}[\rho] + E_{ZZ}. \tag{2.40}$$

The **r**'s and d**r**'s in the functions and integrals have been omitted for clarity. This formulation has several interesting features. Notice first that the second term, the first of the two integrals, is just the Hartree energy *with a negative sign!* This is because the interaction of the electrons with the Hartree potential already appears in the Hamiltonian term, where it counts this part of the electron–electron interaction twice. The second term can therefore be seen as a *double counting correction.* The third term subtracts off the exchange-correlation term in the Hamiltonian and all the exchange-correlation energy reappears in the fourth term E_{xc}.

The double counting correction makes obvious how the electrostatic divergence mentioned before is removed. Suppose that, as a first estimate, the charge density is represented by a superposition of spherically symmetric atomic charge densities. This density, let us call it ρ^{in}, is actually not a bad first approximation to ρ which we will make further use of. Thus if the atomic nuclei are labelled I and the associated charge densities ρ^I we have

$$\rho^{\text{in}}(\mathbf{r}) = \sum_I \rho^I(\mathbf{r}). \tag{2.41}$$

In that case the Hartree double counting correction can be divided into two parts:

$$-\frac{1}{2}\int \rho^{\text{in}}(\mathbf{r})V_{\text{H}}\left(\rho^{\text{in}}(\mathbf{r})\right)\,\mathrm{d}\mathbf{r} = -\sum_{I}\frac{1}{2}\int \rho^{I}V_{HI} - \sum_{I}\sum_{J\neq I}\int \frac{1}{2}\rho^{I}V_{HJ} \tag{2.42}$$

where

$$V_{HI}(\mathbf{r}) = \int \frac{\rho^{I}(\mathbf{r}')}{|\mathbf{r}-\mathbf{r}'|}\mathrm{d}\mathbf{r}'. \tag{2.43}$$

These parts represent the self-energy of the atomic charge densities and the electrostatic interaction between the atomic charge densities respectively. If the atoms are neutral, the interactions between the charges I and J described by (2.42) will, because of the negative sign, exactly cancel the direct Coulomb interaction between the nuclei $Z_I Z_J/|\mathbf{R}_I - \mathbf{R}_J|$ as the distance between the nuclei exceeds the range of overlap of the charge densities. Thus the nucleus–nucleus interaction is screened by the double counting correction. The electron-nucleus interaction is described within the Hamiltonian term where its electrostatic divergence is cancelled by the doubly counted electron–electron interaction!

2.4.4 Self-consistent Solution of the Kohn–Sham Equations

I have mentioned that although the single-particle Kohn–Sham equations (2.34) are a fantastic simplification of the original many-electron problem, there is still a self-consistency problem to be solved. This can be very difficult. It will be useful to sketch how it is done without going into details of the many ways in which it is implemented in practice, partly as in introduction to the approximate functionals that will be described later. The first step in all procedures is to guess the charge density. We will call the first guess at the charge density ρ^{in}. It will commonly be a superposition of computed atomic charge densities, but alternatives are possible, including even the uniform charge density of jellium. The subsequent steps are illustrated in Fig. 2.1.

The Hartree potential V_{H}^{in} must be obtained from ρ^{in} by solving the Poisson equation, a procedure which is usually carried out in k-space. The exchange-correlation potential $V_{\text{xc}}^{\text{in}}$ is also calculated from ρ^{in} directly using the LDA, or a more sophisticated functional such as a *generalized gradient approximation* (GGA) which is expressed in terms of gradients of the density as well as the density itself. These potentials are added to the fixed external potential to build the input effective potential $V_{\text{eff}}^{\text{in}}$. Solving the Schrödinger equation with the effective potential $V_{\text{eff}}^{\text{in}}$ gives a set of eigenstates $\{|n^{\text{out}}\rangle\}$. From these first eigenstates a corresponding output charge density ρ^{out} is constructed in the usual way. The HKS energy $E^{\text{HKS}}[\rho^{\text{out}}]$ can now be evaluated, and at this stage the first iteration is complete.

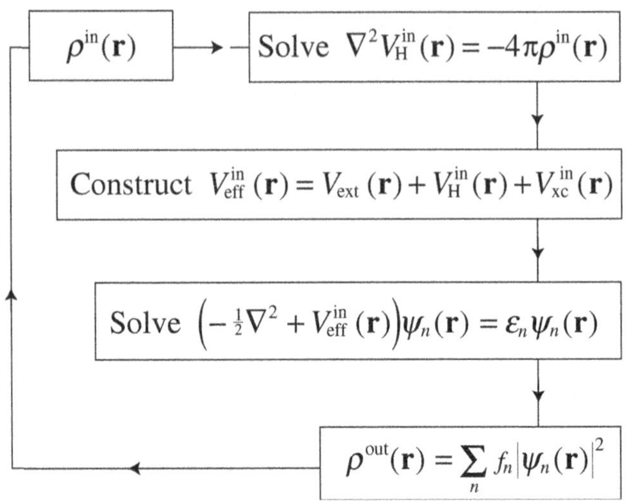

FIG. 2.1 The self-consistency loop.

To proceed, ρ^{out} must now be used to construct a new input charge density which we shall call ρ_1^{in}. The cycle can now be repeated, producing an output charge density ρ_1^{out} and energy $E^{\text{HKS}}[\rho_1^{\text{out}}]$. The third cycle begins when ρ_2^{in} is constructed, and so on. After *it* iterations we have an output charge density ρ_{it}^{out} and an energy $E^{\text{HKS}}[\rho_{it}^{\text{out}}]$. If all is well the charge density and energy will converge in a reasonable number of iterations to their exact HKS values ρ^{ex} and $E^{\text{HKS}}[\rho^{\text{ex}}]$. In practice some tolerances must be set, e.g., err_ρ and err_E, such that the calculation is deemed to have converged when

$$|\rho_{it}^{\text{out}}(\mathbf{r}) - \rho_{it-1}^{\text{out}}(\mathbf{r})| < \text{err}_\rho \tag{2.44}$$

and/or

$$|E^{\text{HKS}}[\rho_{it}^{\text{out}}] - E^{\text{HKS}}[\rho_{it-1}^{\text{out}}]| < \text{err}_E. \tag{2.45}$$

I have glossed over the problem of constructing ρ_{it}^{in} from ρ_{it}^{out}, but the way in which this is done determines whether tens, hundreds or thousands of iterations will be required or if indeed the calculation will converge at all. The simplest option would be to set $\rho_{it}^{\text{in}} = \rho_{it}^{\text{out}}$, but this is unfortunately no good at all; in most cases the iterations never converge. An option which works more often is so-called *linear mixing*. The new input charge density is constructed by mixing ρ_{it}^{out} output with the previous input charge density ρ_{it-1}^{in} according to

$$\rho_{it}^{\text{in}} = \beta\rho_{it}^{\text{out}} + (1-\beta)\rho_{it-1}^{\text{out}}, \tag{2.46}$$

where β is a mixing parameter which may have to be very small to achieve a converging sequence of iterations. But linear mixing is still not very reliable. This seems to be something of a black art. Years of experience in many research groups have led to much more efficient and reliable ways of mixing output and input charge densities, see for example Vanderbilt and Louis (1984) , but it would be beyond the scope of this book to discuss them.

2.4.5 Approximating the Exchange and Correlation: The LDA

The simplest approximation for E_{xc} is the LDA, as in Thomas–Fermi theory, according to which we have:

$$E_{\text{xc}}[\rho] = \int \rho(\mathbf{r})\epsilon_{\text{xc}}\left(\rho(\mathbf{r})\right)\, \mathrm{d}\mathbf{r}. \tag{2.47}$$

The idea with any LDA is that it will be correct in the limit that the charge density is slowly varying. In practical calculations the jellium ϵ_{xc} calculated with quantum Monte Carlo by Ceperley and Alder (1980) is often used in a form parameterized as a function of ρ by Perdew and Zunger (1981). There are alternative treatments favoured by some workers, see for example the review by Jones and Gunnarsson (1989), and a good discussion by Vosko *et al.* (1980). Within the LDA the exchange and correlation potential takes the form

$$V_{\text{xc}}(\mathbf{r}) = \epsilon_{\text{xc}}\left(\rho(\mathbf{r})\right) + \left[\rho \frac{\mathrm{d}\epsilon_{\text{xc}}(\rho)}{\mathrm{d}\rho}\right]_{\rho=\rho(\mathbf{r})}. \tag{2.48}$$

Although the LDA generally gives better results for molecular calculations than simple HF, because it deals with correlation energy, it has a major source of error which HF does not share. Within the LDA the interaction of each electron with itself (called the *self-interaction*) is not properly removed from the total energy. This becomes very obvious in the case of the hydrogen atom for example, for which $V_{\text{ext}}(\mathbf{r}) = -1/r$. The true charge density and energy would only be obtained if $V_{\text{eff}} = V_{\text{ext}}$, in other words V_{xc} should exactly cancel V_{H}. However, the LDA description of exchange and correlation does not achieve this (although a much better job in this case can be done with LDA within the spin-dependent density functional theory). In fact the LDA does surprisingly well, considering how inhomogeneous the electron densities in real systems are. Reasons for its success have been discussed in the literature (Jones and Gunnarsson, 1989).

A helpful way of thinking about the exchange and correlation energy is given by the formula, derived in Parr and Yang (1989, p. 186):

$$E_{\text{xc}}[\rho] = \frac{1}{2} \int \frac{\rho(\mathbf{r})\rho(\mathbf{r}')\rho^{\text{xc}}(\mathbf{r}, \mathbf{r}')}{|\mathbf{r} - \mathbf{r}'|}\, \mathrm{d}\mathbf{r}\, \mathrm{d}\mathbf{r}'. \tag{2.49}$$

This has the form of an electrostatic interaction of each electron with its own *exchange and correlation hole*. The density at $\mathbf{r}'$ of an exchange and correlation hole as seen by electrons at $\mathbf{r}$ is $\rho^{\mathrm{xc}}(\mathbf{r},\mathbf{r}')$ and has the property

$$\int \rho^{\mathrm{xc}}(\mathbf{r},\mathbf{r}')\,\mathrm{d}\mathbf{r}' = -1. \tag{2.50}$$

That is, it represents a depletion of electron density that exactly compensates the charge of an individual electron at $\mathbf{r}$. This depletion is created by the exchange mechanism, but the shape of the exchange and correlation hole is modified by the effect of correlation. An exact treatment of ρ^{xc} would perfectly deal with the self-interaction correction. A clue to the surprising accuracy of the LDA lies in the fact that the *angular* distribution of $\rho^{\mathrm{xc}}(\mathbf{r},\mathbf{r}')$ about $\mathbf{r}$ has no effect on the energy; a result of elementary electrostatics. It is therefore no disadvantage that the LDA exchange and correlation hole contains no information about the angular variation of the true ρ^{xc}.

There have been several attempts to improve on the LDA by adding corrections to ϵ_{xc} depending on the gradient $\nabla\rho$ (GGAs). However, as yet no clear universally applicable improvement of this type is available. More recent attempts to improve it have focussed on bringing explicit orbitals back into the functional, and at the time of writing this is an active area of research.

3

EXPLOITING THE VARIATIONAL PRINCIPLE

In this chapter, the general nature of interatomic forces is described, common to all materials from metals to ionic insulators. The central idea of interatomic forces is expressed by the Hellmann–Feynman theorem (Hellmann, 1937; Feynman, 1939). In its general form this theorem concerns the change in the ground state total energy of the electrons E induced by a change in some parameter λ, which controls the external potential in some way. It states that

$$\frac{\partial E}{\partial \lambda} = \langle \Psi_0 | \frac{\partial V_{\text{ext}}}{\partial \lambda} | \Psi_0 \rangle. \tag{3.1}$$

Amazingly, the variation in the ground state wavefunction due to the variation in λ is not required. The parameter λ may correspond to an atomic (i.e. nuclear) position, in which case we are looking at a formula for the force on an atom. This is the example of the theorem we shall now examine in detail.

3.1 The Hellmann–Feynman Theorem

3.1.1 Statement and Proof

Let us suppose that the distribution of electrons were a classical, continuous negative charge distribution rather than some quantum mechanical entity. Nuclei of charge Z_I are located at positions $\mathbf{R}_I$ within this electron distribution. In this case we know exactly how to calculate the force on a nucleus. The force on nucleus I will be given by elementary electrostatics as

$$\mathbf{F}_I = \sum_{J \neq I} \frac{Z_I Z_J (\mathbf{R}_I - \mathbf{R}_J)}{|\mathbf{R}_I - \mathbf{R}_J|^3} - \int \frac{\rho(\mathbf{r})(\mathbf{R}_I - \mathbf{r})\mathrm{d}\mathbf{r}}{|\mathbf{R}_I - \mathbf{r}|^3}. \tag{3.2}$$

The first term is the electrostatic repulsion of the other nuclei, the second term is the electrostatic attraction of the negative charge distribution.

The amazing thing is that this result is true also for the real system in which $\rho(\mathbf{r})$ is actually a distribution of electrons which depends on the positions of the nuclei according to the solution of a complicated, quantum mechanical many-body problem. All the complexity of quantum mechanics goes into the determination of $\rho(\mathbf{r})$; once we have it, the calculation of the forces is a classical problem. What makes this surprising I think is that we can regard it as a calculation of the derivative of the total energy with respect to the position of a nucleus. Or, more generally, it is the derivative of the total energy with respect to some external potential that happens to be provided by nuclei. So in effect we are able to calculate the small change in total energy for a small change in external potential, which of course changes the motion of all the electrons and their resulting density, *without doing any quantum mechanics!* The reason is actually rather simple: the small change in energy is of *second* order in the change in electron density, because the total energy was variationally mimimum with respect to changes in the density.

The proof from the point of view of DFT runs as follows. Consider the functional $E[\rho]$ with the electrons in the ground state. Let us include the nucleus–nucleus interaction. Now displace a nucleus slightly, or change the external potential in any other way by a small amount δV_{ext}. The potential V_{ext} will later be provided by the nuclei, but for the following equation its origin makes no difference. Make any necessary change $\delta\rho$ in the density to keep the electrons in the ground state corresponding to the new configuration of the nuclei. To first order the change in the energy, from (2.25), is

$$\begin{aligned}\delta E[\rho] &= \delta F[\rho] + \int \delta\rho V_{\text{ext}}\,\mathbf{dr} + \int \rho\delta V_{\text{ext}}\,\mathbf{dr} + \delta E_{ZZ} \\ &= \int \frac{\delta F[\rho]}{\delta\rho}\delta\rho\,\mathbf{dr} + \int \delta\rho V_{\text{ext}}\,\mathbf{dr} + \int \rho\delta V_{\text{ext}}\,\mathbf{dr} + \delta E_{ZZ}. \end{aligned} \tag{3.3}$$

But by the variational property of the functional (2.27) the terms in $\delta\rho$ sum to zero and we are left with

$$\delta E[\rho] = \int \rho\delta V_{\text{ext}}\mathbf{dr} + \delta E_{ZZ} \tag{3.4}$$

or

$$\frac{\delta E[\rho]}{\delta V_{\text{ext}}(\mathbf{r})} = \rho(\mathbf{r}) + \frac{\delta E_{ZZ}}{\delta V_{\text{ext}}(\mathbf{r})} \tag{3.5}$$

which is essentially the Hellmann–Feynman theorem. It is an expression which corresponds purely to classical electrostatics, not involving any change in ρ. To complete the connection to the force (3.2) is just a matter of relating δV_{ext} and δE_{ZZ} to the displacement $\delta\mathbf{R}_I$ of nucleus I. We make use of the result

$$\frac{\partial}{\partial\mathbf{R}_I}|\mathbf{r} - \mathbf{R}_I| = -\frac{\mathbf{r} - \mathbf{R}_I}{|\mathbf{r} - \mathbf{R}_I|}, \tag{3.6}$$

where the derivative $\partial f/\partial\mathbf{R}$ is defined to be the vector with Cartesian components $(\partial f/\partial R_1, \partial f/\partial R_2, \partial f/\partial R_3)$, $\mathbf{R}$ being the Cartesian vector (R_1, R_2, R_3).

The differential (3.4) becomes the derivative

$$\frac{\partial E[\rho]}{\partial \mathbf{R}_I} = \int \rho(\mathbf{r}) \frac{\partial V_{\text{ext}}(\mathbf{r})}{\partial \mathbf{R}_I} \, \mathrm{d}\mathbf{r} + \frac{\partial E_{ZZ}}{\partial \mathbf{R}_I}. \tag{3.7}$$

This is evaluated using (3.6) and the expressions (1.13) and (1.218), for V_{ext} and E_{ZZ}, respectively. The resulting force, $-\partial E[\rho]/\partial \mathbf{R}_I$, is then given by (3.2).

The Hellmann–Feynman theorem is applied whenever forces on atoms need to be calculated, once the self-consistent charge density is available. The self-consistency is a crucial requirement. Any attempt to calculate the electrostatic force on a nucleus with an approximate charge density is very likely to give nonsense. The van der Waals interaction is an interesting case in which the Hellmann–Feynman theorem is usually not applied, although it is of course valid, and it provides an excellent example of the importance of using a self-consistent charge density.

3.1.2 The van der Waals Interaction

The van der Waals interaction (see Fig. 3.1 describes the interaction of atoms which are separated enough that their individual charge clouds do not overlap—or so it seems. The interaction is described conventionally as follows. As the electrons fluctuate in position around one of the nuclei, the atom to which they belong acquires a random, varying dipole moment. Suppose it is instantaneously $\mathbf{P}_1$. This dipole, which remember is very rapidly changing in both magnitude and direction, exerts a field proportional to R^{-3} on another atom a distance R away and induces a dipole $\mathbf{P}_2$ on that atom, where $\mathbf{P}_2 \propto -\mathbf{P}_1/R^3$. The force of one dipole on another varies as $(\mathbf{P}_1 \cdot \mathbf{P}_2)/R^4$ and so they attract each other with a force proportional to $(\mathbf{P}_1 \cdot \mathbf{P}_1)/R^7$. The time average of $\mathbf{P}_1 \cdot \mathbf{P}_1$ gives a constant factor. The force

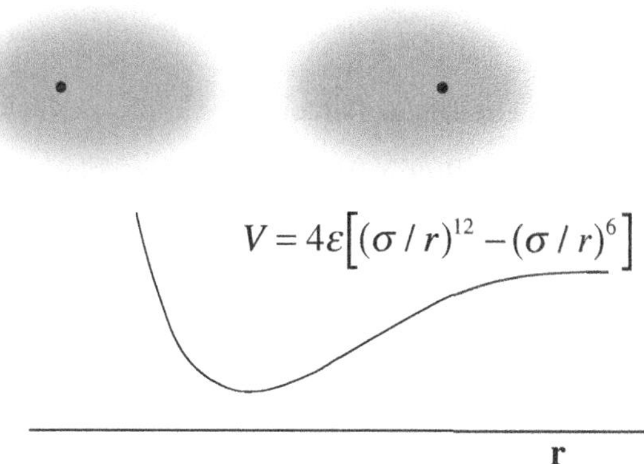

FIG. 3.1 Schematic illustration of the distorted atomic charge densities in a van der Waals interaction.

integrates to give an attractive interaction potential proportional to R^{-6}. A more detailed derivation can be found in Ashcroft and Mermin (1976, Problem 19.1) or in Stone (1996, Chapter 4.3). This picture is modified if the two atoms are so far apart that the speed of light affects the response time (Casimir and Polder, 1948), but that need not concern us. For noble gas atoms, between which there is no chemical bonding, the van der Waals attraction need only be combined with a simple short-ranged repulsion (usually a potential of R^{-12} form) to give a realistic interatomic potential, the well-known Lennard–Jones potential. Within this picture of the interaction the individual atoms remain spherical on average.

However, an alternative description of the interaction follows from the Hellmann–Feynman theorem. There must be some heaping up of electronic charge between the nuclei. If the exact charge density was simply a superposition of the spherical charge densities of the noble gas atoms, then the force between them would be repulsive. The direct repulsive force between the nuclei would exceed the attraction between one nucleus and the electron cloud around the other by exactly $1/R^2$ times the amount of electron charge outside the radius R (Gauss's theorem). The amount of electron charge which heaps up between the nuclei must be more than enough to compensate this net repulsion. Thus there is a *static* polarization of the atoms towards each other, quite distinct from the rapidly fluctuating polarization considered in the standard description of the interaction.

This example illustrates how essential it is to apply the Hellmann–Feynman theorem only if the electron density is self-consistent. A superposition of spherical atomic charge densities, as mentioned before, is an approximation to the true charge density which is useful for certain purposes but definitely not for applying the Hellmann–Feynman theorem. As we can see from Gauss's theorem, the Hellmann–Feynman force from a spherical atomic charge density would always predict a repulsion between the atoms.

3.1.3 Another Example: The Image Potential

At the opposite extreme, it is illuminating to consider a classical electrostatic problem, the energy of interaction of a charge with a planar metal surface (Fig. 3.2). In the classical description, suppose we have a positive charge Z at a distance x from the surface. In the classical description the metal is homogeneous and terminated abruptly. It cannot support an electric field. The charge Z induces a distribution of negative charge $\rho(\mathbf{r})$ which is confined to the surface of the metal. The boundary condition of zero tangential field at the surface and the uniqueness theorem allow the simple solution for the electrostatic potential outside the metal. The distribution $\rho(\mathbf{r})$ acts as if it were a negative charge $-Z$, the image charge, concentrated at a point x below the surface, perpendicularly below Z.

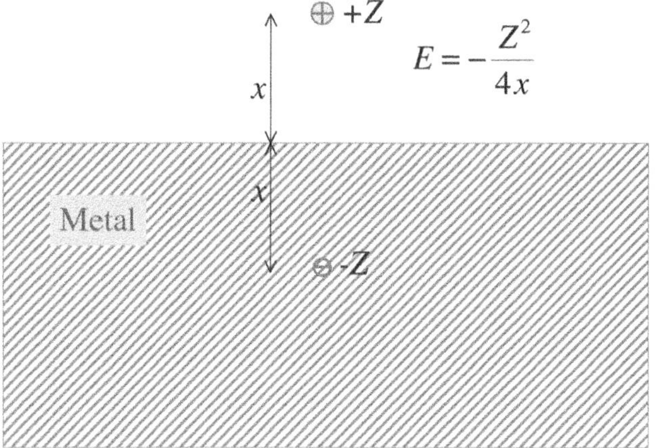

FIG. 3.2 The image interaction between a charge $+Z$ and a metal surface.

The paradox is that the electrostatic energy of the system is $-Z^2/4x$ and not $-Z^2/2x$ as it would be if the image charge were a real point charge. Why is there a factor of a half, when the electric field outside the metal is exactly the same as if the image charge were real? One answer, in terms of field theory, is simply that the energy is proportional to the integral of the square of the electric field, and this is exactly halved in the image problem compared to the case of two real charges. Another answer is that the distribution $\rho(\mathbf{r})$ has a self-energy; the coulomb interaction of the negative charges with each other. This amounts to exactly $Z^2/4x$. The electrostatic interaction of the positive charge with $\rho(\mathbf{r})$ is correctly given by $-Z^2/2x$ as the image charge description would lead us to expect. So the sum of the two gives the correct total electrostatic energy. This might solve the paradox, but it is still a remarkable result, if you are in the mood for it, that all the quantum mechanical complexity of the electronic motion in response to the external charge, the peculiar behaviour of Fermions, gives a total self-energy of this electron distribution which is given by the simple classical formula. This is also another case of the Hellmann–Feynman theorem in action. The induced charge which is equal and opposite to the inducing charge causes a force $\mathbf{F}$ on the inducing charge, where from classical electrostatics (i.e. the Hellmann–Feynman theorem)

$$\mathbf{F} = \frac{Z^2}{(2x)^2}. \tag{3.8}$$

Integrate $\mathbf{F} \cdot \mathrm{d}x$ from infinity to the point x to get the total energy of interaction $-Z^2/4x$.

If we took a more realistic model of the metal, representing the nuclei and electrons explicitly, and treating the electrons by DFT, the above results would be true as long the charge Z is far enough away from the surface; even just 1 nm would be sufficient, as explicit calculations have shown (Finnis *et al.*, 1995). At this distance the discreteness of the surface atomic structure does not have a significant effect on the field at the external charge. Note that the result does not rely on any particular approximation. It would be be found if the calculation were done with the Thomas–Fermi functional as well as with the Kohn–Sham functional, irrespective of the LDA.

The result relies on the linear relation between the induced charge and the inducing charge, and this linear response approximation fails if the inducing charge is too strong. Indeed in extreme cases you could get field emission of electrons for example. Linear response theory always produces a factor of $\frac{1}{2}$ like this in the energy, as we now examine in more detail.

3.2 Perturbation Theory with the Density

3.2.1 Switching on an External Potential

The direct calculation of forces on atoms is not the only use of the Hellmann–Feynman theorem. It also provides some useful formulae equivalent to perturbation theory. The question we ask is this. Suppose we start with a system in which the electrons are in their ground state with density ρ_1, energy $E_1[\rho_1]$, where the external potential is $V_{\text{ext}1}$. Now we let the external potential change from $V_{\text{ext}1}$ to $V_{\text{ext}2}$ and ask what is the change in energy and electron density? The Hellmann–Feynman theorem gives us a way to formulate this change by varying the potential gradually and integrating the differential change in energy (3.4). We introduce a scalar parameter λ which varies between 0 and 1 and which specifies the external potential over the range of interest between $V_{\text{ext}1}$ and $V_{\text{ext}2}$, so that the external potential corresponding to a value λ is

$$V_{\text{ext}}(\mathbf{r}, \lambda) = \lambda V_{\text{ext}2}(\mathbf{r}) + (1 - \lambda) V_{\text{ext}1}(\mathbf{r}). \tag{3.9}$$

You can think of λ as a knob to turn which swiches on the change in potential, so that

$$\delta V_{\text{ext}} = (V_{\text{ext}2} - V_{\text{ext}1})\delta\lambda. \tag{3.10}$$

However, to apply the theorem, the change has to be slow and gentle so that the electron density $\rho(\mathbf{r}, \lambda)$ is always in the ground state corresponding to λ. Under these conditions the Hellmann–Feynman theorem (3.4) becomes

$$\delta E[\rho] = \int \rho(\mathbf{r}, \lambda)\,(V_{\text{ext}2}(\mathbf{r}) - V_{\text{ext}1}(\mathbf{r}))\,\delta\lambda\,\mathrm{d}\mathbf{r} + \delta E_{ZZ} \tag{3.11}$$

The change in energy is given by the integral of (3.11) with respect to λ, giving the expression

$$E[\rho_2, V_{\text{ext2}}] = E[\rho_1, V_{\text{ext1}}] + \int_0^1 \mathrm{d}\lambda \int \rho(\mathbf{r}, \lambda)\,(V_{\text{ext2}}(\mathbf{r}) - V_{\text{ext1}}(\mathbf{r})) + E_{ZZ2} - E_{ZZ1}. \tag{3.12}$$

The difference $(E_{ZZ2} - E_{ZZ1})$ represents the change in the self-energy of the system of charges that are the source of V_{ext}, and this part does not depend on the electron density. This is a pleasing expression but hardly a useful one unless we know something about how ρ depends on λ, which is our next consideration.

3.2.2 First-order Perturbation Theory

The simplest assumption would be to ignore the λ-dependence of ρ, which is the same as saying that to lowest order in λ the charge density does not change. The result is *first-order perturbation theory* which approximates the energy as:

$$E^{\text{fo}}[\rho_2, V_{\text{ext2}}] = E[\rho_1, V_{\text{ext1}}] + \int \rho_1(\mathbf{r})\,(V_{\text{ext2}}(\mathbf{r}) - V_{\text{ext1}}(\mathbf{r}))\,\mathrm{d}\mathbf{r} + E_{ZZ2} - E_{ZZ1}. \tag{3.13}$$

At this level of approximation we only obtain the effect on the energy of the change in potential to first order, and the weaker the change in potential the more accurate this assumption becomes. It is really very convenient if $(V_{\text{ext2}} - V_{\text{ext1}})$ is weak enough for (3.13) to be good, because we do not have to calculate any change in the charge density, so we do not have to think about solving the Schrödinger equation.

3.2.3 Second-order Perturbation Theory

The next level of approximation is *second-order perturbation theory*. This amounts to including just the *linear* dependence of ρ on λ. By the mean value theorem (Whittaker and Watson, 1927), assuming that $\rho(\mathbf{r}, \lambda)$ is a single-valued, continuous and differentiable function of λ:

$$\rho(\mathbf{r}, \lambda) = \rho_1(\mathbf{r}) + \lambda\rho'(\mathbf{r}, 0) + \tfrac{1}{2}\lambda^2\rho''(\mathbf{r}, \theta_\lambda); \quad 0 \le \theta_\lambda \le \lambda. \tag{3.14}$$

We do not have to know the value of the parameter θ_λ, its only importance is to indicate that the error in the linearization of the density is bounded by its second derivative within the range of integration. The primes and double primes denote first and second derivatives with respect to λ. We now neglect the *second* derivative in comparison to the first. Formally, we can do this by simply throwing away the last term in (3.14). An interesting alternative is to apply the mean value theorem again for $\lambda = 1$ to write

$$\rho_2(\mathbf{r}) = \rho_1(\mathbf{r}) + \rho'(\mathbf{r}, 0) + \tfrac{1}{2}\rho''(\mathbf{r}, \theta_1); \quad 0 \leq \theta_1 \leq 1. \tag{3.15}$$

Combining (3.14) and (3.15) then gives

$$\rho(\mathbf{r}, \lambda) = \rho_1(\mathbf{r}) + \lambda\,(\rho_2(\mathbf{r}) - \rho_1(\mathbf{r})) + \tfrac{1}{2}\lambda^2\rho''(\mathbf{r}, \theta_\lambda) - \tfrac{1}{2}\lambda\rho''(\mathbf{r}, \theta_1). \tag{3.16}$$

This equation says explicitly that by neglecting the double primed terms we can make a linear interpolation of ρ between ρ_1 and ρ_2 (see Fig. 3.3):

$$\rho(\mathbf{r}, \lambda) \approx \rho^{(1)}(\mathbf{r}, \lambda) = \rho_1(\mathbf{r}) + \lambda\,(\rho_2(\mathbf{r}) - \rho_1(\mathbf{r}))\,. \tag{3.17}$$

Substituting this approximate form for $\rho(\mathbf{r}, \lambda)$ into (3.12) and integrating over λ gives the result of second-order perturbation theory:

$$\begin{aligned} E^{\mathrm{so}}[\rho_2, V_{\mathrm{ext2}}] = E[\rho_1, V_{\mathrm{ext1}}] &+ \int \rho_1(\mathbf{r})(V_{\mathrm{ext2}}(\mathbf{r}) - V_{\mathrm{ext1}}(\mathbf{r}))\,\mathrm{d}\mathbf{r} + E_{ZZ2} - E_{ZZ1} \\ &+ \frac{1}{2}\int (\rho_2(\mathbf{r}) - \rho_1(\mathbf{r}))\,(V_{\mathrm{ext2}}(\mathbf{r}) - V_{\mathrm{ext1}}(\mathbf{r}))\,\mathrm{d}\mathbf{r}. \end{aligned} \tag{3.18}$$

The result of second-order perturbation theory is simply to replace the ρ_1 in (3.13) by the mean density $\frac{1}{2}(\rho_1 + \rho_2)$. It includes terms up to second order in the

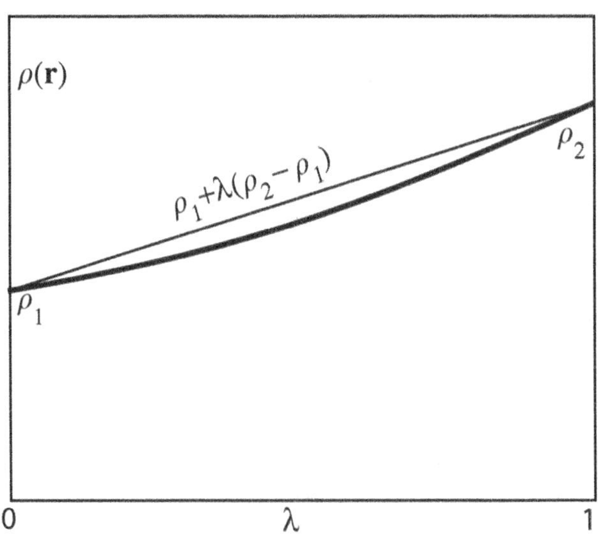

FIG. 3.3 The linear approximation to the induced charge.

change in the potential ΔV_{ext}, where we now define

$$\Delta V_{\text{ext}} = V_{\text{ext2}} - V_{\text{ext1}}. \tag{3.19}$$

The change in the density is a factor which is at least first order in ΔV_{ext}, and it is multiplied by the change ΔV_{ext} and integrated over space to give the second-order term in (3.18). Although the expression (3.18) has been obtained in terms of the exact final density ρ_2 we have seen that any approximation for $\rho(\mathbf{r}, \lambda)$ obtained by a Taylor expansion neglecting ρ'' would be valid, in particular $\rho_2(\mathbf{r}) - \rho_1(\mathbf{r})$ could be replaced by $\rho'(\mathbf{r}, 0)$.

The second-order expression is a very powerful tool, as we shall see later, even though at first sight we still have the problem of calculating the change in density at least to first order in ΔV_{ext}.

Note that the second-order term has exactly the same structure as the energy in the classical image interaction, $-Z^2/4x$, including the factor $\frac{1}{2}$. This is as it should be, since the image charge is assumed to be linearly dependent on the inducing potential of the external charge. In the classical image interaction no other terms than this appear, because E_1 is taken as the zero of energy and the other terms sum to zero. To complete the description of a classical image interaction with eq. (3.18), the terms $\int \rho_1(\mathbf{r})(V_{\text{ext2}}(\mathbf{r}) - V_{\text{ext1}}(\mathbf{r}))d\mathbf{r}$ and $(E_{ZZ2} - E_{ZZ1})$ taken individually describe the interaction energy of a point charge with a uniform half space of negative and positive charge, respectively, which would be infinite. However, taken together they exactly cancel each other.

3.3 The Second-order HKS Functional

3.3.1 Derivation of $E^{(2)}$

We are now going to do a different kind of perturbation theory to obtain what is perhaps the most useful general equation for the purposes of deriving simplified interatomic force models. We start with the HKS functional in the form (2.29) and expand it to second order in deviations of the density from some reference density ρ^{in} which is *not* in general the ground state density. We suppose that the single-particle states we have labelled $|n\rangle$ vary from $|n_{\text{in}}\rangle$ to $|n\rangle$ as the density varies from ρ^{in} to an arbitrary ρ where

$$\begin{aligned}
|n\rangle &= |n^{\text{in}}\rangle + |\delta n\rangle \\
\rho^{\text{in}}(\mathbf{r}) &= \sum f_n \langle \mathbf{r}|n^{\text{in}}\rangle\langle n^{\text{in}}|\mathbf{r}\rangle \\
\rho(\mathbf{r}) &= \rho^{\text{in}}(\mathbf{r}) + \delta\rho(\mathbf{r}) = \sum f_n \langle \mathbf{r}|n\rangle\langle n|\mathbf{r}\rangle.
\end{aligned} \tag{3.20}$$

The next step is to rewrite E_{H} and E_{xc} on the right-hand side of (2.29), in terms of ρ^{in} and $\delta\rho$:

$$E^{\mathrm{HKS}}[\rho] = T_s[\rho] + E_{\mathrm{H}}[\rho^{\mathrm{in}} + \delta\rho] + E_{\mathrm{xc}}[\rho^{\mathrm{in}} + \delta\rho] + \int \rho V_{\mathrm{ext}} + E_{ZZ}. \tag{3.21}$$

The Hartree energy (1.161) can be expanded exactly about ρ^{in} as

$$E_{\mathrm{H}}[\rho] = E_{\mathrm{H}}^{\mathrm{in}} + \int V_{\mathrm{H}}^{\mathrm{in}} \delta\rho + \frac{1}{2}\int \frac{\delta\rho\delta\rho'}{|\mathbf{r}-\mathbf{r}'|}. \tag{3.22}$$

This introduces some simplifying notation. As previously, $\mathbf{r}$ and $\mathbf{dr}$ will not be written explicitly unless it seems useful for clarity, since it should by now be clear when an unspecified integral sign means a single or double volume integral. We also introduce the superscript *in* and the prime to indicate when a function or functional is evaluated at $\rho = \rho^{\mathrm{in}}$ and $\mathbf{r}'$ respectively, so

$$\begin{aligned} E_{\mathrm{H}}^{\mathrm{in}} &\equiv E_{\mathrm{H}}[\rho^{\mathrm{in}}], \\ V_{\mathrm{H}}^{\mathrm{in}}(\mathbf{r}) &= \int \frac{\rho^{\mathrm{in}}(\mathbf{r}')}{|\mathbf{r}-\mathbf{r}'|}\,\mathbf{dr}', \\ \delta\rho' &= \delta\rho(\mathbf{r}'). \end{aligned} \tag{3.23}$$

This notation will be extended as required, so for example we will use the shorthand

$$V_{\mathrm{eff}}^{\mathrm{ex}} \equiv V_{\mathrm{eff}}[\rho]|_{\rho=\rho^{\mathrm{ex}}} \tag{3.24}$$

where ρ^{ex} is the exact density, obtained by minimizing the exact HKS functional (2.29).

Notice that because E_{H} is intrinsically quadratic in ρ, the expression (3.22) is exact, and not a truncated Taylor series. Unfortunately, however, we have to be content with a truncated Taylor series for E_{xc}, which to second order is

$$E_{\mathrm{xc}}^{(2)}[\rho] = E_{\mathrm{xc}}^{\mathrm{in}} + \int V_{\mathrm{xc}}^{\mathrm{in}} \delta\rho + \frac{1}{2}\int \left.\frac{\delta^2 E_{\mathrm{xc}}[\rho]}{\delta\rho\delta\rho'}\right|_{\rho^{\mathrm{in}}} \delta\rho\,\delta\rho'. \tag{3.25}$$

We can now combine (3.22) and (3.25), substituting

$$E_{\mathrm{H}}^{\mathrm{in}} = \frac{1}{2}\int \rho^{\mathrm{in}} V_{\mathrm{H}}^{\mathrm{in}} \tag{3.26}$$

to give

$$\begin{aligned} E_{\mathrm{H}}[\rho] + E_{\mathrm{xc}}^{(2)}[\rho] = &-\frac{1}{2}\int \rho^{\mathrm{in}} V_{\mathrm{H}}^{\mathrm{in}} + E_{\mathrm{xc}}^{\mathrm{in}} - \int \rho^{\mathrm{in}} V_{\mathrm{xc}}^{\mathrm{in}} \\ &+ \int \rho V_{\mathrm{H}}^{\mathrm{in}} + \int \rho V_{\mathrm{xc}}^{\mathrm{in}} + \frac{1}{2}\int C_{\mathrm{in}}(\mathbf{r},\mathbf{r}')\,\delta\rho\,\delta\rho'. \end{aligned} \tag{3.27}$$

where the kernel C_{in} is defined by

$$C_{\text{in}}(\mathbf{r},\mathbf{r}') = \frac{1}{|\mathbf{r}-\mathbf{r}'|} + \left.\frac{\delta^2 E_{\text{xc}}[\rho]}{\delta\rho\delta\rho'}\right|_{\rho^{\text{in}}}. \tag{3.28}$$

Now we define a Hamiltonian corresponding to ρ^{in}

$$\hat{H}^{\text{in}} = \hat{T} + V_{\text{ext}} + V_{\text{H}}^{\text{in}} + V_{\text{xc}}^{\text{in}}, \tag{3.29}$$

which enables us to write

$$T_s[\rho] = \sum f_n \left\langle n|\hat{H}^{\text{in}}|n\right\rangle - \int \rho V_{\text{eff}}^{\text{in}}. \tag{3.30}$$

where

$$V_{\text{eff}}^{\text{in}} = V_{\text{ext}} + V_{\text{H}}^{\text{in}} + V_{\text{xc}}^{\text{in}}. \tag{3.31}$$

Putting this expression for the kinetic energy together with (3.27) into the HKS functional (3.21) results, after cancelling some terms, in a second-order functional:

$$\begin{aligned} E^{(2)}[\rho] = \sum f_n \left\langle n|\hat{H}^{\text{in}}|n\right\rangle + E_{\text{xc}}^{\text{in}} - \int \rho^{\text{in}} V_{\text{xc}}^{\text{in}} \\ - E_{\text{H}}^{\text{in}} + E_{ZZ} + \frac{1}{2}\int C_{\text{in}}(\mathbf{r},\mathbf{r}')\,\delta\rho\,\delta\rho'. \end{aligned} \tag{3.32}$$

The use of this functional will become apparent later, but it has some attractive features which should be mentioned already. Note first that the only approximation so far is the truncation of the expansion of E_{xc} to second order in $\delta\rho = \rho - \rho^{\text{in}}$. Second, minimization of this functional with respect to ρ must still be carried out to get the ground state, and there is still a self-consistency problem to be solved. The states $|n\rangle$ must be regarded as an alternative description of ρ, as defined in (3.20), and included as variables in this minimization. Accordingly, if we seek the stationary value of (3.32) we get a Kohn–Sham equation that has to be solved self-consistently:

$$\left(\hat{T} + \hat{V}_{eff}^{(2)}\right)|n\rangle = \epsilon_n |n\rangle, \tag{3.33}$$

in which the effective potential is

$$V_{\text{eff}}^{(2)}(\mathbf{r}) = V_{\text{eff}}^{\text{in}}(\mathbf{r}) + \int C_{\text{in}}(\mathbf{r},\mathbf{r}')\,\delta\rho(\mathbf{r}')\,\mathrm{d}\mathbf{r}'. \tag{3.34}$$

This equation makes clear the important meaning of the kernel C_{in}, which in general we shall simply write as C when it refers to an arbitrary charge density. It describes the change in the Hartree, exchange and correlation potentials to first order in the change in the charge density.

3.3.2 The Error in $E^{(2)}$

We have defined a new functional $E^{(2)}[\rho]$ which is an approximation to the exact HKS functional in the sense that

$$E^{\text{HKS}}[\rho] = E^{(2)}[\rho] + \Delta E_{\text{xc}}^{\text{in}}[\rho], \tag{3.35}$$

where, in full:

$$\Delta E_{\text{xc}}^{\text{in}}[\rho] = \frac{1}{6}\int \frac{\delta^3 E_{\text{xc}}^{\text{in}}}{\delta\rho\delta\rho'\delta\rho''}(\rho(\mathbf{r}) - \rho^{\text{in}}(\mathbf{r}))(\rho(\mathbf{r}') - \rho^{\text{in}}(\mathbf{r}'))$$
$$\times (\rho(\mathbf{r}'') - \rho^{\text{in}}(\mathbf{r}''))\,d\mathbf{r}\,d\mathbf{r}'\,d\mathbf{r}'' + O[(\rho - \rho^{\text{in}})^4]. \tag{3.36}$$

The question now arises, what is the error we make in the exact energy if we minimize the approximate functional $E^{(2)}$? Let us denote by ρ^{ex} the exact ground state charge density as obtained by minimizing the exact functional E^{HKS}, and by $\rho^{(2)}$ the approximate ground state charge density obtained by minimizing $E^{(2)}$. We want to know the error $\Delta E^{(2)}$ defined by

$$E^{\text{HKS}}[\rho^{\text{ex}}] = E^{(2)}[\rho^{(2)}] + \Delta E^{(2)}. \tag{3.37}$$

We already know from (3.36) that

$$E^{\text{HKS}}[\rho^{\text{ex}}] - E^{(2)}[\rho^{\text{ex}}] = O[(\rho^{\text{in}} - \rho^{\text{ex}})^3].$$

However, it is not yet obvious that $\Delta E^{(2)}$ is of similar order.
Let us write (3.32) in the form

$$E^{(2)}[\rho] = T_s[\rho] + \int \rho V_{\text{eff}}^{\text{in}} + E_{\text{xc}}^{\text{in}} - \int \rho^{\text{in}} V_{\text{xc}}^{\text{in}} - E_{\text{H}}^{\text{in}} + E_{ZZ}$$
$$+ \frac{1}{2}\int C_{\text{in}}(\mathbf{r}, \mathbf{r}')\delta\rho\delta\rho'. \tag{3.38}$$

Then by expanding about $\rho = \rho^{\text{ex}}$:

$$E^{(2)}[\rho] = E^{(2)}[\rho^{\text{ex}}]$$
$$+ \int \left\{ \left.\frac{\delta T_s}{\delta\rho}\right|_{\rho^{\text{ex}}} + V_{\text{eff}}^{\text{in}} + C_{\text{in}}(\rho^{\text{ex}\prime} - \rho^{\text{in}\prime}) \right\} (\rho - \rho^{\text{ex}})$$
$$+ \frac{1}{2}\int \left.\frac{\delta^2 T_s}{\delta\rho\delta\rho'}\right|_{\rho^{\text{ex}}} (\rho - \rho^{\text{ex}})(\rho' - \rho^{\text{ex}\prime})$$
$$+ \frac{1}{2}\int C_{\text{in}}(\rho - \rho^{\text{ex}})(\rho' - \rho^{\text{ex}\prime}) + O[(\rho - \rho^{\text{ex}})^3]. \tag{3.39}$$

The algebra is getting cumbersome, so I have extended the shorthand by also omitting the $\mathbf{r}, \mathbf{r}'$ arguments of C_{in}. At this point, if we set $\rho = \rho^{(2)}$, it looks as if the error might be of second order or worse! The variational principle will save us here; we have yet to use the fact that $E^{(2)}[\rho]$ is a mimimum when $\rho = \rho^{(2)}$. First, we note that from the variational property (2.32) the first-order term *almost* vanishes. To see this we make use of the expansion:

$$V_{\text{eff}}^{\text{ex}} = V_{\text{eff}}^{\text{in}} + \int C_{\text{in}}(\rho^{\text{ex}\prime} - \rho^{\text{in}\prime}) + O[(\rho^{\text{in}} - \rho^{\text{ex}})^2], \tag{3.40}$$

together with the variational condition which defines $E^{\text{HKS}}[\rho^{\text{ex}}]$, namely from (2.32)

$$\left.\frac{\delta T_s}{\delta \rho}\right|_{\rho^{\text{ex}}} + V_{\text{eff}}|_{\rho^{\text{ex}}} = \mu. \tag{3.41}$$

If we substitute for $\delta T_s/\delta\rho$ in the first-order term of (3.39) using (3.41) and (3.40) that term becomes:

$$\int \left\{ \left.\frac{\delta T_s}{\delta \rho}\right|_{\rho^{\text{ex}}} + V_{\text{eff}}^{\text{in}} + C_{\text{in}}(\rho^{\text{ex}\prime} - \rho^{\text{in}\prime}) \right\} (\rho - \rho^{\text{ex}}) = O[(\rho^{\text{in}} - \rho^{\text{ex}})^2(\rho - \rho^{\text{ex}})]. \tag{3.42}$$

We have made use of the fact that the integral of any constant times $\rho - \rho^{\text{ex}}$ vanishes by the constraint of charge conservation. This term is already looking a lot smaller than it did, and how small it is depends on the error in the initial guess ρ^{in}. Now we must apply the variational condition for $E^{(2)}$, namely

$$\left.\frac{\delta E^{(2)}[\rho]}{\delta \rho}\right|_{\rho^{(2)}} = \mu^{(2)}. \tag{3.43}$$

Applying it to (3.39), making use of (3.42), gives

$$\int \left\{ \left.\frac{\delta^2 T_s}{\delta \rho \delta \rho'}\right|_{\rho^{\text{ex}}} + C_{\text{in}} \right\} (\rho^{(2)\prime} - \rho^{\text{ex}\prime}) = O[(\rho^{\text{in}} - \rho^{\text{ex}})^2] + O[(\rho^{(2)} - \rho^{\text{ex}})^2]. \tag{3.44}$$

Finally, we need only to expand $\rho^{(2)} - \rho^{\text{ex}}$ in powers of $\rho^{\text{in}} - \rho^{\text{ex}}$ to see that, provided the kernel in the braces has an inverse (an assumption akin to v-representability, as we shall see):

$$\rho^{(2)} - \rho^{\text{ex}} = O[(\rho^{\text{in}} - \rho^{\text{ex}})^2]. \tag{3.45}$$

This is a gratifying result, because we can substitute it into (3.42), and then substitute it together with (3.42) into (3.39) to give

$$E^{(2)}[\rho^{(2)}] = E^{(2)}[\rho^{\text{ex}}] + O[(\rho^{\text{in}} - \rho^{\text{ex}})^4]. \tag{3.46}$$

This result in turn means that we can neglect the difference between $E^{(2)}[\rho^{(2)}]$ and $E^{(2)}[\rho^{\text{ex}}]$ in comparison to the third-order error already noted in $E^{(2)}[\rho^{\text{ex}}]$, so that the final result for the error can be written:

$$E^{(2)}[\rho^{(2)}] = E^{\text{HKS}}[\rho^{\text{ex}}] + O[(\rho^{\text{in}} - \rho^{\text{ex}})^3]. \tag{3.47}$$

3.4 The Harris–Foulkes Functional and its Generalizations

3.4.1 The First-order Functional and the Harris–Foulkes Functional

A very fruitful idea was developed independently by Harris (1985) and Foulkes (1987). They studied the first-order functional, which we can define simply by dropping the second-order terms in (3.32):

$$\begin{aligned} E^{(1)}[\rho] &= \sum f_n \langle n|\hat{H}^{\text{in}}|n\rangle + E^{\text{in}}_{\text{xc}} - \int \rho^{\text{in}} V^{\text{in}}_{\text{xc}} - E^{\text{in}}_{\text{H}} + E_{ZZ} \\ &\equiv T_s[\rho] + \int \rho V^{\text{in}}_{\text{eff}} + E^{\text{in}}_{\text{xc}} - \int \rho^{\text{in}} V^{\text{in}}_{\text{xc}} - E^{\text{in}}_{\text{H}} + E_{ZZ}. \end{aligned} \tag{3.48}$$

Minimizing this with respect to ρ is much easier than in the case of the exact functional or $E^{(2)}$, because now the Kohn–Sham equation

$$\hat{H}^{\text{in}}|n\rangle = \epsilon_n|n\rangle \tag{3.49}$$

does not have to be solved self-consistently. The potential $V^{\text{in}}_{\text{eff}}$ is constructed once and for all for a given set of atomic positions using the guessed charge density ρ^{in}. The eigenstates $|n^{\text{out}}\rangle$ that are found could be used to construct a first-order charge density according to

$$\rho^{\text{out}}(\mathbf{r}) = \sum f_n \left|\langle \mathbf{r}|n^{\text{out}}\rangle\right|^2. \tag{3.50}$$

In a standard self-consistency procedure, ρ^{out} would then be used to calculate a new value of V_{eff} and the cycle repeated until convergence is achieved, as outlined in Section 2.4.4. If $E^{(1)}[\rho^{\text{out}}]$ is already accurate enough, the self-consistency cycle is not necessary, and an enormous amount of computation could be saved. The question then arises, just as in the case of $E^{(2)}$, as to what is the magnitude of the error $\Delta E^{(1)}$ defined by

$$E^{\text{HKS}}[\rho^{\text{ex}}] = E^{(1)}[\rho^{\text{out}}] + \Delta E^{(1)}. \tag{3.51}$$

As in the case of $E^{(2)}$, we expect the error to be less the better our guess at ρ^{in}, but not as small as $\Delta E^{(2)}$.

Before addresssing this question in detail, to be strict about definitions the *Harris–Foulkes functional* is nearly but not quite the same thing as $E^{(1)}[\rho]$. Remember ρ in this functional is equivalent to a set of single particle states (or a single

Slater determinant of states), which become the set $\{|n^{\text{out}}\rangle\}$ when $\rho = \rho^{\text{out}}$. Let us include as an argument of $E^{(1)}$ the input charge density and write the former functional as $E^{(1)}[\rho, \rho^{\text{in}}]$. We now fix ρ at ρ^{out}, which depends on ρ^{in}, and define the Harris–Foulkes functional $E^{\text{HF}}[\rho^{\text{in}}]$, a functional of the *input* charge density:

$$E^{\text{HF}}[\rho^{\text{in}}] = E^{(1)}[\rho^{\text{out}}[\rho^{\text{in}}], \rho^{\text{in}}]. \tag{3.52}$$

If $\rho^{\text{in}} = \rho^{\text{ex}}$ clearly $\Delta E^{(1)} = 0$, and the Harris–Foulkes functional gives the exact energy

$$E^{\text{HF}}[\rho^{\text{ex}}] = E^{\text{HKS}}[\rho^{\text{ex}}]. \tag{3.53}$$

Harris and Foulkes' discovery was that the error $\Delta E^{(1)}$ is of *second order*, in the sense we shall define precisely in Section 4.5. It is convenient to defer a discussion of the error in the Harris–Foulkes functional until after linear response functions have been introduced.

3.4.2 Generalization of the Harris–Foulkes Functional

The first-order functional has additional surprising and useful properties, which were pointed out by Methfessel (1995), Jacobsen *et al.* (1987) and Haydock (1998). We have observed that, since ρ^{out} is obtained from a single-particle Schrödinger equation with the potential $V_{\text{eff}}^{\text{in}}$, $E^{(1)}$ is variationally minimum with respect to the single-particle wavefunctions, or, equivalently, with respect to ρ^{out} at fixed ρ^{in}. We have used this property to derive the variational behaviour of E^{HF}. So $E^{(1)}$ has *two* variational properties. Now we shall see that there is even a third! In a certain sense, $E^{(1)}$ is variationally maximum with respect to the effective potential $V_{\text{eff}}^{\text{in}}$. To see this we write the functional dependence of $E^{(1)}$ as follows:

$$E^{(1)}[\rho, \rho^{\text{in}}] = T_s[\rho] + \rho V_{\text{eff}}^{\text{in}} - \rho^{\text{in}} V_{\text{eff}}^{\text{in}} + \rho^{\text{in}} V_{\text{ext}} + E_{\text{xc}}^{\text{in}} + E_{\text{H}}^{\text{in}} + E_{ZZ}. \tag{3.54}$$

This equation is really just (3.48), rewritten by adding to the right-hand side the following terms which sum to zero:

$$-\rho^{\text{in}} V_{\text{eff}}^{\text{in}} + \rho^{\text{in}} V_{\text{ext}} + \rho^{\text{in}} V_{\text{xc}}^{\text{in}} + \rho^{\text{in}} V_{\text{H}}^{\text{in}} \equiv 0, \tag{3.55}$$

and substituting the identity

$$\rho^{\text{in}} V_{\text{H}}^{\text{in}} \equiv 2E_{\text{H}}^{\text{in}}. \tag{3.56}$$

Now let us replace $V_{\text{eff}}^{\text{in}}$ by an arbitrary potential V_{eff} which is independent of ρ^{in} and ρ, thereby defining the following new functional of three functions:

$$E^{\text{HFG}}[\rho, \rho^{\text{in}}, V_{\text{eff}}] = T_s[\rho] + \rho V_{\text{eff}} - \rho^{\text{in}} V_{\text{eff}} + \rho^{\text{in}} V_{\text{ext}} + E_{\text{xc}}^{\text{in}} + E_{\text{H}}^{\text{in}} + E_{ZZ}. \tag{3.57}$$

Consider first the variation with respect to V_{eff}:

$$\frac{\delta E^{\text{HFG}}}{\delta V_{\text{eff}}} = \rho - \rho^{\text{in}}. \tag{3.58}$$

This vanishes if and only if $\rho = \rho^{\text{in}}$. Second, consider the variation of E^{HFG} with respect to ρ, subject to charge conservation. The Euler–Lagrange equation is the familiar condition that ρ is the ground state density of non-interacting electrons in the potential V_{eff}, eq. (2.32). When it is satisfied E^{HFG} is minimized with respect to ρ, at an output charge density ρ^{out} and we have from (2.39) that

$$T_s + \rho^{\text{out}} V_{\text{eff}} = \sum_n f_n \epsilon_n. \tag{3.59}$$

Third, consider the variation of E^{HFG} with respect to ρ^{in}:

$$\frac{\delta E^{\text{HFG}}}{\delta \rho^{\text{in}}} = -V_{\text{eff}} + V_{\text{ext}} + V_{\text{xc}}^{\text{in}} + V_{\text{H}}^{\text{in}}. \tag{3.60}$$

From the definition of V_{eff} this variation vanishes if and only if $V_{\text{eff}} = V_{\text{eff}}^{\text{in}}$. Under what conditions will all three variations vanish? Well, this requires both $\rho = \rho^{\text{out}} = \rho^{\text{in}}$ and $V_{\text{eff}} = V_{\text{eff}}^{\text{in}}$. But if the ground state density ρ^{out} in the potential $V_{\text{eff}}^{\text{in}}$, which you remember was constructed from ρ^{in}, is itself ρ^{in}, we must be at the exact, self-consistent solution. So the answer is when $\rho = \rho^{\text{in}} = \rho^{\text{ex}}$, and $V_{\text{eff}} = V_{\text{eff}}^{\text{in}} = V_{\text{eff}}^{\text{ex}}$, that is at the exact ground state density.

The significance of this discovery that $E^{\text{HFG}}[\rho, \rho^{\text{in}}, V_{\text{eff}}]$ is stationary with respect to all three functions at the exact ground state density and effective potential is the following. If we make a guess *independently* not only of the charge density ρ^{in} but also of the effective potential V_{eff} and solve the HKS equation to obtain the eigenvalues ϵ_n, the resulting estimate of the energy given by:

$$E^{\text{HFG}}[\rho^{\text{out}}, \rho^{\text{in}}, V_{\text{eff}}] = \sum_n f_n \epsilon_n - \rho^{\text{in}} V_{\text{eff}} + \rho^{\text{in}} V_{\text{ext}} + E_{\text{xc}}^{\text{in}} + E_{\text{H}}^{\text{in}} + E_{ZZ}. \tag{3.61}$$

will differ from the exact ground state energy only by terms of second order in the deviations $(\rho^{\text{in}} - \rho^{\text{ex}})$ and $(V_{\text{eff}}^{\text{in}} - V_{\text{eff}}^{\text{ex}})$. The advantage of the standard Harris–Foulkes functional that it is not necessary to evaluate ρ^{out} explicitly is preserved in the generalized functional. It has the further advantage that it does not require V_{eff} to be computed from ρ^{in}. The significance of this is not merely practical. Because V_{xc} is not tied to ρ^{in}, one could use a model for V_{eff} which implicitly takes account of the non-locality of E_{xc}. Thus, paradoxically it may be possible to get *better* results with the approximate functional E^{HFG} than with the exact functional E^{HKS} in which the LDA or GGA must be applied!

An alternative expression for E^{HFG} is obtained by rearranging the Hartree terms as before for E^{HF}:

$$E^{\mathrm{HFG}}[\rho^{\mathrm{out}}, \rho^{\mathrm{in}}, V_{\mathrm{eff}}] = \sum_{n} f_n \epsilon_n - \rho^{\mathrm{in}} V_{\mathrm{xc}} + E_{\mathrm{xc}}^{\mathrm{in}} - E_{\mathrm{H}}^{\mathrm{in}} + E_{ZZ}. \tag{3.62}$$

This formulation has the feature discussed following eq. (2.40), that, if the input charge density is a superposition of spherical atomic-like densities, the last two terms take the form of a constant (the Hartree energies of the separate atomic densities, with a negative sign!), plus a sum of rapidly screened Coulomb repulsions between the ions.

4

LINEAR RESPONSE THEORY

There are two situations in which we need to know about the linear response of the electron density to a perturbation in the external potential. One, which will be described in detail in Chapter 7, is the theory of the energy and interatomic forces in simple metals. These 'simple' metals are the nearly free electron metals, in which the electron density and hence (from the Hellman–Feynman theorem) the interatomic forces can be described in terms of the linear response of jellium to the perturbation induced by discrete ions. The other situation is in the calculation of phonon frequencies, which can be formulated in terms of the response of the electron density to a small displacement of a nucleus. In this chapter, we shall examine the concepts and the formulae of linear response theory that are useful for these purposes and see how they are related to the ideas of previous chapters. For a general view of the subject, an excellent reference is Inkson (1984).

4.1 Definition of the Response Function $\chi_e(\mathbf{r}, \mathbf{r}')$

The basic idea of static linear response theory is as follows. We start with a system of electrons in its ground state with an external potential V_{ext1} and density ρ_1 and apply the perturbation ΔV_{ext} as before, where we would like ΔV_{ext} to be small enough for change in the density to be accurately given by the first term in the expansion:

$$\rho_2 - \rho_1 = \Delta\rho(\mathbf{r}) = \int \chi_e(\mathbf{r}, \mathbf{r}')\Delta V_{ext}(\mathbf{r}')\,d\mathbf{r}' + O[(\Delta V_{ext})^2]. \tag{4.1}$$

The function χ_e will be referred to as the *density response function* and linear response theory is concerned with its functional form and properties. The response function here is a function of the electron density ρ_1, evaluated at the ground state in the potential V_{ext1}. χ_e will be different for each system, and a functional of the ground state density, but in the interests of light notation we shall not explicitly show the density-dependence of χ_e unless we need it. As before, we imagine the perturbation to be applied sufficiently gently that the electrons are always in their ground state.

A full linear response theory also deals with time-dependent perturbations, which usually means situations in which this *adiabatic* condition is not a valid approximation. One common way to make non-adiabatic changes for example would be to shine light on a material, which is a simple way of applying a rapidly oscillating perturbation to the external potential. However, non-adiabatic response is a major topic in itself and beyond the scope of this book.

If the external potential simply changes by a constant, independent of $\mathbf{r}$, there will be no change induced in the electron density. The potential is in any case measured with respect to a completely arbitrary value which is taken as the zero of potential. Formally, this implies that for any response function χ_e

$$\int \chi_e(\mathbf{r}, \mathbf{r}')\,d\mathbf{r}' = 0. \tag{4.2}$$

The reader should be aware of a common source of confusion when comparing formulae between sources. Some authors adopt the convention that the response function gives the response to a change ΔV_H in V_H rather than to the change ΔV_{ext}.

The relevance of linear response to interatomic forces lies in the fact that the response function is closely related to the second order term in the density functional. Hence a knowledge of χ_e helps us to build approximate density functionals. Having adopted a density functional, we may want to study the vibration frequencies or phonons in a system. They depend on the force induced on each atom when another atom is displaced. These forces are directly related via the Hellmann–Feynman theorem to the induced charges, which in turn are determined by the response function.

4.2 Relationship to HKS Density Functional

To see the connection between χ_e and the HKS functional, we expand the exact energy functional about the exact charge density ρ_1 in the same way as we did to derive the second-order functional $E^{(2)}[\rho]$. It is convenient to separate the part of E^{HKS} defined as F^{HKS} in (2.29), which does not depend on the external potential, the expansion of which is easily shown to be:

$$\begin{aligned} F^{\mathrm{HKS}}[\rho] = F^{\mathrm{HKS}}[\rho_1] &+ \int \left(\left.\frac{\delta T_s}{\delta \rho}\right|_1 + V_{\mathrm{H}1} + V_{\mathrm{xc}1} \right) (\rho - \rho_1) \\ &+ \frac{1}{2} \int K_1 (\rho - \rho_1)(\rho' - \rho_1') + O[(\rho - \rho_1)^3]. \end{aligned} \tag{4.3}$$

We have defined the second derivative

$$K(\mathbf{r}, \mathbf{r}') = \frac{\delta^2 T_s}{\delta\rho\delta\rho'} + C(\mathbf{r}, \mathbf{r}') \tag{4.4}$$

and a quantity bearing the suffix 1 is evaluated at $\rho = \rho_1$.

Recall from (2.29) that the HKS total energy functional, including all terms in the external potential, is

$$E^{\text{HKS}}[\rho] = F^{\text{HKS}}[\rho] + \int \rho V_{\text{ext}} + E_{ZZ}. \tag{4.5}$$

Hence we have the following expansion for E^{HKS} about ρ_1 and $V_{\text{ext}1}$:

$$\begin{aligned} E^{\text{HKS}}[\rho, V_{\text{ext}2}] = E^{\text{HKS}}[\rho_1, V_{\text{ext}1}] + \int \rho \Delta V_{\text{ext}} + \Delta E_{ZZ} \\ + \frac{1}{2} \int K_1(\rho - \rho_1)(\rho' - \rho'_1) + O[(\rho - \rho_1)^3]. \end{aligned} \tag{4.6}$$

The first-order term has vanished by the variational property; since ρ_1 is the ground state density in the external potential $V_{\text{ext}1}$, eq. (2.30) tells us that

$$\left.\frac{\delta T_s}{\delta \rho}\right|_1 + V_{\text{H}1} + V_{\text{xc}1} + V_{\text{ext}1} = \mu_1. \tag{4.7}$$

Notice that the exact HKS functional (2.29) depends on the external potential only linearly through the term $\int \rho V_{\text{ext}}$, so V_{ext} does not appear in any quadratic or higher order terms in its expansion (3.35).

In the presence of ΔV_{ext} the minimum of E^{HKS} is no longer at $\rho = \rho_1$ but at $\rho = \rho_1 + \Delta\rho$. The density change must satisfy the Euler–Lagrange equation obtained by differentiating (4.6), namely to first order:

$$\Delta V_{\text{ext}} + \int K_1 \Delta \rho' = \mu. \tag{4.8}$$

Comparing this with (4.1), recalling that any constant can be added to ΔV_{ext}, we see that χ_e looks like the inverse of K_1, with a minus sign. Formally the relationship can be established as follows. First, consider (4.8) evaluated explicitly at $\mathbf{r}'$:

$$\Delta V_{\text{ext}}(\mathbf{r}') + \int K_1(\mathbf{r}', \mathbf{r}'')\Delta\rho(\mathbf{r}'')\, d\mathbf{r}'' = \mu. \tag{4.9}$$

Now multiply by $\chi_{e1}(\mathbf{r}, \mathbf{r}')$ and integrate over $\mathbf{r}'$, using (4.2), to give

$$\Delta\rho(\mathbf{r}) + \int \chi_{e1}(\mathbf{r}, \mathbf{r}')K_1(\mathbf{r}', \mathbf{r}'')\Delta\rho(\mathbf{r}'')\, d\mathbf{r}''\, d\mathbf{r}' = 0. \tag{4.10}$$

If **v**-representability holds, then any $\Delta\rho$ we choose must correspond to some ΔV_{ext} for which (4.10) must be true. It follows that (4.10) is true for arbitrary $\Delta\rho$. Hence

$$-\int \chi_{e1}(\mathbf{r}, \mathbf{r}')K_1(\mathbf{r}', \mathbf{r}'')\, d\mathbf{r}' = \delta(\mathbf{r} - \mathbf{r}''). \tag{4.11}$$

4.2.1 Matrix and Vector Algebra Notation for Integrals

The form of (4.11) is reminiscent of the product of two infinite dimensional matrices, or the product of two operators in the **r**-represention (see page 16). We could think of the operator $\hat{\chi}_e$ for example as having matrix elements $\langle \mathbf{r}|\hat{\chi}_e|\mathbf{r}'\rangle$. With this way of thinking, $-K(\mathbf{r}', \mathbf{r}'')$ is the matrix element of the operator which is inverse to $\hat{\chi}_e$, which we can denote symbolically by $\hat{\chi}_e{}^{-1}$. These operators satisfy the relation

$$\hat{\chi}_e \hat{\chi}_e{}^{-1} = \hat{\mathbf{1}} \tag{4.12}$$

and the matrices satisfy the relation

$$\int \chi_e(\mathbf{r}, \mathbf{r}'')\chi_e^{-1}(\mathbf{r}'', \mathbf{r}')\,d\mathbf{r}'' = \delta(\mathbf{r} - \mathbf{r}'). \tag{4.13}$$

It should be clear that $\chi_e^{-1}(\mathbf{r}'', \mathbf{r}')$ does *not* mean the inverse or reciprocal of the function $\chi_e(\mathbf{r}'', \mathbf{r}')$, it means the $(\mathbf{r}'', \mathbf{r}')$ element of the matrix χ_e^{-1}. An appreciation that these objects with two spacial coordinates are matrices also allows us to drop the hat from them in equations such as (4.12). Furthermore if we also understand quantities with one position argument, such as local potentials, to be vectors, we can further simplify our notation by dropping integral signs where the meaning is obvious. This matrix and vector convention will be adopted henceforth as a convenient shorthand.

In terms of χ_{e1} we can write the HKS functional to second order (4.6) as

$$\begin{aligned} E^{\mathrm{HKS}}[\rho, V_{\mathrm{ext2}}] &= E^{\mathrm{HKS}}[\rho_1, V_{\mathrm{ext1}}] + \rho\Delta V_{\mathrm{ext}} + \Delta E_{ZZ} \\ &\quad - \tfrac{1}{2}(\rho - \rho_1)\chi_{e1}^{-1}(\rho - \rho_1) + O[(\rho - \rho_1)^3]. \end{aligned} \tag{4.14}$$

We see immediately that if we minimize this with respect to ρ we recover directly (4.1), or in our more compact notation

$$\rho - \rho_1 \to \Delta\rho = \chi_{e1}\Delta V_{\mathrm{ext}}. \tag{4.15}$$

The relationship between (4.14) and the second-order energy E^{so} also follows directly. If we substitute the ground state final density ρ_2 given by (4.1) into (4.14) we find:

$$\begin{aligned} E^{\mathrm{HKS}}[\rho_2, V_{\mathrm{ext2}}] &= E^{\mathrm{HKS}}[\rho_1, V_{\mathrm{ext1}}] + \int \rho_1(\mathbf{r})\Delta V_{\mathrm{ext}}(\mathbf{r})\,d\mathbf{r} + \Delta E_{ZZ} \\ &\quad + \frac{1}{2}\int \Delta\rho(\mathbf{r})\Delta V_{\mathrm{ext}}(\mathbf{r})\,d\mathbf{r} + O[(\Delta\rho)^3], \end{aligned} \tag{4.16}$$

which up to second order reproduces E^{so} from (3.18). The first two terms in this expression, truncating it to first order, are the Hellmann–Feynman theorem. Notice how the factor of $\frac{1}{2}$ has appeared again in the second-order term. To obtain (4.16) we made these substitutions in (4.14):

$$\rho \Delta V_{\text{ext}} \to \rho_1 \Delta V_{\text{ext}} + \Delta\rho \Delta V_{\text{ext}}$$

and

$$-\tfrac{1}{2}\Delta\rho \chi_{\text{e}1}^{-1} \Delta\rho \to -\tfrac{1}{2}\Delta\rho\Delta V_{\text{ext}}.$$

The cancellation of half of the term $\Delta\rho\Delta V_{\text{ext}}$ has exactly the same physical origin as the way the factor of $\frac{1}{2}$ appears in the classical image interaction. The direct negative energy of interaction of $\Delta\rho$ with ΔV_{ext} is exactly half compensated by the self-energy of $\Delta\rho$, which is described by the term $-\frac{1}{2}\Delta\rho\chi_{\text{e}1}^{-1}\Delta\rho$. That term is actually positive in sign, whatever the density change $\Delta\rho$. We can express this last property formally by saying that the matrix $-\chi_{\text{e}}^{-1}$ is *positive definite*, where we now drop the suffix 1 because these considerations must apply to *any* ground state density. For convenience let us write this matrix explicitly here in the **r**-representation:

$$\begin{aligned} -\chi_{\text{e}}^{-1}(\mathbf{r},\mathbf{r}') &= \frac{\delta^2 T_{\text{s}}}{\delta\rho(\mathbf{r})\delta\rho(\mathbf{r}')} + C(\mathbf{r},\mathbf{r}') \\ &\equiv \frac{\delta^2 T_{\text{s}}}{\delta\rho(\mathbf{r})\delta\rho(\mathbf{r}')} + \frac{1}{|\mathbf{r}-\mathbf{r}'|} + \frac{\delta^2 E_{\text{xc}}[\rho]}{\delta\rho(\mathbf{r})\delta\rho(\mathbf{r}')}. \end{aligned} \tag{4.17}$$

4.3 The Non-interacting Response Function

We have discussed how the first-order density response is given by the change in external potential by $\Delta\rho = \chi_{\text{e}}\Delta V_{\text{ext}}$. Recall that the real system is associated with a fictitious system of non-interacting electrons. Non-interacting electrons have their own response function which we can call χ_{s}. In order for the non-interacting system to always faithfully reproduce the density of the real, interacting system, the density change $\Delta\rho$ must in addition be given by

$$\Delta\rho = \chi_{\text{s}}\Delta V_{\text{eff}}. \tag{4.18}$$

Since $C = 0$ for a system of non-interacting electrons, from (4.17) it must be true that

$$-\chi_{\text{s}}^{-1}(\mathbf{r},\mathbf{r}') = \frac{\delta^2 T_{\text{s}}}{\delta\rho(\mathbf{r})\delta\rho(\mathbf{r}')}. \tag{4.19}$$

From the definitions of χ_{e}^{-1} and χ_{s}^{-1} in (4.17) and (4.19) respectively, these matrices are symmetric in their 'indices' $(\mathbf{r},\mathbf{r}')$. It follows that χ_{e} and χ_{s} are also symmetric matrices.

We can also express the interaction function C in terms of these response functions using (4.17) and (4.19):

$$C = \chi_s^{-1} - \chi_e^{-1}. \tag{4.20}$$

4.4 The Dielectric Function

Another closely related function is sometimes used, the electron dielectric function ϵ_e, which is first defined in terms of its inverse:

$$\Delta V_{\text{eff}} = \epsilon_e^{-1} \Delta V_{\text{ext}}. \tag{4.21}$$

It describes how within the model of independent particles a 'bare' potential is screened to give an effective potential. From the relation

$$\Delta V_{\text{eff}} = \Delta V_{\text{ext}} + C \Delta\rho \tag{4.22}$$

and using (4.1) we have that

$$\epsilon_e^{-1} \Delta V_{\text{ext}} = \Delta V_{\text{ext}} + C \chi_e \Delta V_{\text{ext}}. \tag{4.23}$$

This has to be true for all ΔV_{ext}, therefore in matrix notation:

$$\epsilon_e^{-1} = 1 + C\chi_e. \tag{4.24}$$

If we substitute C from (4.20) we have

$$\epsilon_e^{-1} = \chi_s^{-1} \chi_e \tag{4.25}$$

from which

$$\epsilon_e = \chi_e^{-1} \chi_s \equiv 1 - C\chi_s. \tag{4.26}$$

Each term in a matrix equation is a matrix. So the simple 1 in eqs. (4.24) or (4.26) now stands for the unit or identity matrix, the matrix elements of which are (in the $\mathbf{r}$-representation) $\delta(\mathbf{r} - \mathbf{r}')$. The usual rules of matrix algebra apply even to these infinite dimensional matrices. If in doubt, translate the equations into the standard integral notation and try them out.

If there were no exchange and correlation, C would reduce to the familiar Coulomb interaction $v_C = 1/\left|\mathbf{r} - \mathbf{r}'\right|$. The exchange and correlation is sometimes regarded as a correction to the Coulomb potential due to the local field, and C is then written in the form

$$C = v_C(1 - \mathcal{G}), \tag{4.27}$$

where $\mathcal{G}$ is called the *local field correction*. Note again that this is a multiplication of infinite matrices, which can be represented in $\mathbf{r}$- or $\mathbf{k}$-space. The $\mathbf{k}$-space representation has certain advantages, as we will discuss in more detail.

Whereas the dielectric function ϵ_e describes how a bare external potential is screened to give the effective potential as seen by an *electron* there is a classical dielectric function that describes how the bare external potential is screened to give the potential as seen by a *test charge*. The difference is that a test charge sees the electrostatic potential without the exchange and correlation part. Let us call this dielectric function simply ϵ, since there will be no likelihood of confusion with the energies that take this symbol. It is defined by the relation:

$$\Delta V_H + \Delta V_{ext} = \epsilon^{-1} \Delta V_{ext}. \tag{4.28}$$

We can express ϵ in terms of the response function defined previously; using

$$\Delta V_H = v_C \Delta \rho. \tag{4.29}$$

We find

$$\epsilon^{-1} = 1 + v_C \chi_e, \tag{4.30}$$

which can be written out in full as

$$\epsilon^{-1}(\mathbf{r}, \mathbf{r}') = \delta(\mathbf{r} - \mathbf{r}') + \int \frac{1}{|\mathbf{r} - \mathbf{r}''|} \chi_e(\mathbf{r}'', \mathbf{r}) \, d\mathbf{r}''. \tag{4.31}$$

4.5 The Error in the Harris–Foulkes Functional

With the help of the response functions we now return to the still open question of the size of the error in the Harris–Foulkes functional. To study the error, we can follow the same approach as we did to study the error in the second order functional. In addition we will apply the ideas of linear response theory. Again to lighten the equations, I will abbreviate the notation a little more, by omitting, when there is no ambiguity, the integral sign before the product of a charge density and a potential. It is understood that such products are like matrix products, or traces to be more precise. The first step is to expand $E^{HF}[\rho^{in}]$ to second order about ρ^{ex} (we could equally well expand $E^{HF}[\rho^{ex}]$ about ρ^{in}). We do this by first writing it in the form:

$$\begin{aligned} E^{HF}[\rho^{in}] = {} & E^{HF}[\rho^{ex}] + T_s[\rho^{out}] - T_s[\rho^{ex}] + \left(\rho^{out} V_{eff}^{in} - \rho^{ex} V_{eff}^{ex}\right) \\ & + (E_{xc}^{in} - E_{xc}^{ex}) - \left(\rho^{in} V_{xc}^{in} - \rho^{ex} V_{xc}^{ex}\right) - \left(E_H^{in} - E_H^{ex}\right), \end{aligned} \tag{4.32}$$

in order to expose the changes in all the ingredients of E^{HF} as we change the density from ρ^{ex} to ρ^{in}. It is these changes which we then expand to second order, using the relationshps we used before. This is an exercise for the interested reader. Having done that, it is necessary to exploit the two variational conditions, namely the one for the interacting electrons (3.41) and the one for the non-interacting electrons analogous to (3.43):

$$\left.\frac{\delta T_s[\rho]}{\delta \rho}\right|_{\rho^{\text{out}}} + V_{\text{eff}}^{\text{in}} = \mu^{(1)}. \tag{4.33}$$

The first condition (3.41) makes the first-order terms vanish, reducing the expansion to

$$\begin{aligned} E^{\text{HF}}[\rho^{\text{in}}] = E^{\text{HF}}[\rho^{\text{ex}}] &- \tfrac{1}{2}(\rho^{\text{out}} - \rho^{\text{ex}})\chi_s^{-1}(\rho^{\text{out}} - \rho^{\text{ex}}) \\ &+ (\rho^{\text{in}} - \rho^{\text{ex}})K_{\text{ex}}(\rho^{\text{out}} - \rho^{\text{ex}}) \\ &- \tfrac{1}{2}(\rho^{\text{in}} - \rho^{\text{ex}})K_{\text{ex}}(\rho^{\text{in}} - \rho^{\text{ex}}). \end{aligned} \tag{4.34}$$

It clearly looks as if the error is second order, but we want to be more precise than that, we want to be sure that it is second order in $\rho^{\text{in}} - \rho^{\text{ex}}$. In other words we would like to convert the factor of $\rho^{\text{out}} - \rho^{\text{ex}}$ in the second line of the above equation into a factor of $\rho^{\text{in}} - \rho^{\text{ex}}$. This is where linear response theory comes in useful.

The relation (4.33) can also be expanded about ρ^{ex} to give the useful equation to first order:

$$\begin{aligned} &\int \frac{\delta^2 T_s}{\delta\rho(\mathbf{r})\delta\rho(\mathbf{r}')}(\rho^{\text{out}}(\mathbf{r}') - \rho^{\text{ex}}(\mathbf{r}'))\,d\mathbf{r}' + \int \frac{1}{|\mathbf{r} - \mathbf{r}'|}(\rho^{\text{in}}(\mathbf{r}') - \rho^{\text{ex}}(\mathbf{r}'))\,d\mathbf{r}' \\ &+ \int \frac{\delta^2 E_{\text{xc}}}{\delta\rho(\mathbf{r})\delta\rho(\mathbf{r}')}(\rho^{\text{in}}(\mathbf{r}') - \rho^{\text{ex}}(\mathbf{r}'))\,d\mathbf{r}' = \text{constant}. \end{aligned} \tag{4.35}$$

In terms of the matrices we have defined before, if we multiply this relation by χ_s we obtain to first order

$$\rho^{\text{out}} - \rho^{\text{ex}} = \chi_s C(\rho^{\text{in}} - \rho^{\text{ex}}). \tag{4.36}$$

Equation (4.36) has a simple interpretation; indeed, some people might find it obvious. If we start with the ground state density ρ^{ex} and distort it by means of some hypothetical force field to ρ^{in}, this change in ρ induces an effective potential $C(\rho^{\text{in}} - \rho^{\text{ex}})$. This potential perturbs the density of the electrons, and their response is given by (4.36). Equation (4.36) can be substituted into the expansion of $E^{\text{HF}}[\rho^{\text{in}}]$ to give it the simple form, entirely in terms of $\rho^{\text{in}} - \rho^{\text{ex}}$:

$$E^{\text{HF}}[\rho^{\text{in}}] = E^{\text{HF}}[\rho^{\text{ex}}] - \tfrac{1}{2}(\rho^{\text{in}} - \rho^{\text{ex}})\epsilon_e C(\rho^{\text{in}} - \rho^{\text{ex}}). \tag{4.37}$$

This shows explicitly that the error is second order, just as the error in the HKS functional at a density close to ρ^{ex}. Note however that $E^{\text{HF}}[\rho^{\text{in}}]$ is not an upper bound to the exact energy, as for example $E^{\text{HKS}}[\rho^{\text{in}}]$ would be. In fact it was conjectured at one time that it might be a *lower* bound to the energy (Finnis, 1990; Zaremba, 1990). As Zaremba (1990) showed, for this to be true the matrix $\epsilon_e C$ would have to be positive definite. He showed that a necessary and sufficient condition for this is that the matrix C be positive definite and that this is equivalent to the matrix $\epsilon_e - 1$ being positive definite. At first sight these conditions seemed rather likely to be satisfied. After all, C is just the Coulomb interaction v_e, which is obviously positive definite, reduced by a local field correction. If $E^{\mathrm{HF}}[\rho^{\mathrm{in}}]$ were an upper bound just as $E^{\mathrm{HKS}}[\rho]$ is a lower bound to the exact energy $E^{\mathrm{HKS}}[\rho^{\mathrm{ex}}]$ there would be interesting possibilities for bracketing the exact energy, as well as for algorithms which sought the extremum of E^{HF}. Could the local field correction $\mathcal{G}$ in (4.27) really turn v_{C} into an attraction? Sadly, it was soon proved (Robertson and Farid, 1991; Farid *et al.*, 1993) that these conditions for a maximum in E^{HF} are certainly not satisfied. The local field corrections do, for short wavelength disturbances of the electron density, outweigh the repulsive Coulomb interaction, since the latter goes to zero as the square of the wavelength. This became evident in practical calculations.

As a consequence we must accept that the Harris–Foulkes functional has a saddle-point in the space of variation of ρ^{in} when it is equal to the exact charge density, rather than a maximum. Nevertheless, the error is of second order in $(\rho^{\mathrm{in}} - \rho^{\mathrm{ex}})$ and it has been shown that the Harris–Foulkes functional can therefore be remarkably accurate (Polatoglou and Methfessel, 1988; Paxton *et al.*, 1990; Polatoglou and Methfessel, 1990). Taking free atom charge densities and rigidly overlapping them is the standard procedure for constructing ρ^{in}. Polatoglou and Methfessel were the first to show that even in the case of the ionic material NaCl, by overlapping *atomic* charge densities the cohesive energy, lattice parameter and bulk modulus could be well reproduced by E^{HF} (see Table 4.1).

A trickier case arises at surfaces in metals, where the relaxation of the charge becomes more important than in the bulk. It was shown (Finnis, 1990) that for E^{HF} to give reasonably good results, ρ^{in} has to be better than superimposed atomic charges. A small contraction of the atomic charge turns out to be sufficient to reproduce the surface energy very well. It was surprising to discover that the exact

TABLE 4.1 The cohesive energy (E^{coh}) lattice parameter (a) and bulk modulus (B) of NaCl calculated with the Harris–Foulkes functional and overlapping atomic charge densities (Polatoglou and Methfessel, 1988)

	E^{coh} (eV/atom)	a (a.u.)	B (Mbar)
Harris–Foulkes	3.56	10.72	0.25
Experiment	3.25	10.65	0.25

atomic charge densities in this case are simply too crude a starting point, when they had given such startlingly good results for bulk energies versus volume for a range of materials, even comparing the small energy differences between crystal structures of the same material. The explanation for this lies in the sensitivity of the surface energy to the electrostatic dipole across the surface; it is necessary to use input atomic-like densities which get this at least approximately right, which exact atomic charge densities notoriously fail to do. These leads to a large error in ρ^{out}, and the second-order error becomes comparable to the surface energy itself. The same contracted atomic charge densities also do a rather better job of the vacancy formation energy than free atom densities. Alternative improvements to free atomic charge densities have been discussed by other authors (Chetty *et al.*, 1991; Stokbro *et al.*, 1994; Hartford *et al.*, 1996).

4.6 Linear Response and the Green Function

We have seen in eq. (1.212) how the electron density can be expressed in terms of the single-particle Green function operator $\hat{G}$, which in the limit $\eta \to 0$ is the inverse of the operator $(\epsilon + i\eta - \hat{H})$. It must therefore be possible to express a change in ρ in terms of a change in $\hat{G}$, and this in terms of a change in $\hat{H}$. By exploring this relation we obtain a useful expression for the response function in terms of $\hat{G}$.

Let us change the Hamiltonian by ΔV_{eff}. This sounds a peculiar strategy, since we really have no direct control over ΔV_{eff}, nevertheless you can imagine that we actually create the change in V_{eff} by changing V_{ext}, which we suppose we *can* control. We consider the change ΔV_{eff} rather than ΔV_{ext}, at least in the first instance, because the resulting $\Delta\rho$ is related to ΔV_{eff} directly via the non-interacting response function χ_s, as in eq. (4.18). The Green operator changes to

$$\begin{aligned}
\hat{G} + \Delta\hat{G} &= (\epsilon - \hat{H} - \Delta V_{\text{eff}})^{-1} \\
&= (\epsilon - \hat{H} - (\epsilon - \hat{H})\hat{G}\Delta V_{\text{eff}})^{-1} \\
&= [(\epsilon - \hat{H})(1 - \hat{G}\Delta V_{\text{eff}})]^{-1} \\
&= (1 - \hat{G}\Delta V_{\text{eff}})^{-1}(\epsilon - \hat{H})^{-1} \\
&= (1 - \hat{G}\Delta V_{\text{eff}})^{-1}\hat{G}.
\end{aligned} \tag{4.38}$$

Now we can expand the first factor in powers of ΔV_{eff} and ignore all but the first-order term to get:

$$\Delta\hat{G} = \hat{G}\Delta V_{\text{eff}}\hat{G} + O\big[\Delta V_{\text{eff}}^2\big]. \tag{4.39}$$

Turning next to the induced electron density, from (1.212) we can write

$$\Delta\rho(\mathbf{r}) = -\frac{2}{\pi}\mathrm{Im}\int_{-\infty}^{+\infty} \mathrm{d}\epsilon f_F(\epsilon)\langle\mathbf{r}|\Delta\hat{G}|\mathbf{r}\rangle, \tag{4.40}$$

where we have assumed that the perturbation is such as not to change the Fermi energy. This will be the case if it is a perturbation on the atomic scale somewhere in a system which we can suppose to be arbitrarily large. By making the system larger, any change in the Fermi energy becomes smaller. From (4.39) and (4.40) we have, in the linear response limit:

$$\Delta\rho(\mathbf{r}) = -\frac{2}{\pi}\mathrm{Im}\int_{-\infty}^{+\infty} \mathrm{d}\epsilon f_F(\epsilon)\int G(\mathbf{r},\mathbf{r}')\Delta V_{\mathrm{eff}}(\mathbf{r}',\mathbf{r}'')G(\mathbf{r}'',\mathbf{r})\,\mathrm{d}\mathbf{r}'\,\mathrm{d}\mathbf{r}''. \tag{4.41}$$

If the change in potential is local, then we can write

$$\Delta V_{\mathrm{eff}}(\mathbf{r}',\mathbf{r}'') = \Delta V_{\mathrm{eff}}(\mathbf{r}')\delta(\mathbf{r}-\mathbf{r}'') \tag{4.42}$$

and identify the response function as

$$\chi_s(\mathbf{r},\mathbf{r}') = -\frac{2}{\pi}\mathrm{Im}\int_{-\infty}^{+\infty} \mathrm{d}\epsilon f_E(\epsilon)G(\mathbf{r},\mathbf{r}')G(\mathbf{r}',\mathbf{r}). \tag{4.43}$$

A useful expression is obtained by representing $\Delta\hat{G}$ in (4.40) in the basis of eigenstates, in which G is diagonal. By inserting the identity operator we have

$$\Delta\rho(\mathbf{r}) = -\frac{2}{\pi}\mathrm{Im}\int_{-\infty}^{+\infty} \mathrm{d}\epsilon f_F(\epsilon)\sum_{n,n'}\langle\mathbf{r}|n\rangle\langle n|\hat{G}|n\rangle\langle n|\hat{\Delta}V_{\mathrm{eff}}|n'\rangle\langle n'|\hat{G}|n'\rangle\langle n'|\mathbf{r}\rangle. \tag{4.44}$$

The integral over energy can be taken further if we write the Green function matrix elements explicitly, using eq. (1.202), as

$$\begin{aligned}\langle n|\hat{G}|n\rangle\langle n'|\hat{G}|n'\rangle &= \frac{1}{(\epsilon+\mathrm{i}\eta-\epsilon_n)}\frac{1}{(\epsilon+\mathrm{i}\eta-\epsilon_{n'})}\\ &= \frac{1}{(\epsilon_n-\epsilon_{n'})}\left[\frac{1}{(\epsilon+\mathrm{i}\eta-\epsilon_n)}-\frac{1}{(\epsilon+\mathrm{i}\eta-\epsilon_{n'})}\right].\end{aligned} \tag{4.45}$$

Substituting this form into (4.44) and using the definition (1.197) of the delta function enables us to do the integral over ϵ to obtain

$$\Delta\rho(\mathbf{r}) = 2\sum_{n,n'}\langle\mathbf{r}|n\rangle\langle n'|\mathbf{r}\rangle\frac{(f_F(\epsilon_n)-f_F(\epsilon_{n'}))}{(\epsilon_n-\epsilon_{n'})}\langle n|\hat{\Delta}V_{\mathrm{eff}}|n'\rangle. \tag{4.46}$$

This represents a general form of the linear response equation which offers a practical route for calculations. In particular at $T = 0\,K$ it has a particularly simple structure, because the Fermi factors are either 1 or 0, depending on whether ϵ_n and $\epsilon_{n'}$ are below or above ϵ_F. So the only contributing terms from the pairs n and n' are those when one state is occupied and the other is unoccupied. This also has a physical interpretation, in that the density perturbation arises from electrons that are scattered from below to above the Fermi energy.

4.7 Linear Response in Jellium

In particular, we can obtain the jellium linear response function from (4.46) by inserting the eigenstates and energies of plane waves. Equation (4.46) becomes:

$$\Delta\rho(\mathbf{r}) = 2\sum_{\mathbf{k},\mathbf{k}'}\langle\mathbf{r}|\mathbf{k}\rangle\langle\mathbf{k}'|\mathbf{r}\rangle\frac{(f_F(\epsilon_{\mathbf{k}}) - f_F(\epsilon_{\mathbf{k}'}))}{(\epsilon_{\mathbf{k}} - \epsilon_{\mathbf{k}'})}\langle\mathbf{k}|\Delta\hat{V}_{\text{eff}}|\mathbf{k}'\rangle, \tag{4.47}$$

where $\epsilon_{\mathbf{k}} = k^2/2$ and $\epsilon_{\mathbf{k}'} = k'^2/2$.

To obtain an explicit form for the response function we now consider the effect of a sinusoidal perturbation to the effective potential, a local potential given by

$$\Delta V_{\text{eff}}(\mathbf{r}) = \Delta V_q \exp(-\mathbf{q}\cdot\mathbf{r}), \tag{4.48}$$

whose matrix elements in $\boldsymbol{r}$-space are

$$\langle\mathbf{r}|\Delta\hat{V}_{\text{eff}}|\mathbf{r}'\rangle = \Delta V_q \exp(-\mathbf{q}\cdot\mathbf{r})\delta(\mathbf{r}-\mathbf{r}'). \tag{4.49}$$

Its matrix elements in $\boldsymbol{k}$-space are easily evaluated, and are

$$\langle\mathbf{k}|\Delta\hat{V}_{\text{eff}}|\mathbf{k}'\rangle = \frac{(2\pi)^3}{V}\Delta V_q\delta(\mathbf{k}'-\mathbf{q}-\mathbf{k}). \tag{4.50}$$

We can now start with the $\mathbf{k}'$ summation in (4.47), converting it to an integral in the usual way by substituting $\sum_{\mathbf{k}'} \to V/(2\pi)^3\int d\mathbf{k}'$. The result is easily found to be

$$\begin{aligned}\Delta\rho(\mathbf{r}) &= 2\sum_{\mathbf{k}}\langle\mathbf{r}|\mathbf{k}\rangle\langle\mathbf{k}+\mathbf{q}|\mathbf{r}\rangle\frac{(f_F(\epsilon_{\mathbf{k}}) - f_F(\epsilon_{\mathbf{k}+\mathbf{q}}))}{(\epsilon_{\mathbf{k}} - \epsilon_{\mathbf{k}+\mathbf{q}})}\Delta V_q \\ &= \frac{2}{V}\sum_{\mathbf{k}}\frac{(f_F(\epsilon_{\mathbf{k}}) - f_F(\epsilon_{\mathbf{k}+\mathbf{q}}))}{(\epsilon_{\mathbf{k}} - \epsilon_{\mathbf{k}+\mathbf{q}})}\Delta V_q\exp(-i\mathbf{q}\cdot\mathbf{r}).\end{aligned} \tag{4.51}$$

The remaining summation over $\mathbf{k}$ can also be done by expressing it as an integral, but this one is a bit tricky. Let us confine ourselves to the $T = 0$ K case, in which the Fermi function $f_E(\epsilon)$, defined in eqn (1.168), is unity for $\epsilon \leq \epsilon_F$ and zero for energies $\epsilon > \epsilon_F$. The effect of finite temperature is to smear out the integrand close to the Fermi surface, which would make the integral even more difficult. As it is

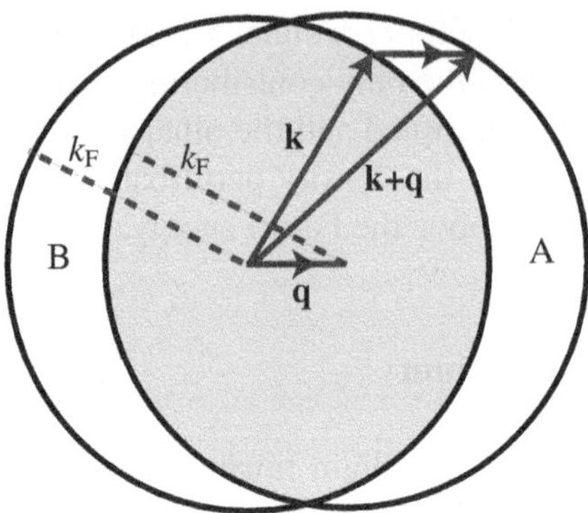

FIG. 4.1 Fermi spheres displaced by $\mathbf{q}$.

there are two cases to consider. First, consider the case $q < 2k_F$. Let the vector $\mathbf{k}$ range over the interior of the Fermi sphere. As it does so $\mathbf{k}+\mathbf{q}$ is the vector measured from the centre of the sphere shifted by $-\mathbf{q}$ as in Fig. 4.1. The only terms which do not vanish when $k < k_F$ are those for which the vector $|\mathbf{k}+\mathbf{q}| > k_F$. This only occurs when $\mathbf{k}$ is within the shaded moon-shaped segment to the right of the sphere, let us call it region A. Now let $\mathbf{k}$ range outside the Fermi sphere ($k > k_F$). Exactly the same contributions come from the moon-shaped segment to the left of the sphere, let us call it region B. Hence there is an integration over region A to be done, which we can multiply by two to take care of the integration over region B. Equation (4.51) becomes:

$$\Delta\rho(\mathbf{r}) = \frac{1}{2\pi^3}\int_A \frac{1}{(\epsilon_{\mathbf{k}} - \epsilon_{\mathbf{k}+\mathbf{q}})}\,\mathrm{d}\mathbf{k}\Delta V_q \exp(-\mathrm{i}\mathbf{q}\cdot\mathbf{r}). \tag{4.52}$$

This reduces to a two-dimensional integral because of the axial symmetry about $\mathbf{q}$. It is straightforward to evaluate it by expressing $\mathrm{d}\mathbf{k}$ as $2\pi k^2 \sin\theta \mathrm{d}k\mathrm{d}\theta$; details are left as an exercise for the reader. The result is

$$\Delta\rho(\mathbf{r}) = -\frac{k_F}{\pi^2}\left[\frac{1}{2} + \frac{4k_F^2 - q^2}{8k_F q}\ln\left(\frac{2k_F + q}{2k_F - q}.\right)\right]\Delta V_q \exp(-\mathrm{i}\mathbf{q}\cdot\mathbf{r}). \tag{4.53}$$

The complication is now apparent in the negative argument of the logarithm when $q > 2k_F$; the Fermi spheres centred at the origin and at q no longer overlap, so the moon-shaped region of integration is no longer appropriate, although the procedure is very similar. The result of the integration is almost the same as before except the argument of the logarithm has the opposite sign, with the effect that it remains positive for all q. The results of both regimes can be summarized then as follows:

$$\Delta\rho(\mathbf{r}) = \chi_s(q)\Delta V_q \exp(-\mathrm{i}\mathbf{q}\cdot\mathbf{r}), \tag{4.54}$$

where

$$\chi_s(q) = \frac{2}{V}\sum_{\mathbf{k}} \frac{(f_F(\epsilon_{\mathbf{k}}) - f_F(\epsilon_{\mathbf{k}+\mathbf{q}}))}{(\epsilon_{\mathbf{k}} - \epsilon_{\mathbf{k}+\mathbf{q}})}$$

$$= -\frac{k_F}{\pi^2}\left[\frac{1}{2} + \frac{4k_F^2 - q^2}{8k_F q}\ln\left|\frac{2k_F + q}{2k_F - q}\right|\right] \tag{4.55}$$

is a third represention of the response function χ_s, and the one most commonly used. It is plotted in Fig. 4.2.

Notice that in the long wavelength (small q) limit, χ_s tends to $-k_F/\pi^2$. This is exactly the value χ_s takes for all q in the Thomas–Fermi approximation. It corresponds to completely local screening in **r**-space. The reduction in magnitude with increasing q makes sense if we think that shorter wavelength disturbances of the electron gas are associated with plane waves of higher kinetic energy, which will be harder to excite. The singularity at $q = 2k_F$ is interesting, because although it is weak as singularities go, it has measurable effects for example on vibration spectra (so-called *Kohn-anomalies*). Remember the singularity only arises because we took the $T = 0$ limit, so there is a sharp transition in the occupancies $f_F(\epsilon_k)$ from 1 to 0 as **k** goes through the Fermi surface.

Equation (4.54) is telling us that each Fourier component of the perturbing potential in jellium is screened independently simply by a multiplicative factor which depends on its wavelength. A Fourier analysis of any perturbing potential can be made, and the Fourier components of the induced charge can be generated by multiplying each by $\chi_s(q)$. This is a simple result which is possible only because of the

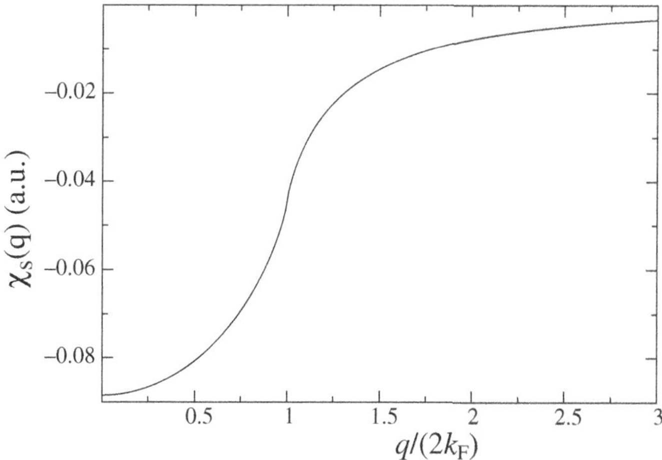

FIG. 4.2 The response function χ_s for independent electrons at the density of aluminium.

translational invariance of jellium, as we shall now study. Formally, the connection between this new (q) representation and the ($\mathbf{k}, \mathbf{k}'$) and ($\mathbf{r}, \mathbf{r}'$) representations can be derived by expressing the original linear response equation in **k**-space:

$$\Delta\rho(\mathbf{k}) = \sum_{\mathbf{k}} \chi_s(\mathbf{k}, \mathbf{k}')\Delta V_{\text{eff}}(\mathbf{k}') = \frac{V}{2\pi^3} \int \chi_s(\mathbf{k}, \mathbf{k}')\Delta V_{\text{eff}}(\mathbf{k}')\, \mathrm{d}\mathbf{k}'. \qquad (4.56)$$

In order to make the connection we need to explicitly assign the precise meaning to these quantities, which are Fourier transforms of their real space equivalents. Any of the matrices defined previously can be represented in **k**-space as follows. Taking the operator $\hat{\chi}_e$ as an example we can express it in **k**-space by analogy with the Coulomb potential in eqn (1.64):

$$\langle\mathbf{k}|\hat{\chi}_e|\mathbf{k}'\rangle = \int \langle\mathbf{k}|\mathbf{r}\rangle\langle\mathbf{r}|\hat{\chi}_e|\mathbf{r}'\rangle.\langle\mathbf{r}'|\mathbf{k}'\rangle\, \mathrm{d}\mathbf{r}\, \mathrm{d}\mathbf{r}'. \qquad (4.57)$$

For simplicity we will adopt a conventional notation in which $\mathbf{q}$ is a vector in **k**-space and a function with arguments $\mathbf{q}$ or $\mathbf{g}$ (a reciprocal lattice vector) or $\mathbf{k}$ is understood to be the Fourier transform of the function with arguments $\mathbf{r}$ or $\mathbf{r}'$, without introducing extra notation such as a bar or a twiddle over the function, as is sometimes done. The matrix $\chi_s(\mathbf{k}, \mathbf{k}')$ for example is then the Fourier transform of the matrix $\chi_s(\mathbf{r}, \mathbf{r}')$. Although one might want to distinguish it, perhaps by writing $\tilde{\chi}_s(\mathbf{q}, \mathbf{q}')$, I shall not do so here. By a similar convention $\mathbf{u}$ and $\mathbf{r}$ will always mean vectors in $\boldsymbol{r}$-space. The explicit Fourier transform is then

$$\chi_s(\mathbf{k}, \mathbf{k}') = \frac{1}{V} \int \chi_s(\mathbf{r}, \mathbf{r}') \exp(\mathrm{i}(\mathbf{k} \cdot \mathbf{r} - \mathbf{k}' \cdot \mathbf{r}'))\, \mathrm{d}\mathbf{r}\, \mathrm{d}\mathbf{r}', \qquad (4.58)$$

with its inverse

$$\chi_s(\mathbf{r}, \mathbf{r}') = \frac{V}{(2\pi)^6} \int \chi_s(\mathbf{k}, \mathbf{k}') \exp(-\mathrm{i}(\mathbf{k} \cdot \mathbf{r} - \mathbf{k}' \cdot \mathbf{r}'))\, \mathrm{d}\mathbf{k}\, \mathrm{d}\mathbf{k}'. \qquad (4.59)$$

Equations (4.58) and (4.59) would apply to any system. In the special case of jellium the absolute location in space of a perturbation is irrelevant. If the perturbation is shifted by $\mathbf{u}$, the response is also shifted by $\mathbf{u}$. In this case, there is no need to express the response functions as functions of $\mathbf{r}$ and $\mathbf{r}'$ independently, since they can be expressed as functions of $|\mathbf{r} - \mathbf{r}'|$, which also makes their Fourier transforms simpler. Let $\mathbf{u} = \mathbf{r} - \mathbf{r}'$ and replace the variables $\mathbf{r}$ and $\mathbf{r}'$ by $\mathbf{r}$ and $\mathbf{u}$. Equation (4.58) becomes:

$$\begin{aligned}
\chi_s(\mathbf{k}, \mathbf{k}') &= \frac{1}{V} \int \chi_s(u) \exp(\mathrm{i}(\mathbf{k} - \mathbf{k}') \cdot \mathbf{r}) \exp(\mathrm{i}\mathbf{k}' \cdot \mathbf{u}))\, \mathrm{d}\mathbf{r}\, \mathrm{d}\mathbf{u} \\
&= \frac{(2\pi)^3}{V} \int \chi_s(u)\delta(\mathbf{k} - \mathbf{k}') \exp(\mathrm{i}\mathbf{k}' \cdot \mathbf{u})\, \mathrm{d}\mathbf{u} \\
&= \frac{(2\pi)^3}{V} \chi_s(k')\delta(\mathbf{k} - \mathbf{k}'), \qquad (4.60)
\end{aligned}$$

where $u = |\mathbf{u}|$, $k' = |\mathbf{k}'|$ and we have defined the Fourier transform of $\chi_s(u)$ by

$$\chi_s(q) = \int \chi_s(u) \exp(\mathrm{i}\mathbf{q} \cdot \mathbf{u}) \, \mathrm{d}\mathbf{u}. \tag{4.61}$$

Similarly the inverse Fourier transform (4.59) becomes

$$\chi_s(\mathbf{r}, \mathbf{r}') = \chi_s(|\mathbf{r} - \mathbf{r}'|) = \chi_s(u) = \frac{1}{(2\pi)^3} \int \chi_s(q) \exp(-\mathrm{i}\mathbf{q} \cdot \mathbf{u}) \, \mathrm{d}\mathbf{q}. \tag{4.62}$$

Equation (4.60) shows explicitly that the response function in jellium is diagonal in **k**-space. Being fanciful we could say that linear response is local in **k**-space, rather as the potential operators we have met are local in **r**-space.

It is easy to show that the matrix element $\chi_s^{-1}(\mathbf{k}, \mathbf{k}') = \langle \mathbf{k} | \chi_s^{-1} | \mathbf{k}' \rangle$ (not to be confused with $\chi_s(\mathbf{k}, \mathbf{k}')^{-1}$, which is not a particularly useful quantity!) is given similarly to (4.60) by

$$\chi_s^{-1}(\mathbf{k}, \mathbf{k}') = \frac{(2\pi)^3}{V} \chi_s(k')^{-1} \delta(\mathbf{k} - \mathbf{k}'), \tag{4.63}$$

and hence by analogy with (4.60) we have the unsurprising but satisfying result

$$\chi_s^{-1}(q) = \chi_s(q)^{-1}. \tag{4.64}$$

Finally, we still have to show that the function $\chi_s(q)$ defined by (4.61) is the same as the function of the same name which we defined by (4.54). If we substitute the diagonal form (4.60) into (4.56) we find the simple proportionality

$$\Delta\rho(\mathbf{k}) = \chi_s(k) \Delta V_{\text{eff}}(\mathbf{k}). \tag{4.65}$$

Now if $V_{\text{eff}}(\mathbf{r}) = \Delta V_q \exp(-\mathrm{i}\mathbf{q} \cdot \mathbf{r})$ its Fourier transform is

$$V_{\text{eff}}(\mathbf{k}) = (2\pi)^3 \Delta V_q \delta(\mathbf{k} - \mathbf{q}). \tag{4.66}$$

Inserting this into (4.65) and taking the inverse Fourier transform gives precisely (4.54).

4.8 Electron–Electron Interactions in the Jellium Response

Based on the above expression for χ_s we can now build the other response functions for jellium , χ_e and ϵ_e. However, we do need to know the local field correction $\mathcal{G}$ defined by (4.27) in order to take us from the effect of a perturbation ΔV_{eff} to the effect of the perturbation which we actually control, namely ΔV_{ext}. There is a long history of approximations for $\mathcal{G}$ in the same representation $\mathcal{G}(q)$ as $\chi_s(q)$ above. The earliest and simplest was that of Hubbard (1958), who suggested

$$\mathcal{G}_{\text{Hubbard}}(q) = \frac{q^2}{2(q^2 + k_F^2)}, \tag{4.67}$$

and the subsequent proposals have in common that they increase from 0 at $q = 0$, but not always monotonically. They have been reviewed by Ichimaru (1982) who recommends the function given in Ichimaru and Utsumi (1981) and by Farid *et al.* (1993), who also recommend a function of their own. It would be beyond the scope of this book to enter into details, except to say that form of $\mathcal{G}(q)$ as $q \to 0$ is very important because it is related to the compressibility of the electron gas, the so-called *compressibility sum rule*, to which we shall return. The latter approximations to $\mathcal{G}(q)$ treat this limit exactly, and also take into account other constraints on $\mathcal{G}(q)$, notably the positivity of the pair correlation function between electrons, which were violated by some of the earlier theories. However, the original and simplest response function, which ignores $\mathcal{G}$ completely and therefore treats the electrons by a mean field description, is still the basis for the exact theory. In this approximation $C = v_C$. Let us now see how it looks.

The dielectric function ϵ_e in the $\mathcal{G} = 0$ approximation is called the Lindhard or Hartree function, which we will denote ϵ_H. From the relation (4.26) we can write

$$\epsilon_H = 1 - v_C \chi_s \tag{4.68}$$

or in $\boldsymbol{k}$-space, a matrix element of the above equation is:

$$\langle \mathbf{k} | \epsilon_H | \mathbf{k}' \rangle = \langle \mathbf{k} | \mathbf{k}' \rangle - \sum_{\mathbf{k}''} \langle \mathbf{k} | v_C | \mathbf{k}'' \rangle \langle \mathbf{k}'' | \chi_s | \mathbf{k}' \rangle. \tag{4.69}$$

It is straightforward to show along the lines of the derivation of (1.68) that the matrix element of the Coulomb interaction in $\boldsymbol{k}$-space is

$$v_C(\mathbf{k}, \mathbf{k}') = \frac{(2\pi)^3}{V} \frac{4\pi}{k^2} \delta(\mathbf{k} - \mathbf{k}'). \tag{4.70}$$

Also, in a similar way to $\chi_s(\mathbf{k}, \mathbf{k}')$, the matrix element of ϵ_H must have the form

$$\epsilon_H(\mathbf{k}, \mathbf{k}') = \frac{(2\pi)^3}{V} \epsilon_H(k') \delta(\mathbf{k} - \mathbf{k}'). \tag{4.71}$$

Putting these ingredients together into (4.69) and integrating over $\mathbf{k}'$ to remove the delta function gives us the standard expression for the Lindhard dielectric function:

$$\epsilon_H(q) = 1 - \frac{4\pi}{q^2} \chi_s(q) \tag{4.72}$$

which is very similar in form to the general matrix expression (4.26). Evidently the screening falls off in strength with increasing q, in other words the electrons screen less effectively at shorter wavelength.

By analogy with (4.61) the Fourier transformed Coulomb interaction, from (4.70) is

$$v_{\mathrm{C}}(q) = \frac{4\pi}{q^2}. \tag{4.73}$$

With it we can construct the other response functions in jellium as a function of a single wavelength, making use of equations (4.26) and (4.27). For future reference the results are

$$\epsilon_{\mathrm{e}}(q) = 1 - v_{\mathrm{C}}(q)(1 - \mathcal{G}(q))\chi_{\mathrm{s}}(q) \tag{4.74}$$

and

$$\chi_{\mathrm{e}}(q) = \frac{\chi_{\mathrm{s}}(q)}{\epsilon_{\mathrm{e}}(q)}. \tag{4.75}$$

The latter formula shows clearly how the response function of non interacting electrons is 'screened' by the dielectric function to give the response function of real electrons; it is illustrated in Fig. 4.3.

It is very convenient that the matrix (or operator) inverses in the **k**-space representation have now become simple algebraic inverses, but I emphasize that this is only possible because of the translational variance of jellium.

This special **k**-space representation in terms of only the *modulus* q or k of the **k**-vector is so common that it deserves a special name. Let us call it the *isotropic* **k**-space representation. Use will be made of these formula in the later chapter deriving pair-potentials in simple metals.

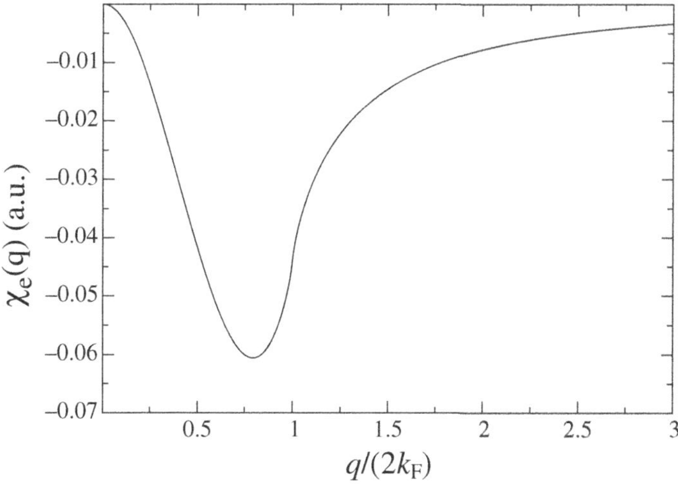

FIG. 4.3 The response function $\chi_{\mathrm{e}}(q)$ of the interacting electron gas in jellium, eq. (4.75), using the local field correction $\mathcal{G}(q)$ of Ichimaru and Utsumi (1981).

4.9 The Long Wavelength Limit of Response Functions in Jellium

Response functions in the long wavelength limit ($q \to 0$) conceal a surprising secret I have mentioned before, which is known as the *compressibility sum rule* (Pines and Nozieres, 1966). A sum rule is something that physicists are fond of, because it is a constraint on the behaviour or form of an imperfectly known function that enables the function to be pinned down with greater accuracy. This has been useful in making better approximations to χ_e, which in turn have led to a more accurate description of interatomic forces in simple metals, as we shall see.

4.9.1 The Thomas–Fermi Response Function

It is instructive to start with Thomas–Fermi theory and its Euler–Lagrange equation (2.21). If we make a small change ΔV_{eff}, the corresponding change in ρ is given by:

$$\tfrac{1}{3}(3\pi^2)^{2/3}\rho(\mathbf{r})^{-1/3}\Delta\rho(\mathbf{r}) = -\Delta V_{\text{eff}}(\mathbf{r}) + \Delta\mu_{\text{TFD}}. \tag{4.76}$$

We now discard the term $\Delta\mu_{\text{TFD}}$, because we are considering a peturbation in a finite region of space within a very large system of volume V, which itself is periodically repeated remember. In the limit of $V \to \infty$ the change in μ tends to zero. On the other hand if we insist that the perturbation is infinite in extent, let us also insist that the zero of potential is adjusted automatically so as to ensure $\Delta\mu_{\text{TFD}} = 0$. Since we are conserving total charge, a constant shift in the potential can have no effect on the charge distribution. With that little matter aside we can write

$$\Delta\rho(\mathbf{r}) = -\frac{3\rho(\mathbf{r})^{1/3}}{(3\pi^2)^{2/3}}\Delta V_{\text{eff}}(\mathbf{r}) \tag{4.77}$$

and hence

$$\chi_s^{\text{TF}}(\mathbf{r}, \mathbf{r}') = -\frac{3\rho(\mathbf{r})^{1/3}}{(3\pi^2)^{2/3}}\delta(\mathbf{r} - \mathbf{r}'). \tag{4.78}$$

Notice that this non-interacting response function is the same for simple Thomas–Fermi as for the Thomas–Fermi–Dirac or any more elaborate functionals, as long as they have a local kinetic energy part. In jellium, where ρ is a constant and related to the Fermi wavevector k_{F} by eq. (1.227), this becomes

$$\chi_s^{\text{TF}}(\mathbf{r}, \mathbf{r}') = -\frac{k_{\text{F}}}{\pi^2}\delta(\mathbf{r} - \mathbf{r}'). \tag{4.79}$$

The **k**-space representation, from (4.61), is

$$\chi_s^{\text{TF}}(q) = -\int \frac{k_{\text{F}}}{\pi^2}\delta(\mathbf{u})\,d\mathbf{u} = -\frac{k_{\text{F}}}{\pi^2}. \tag{4.80}$$

Hence in the plain Thomas–Fermi approximation, without any account of exchange and correlation, the dielectric constant, from (4.74), is given by

$$\epsilon_{\rm e}^{\rm TF}(q) = 1 + \frac{4k_{\rm F}}{\pi q^2} \equiv 1 + \frac{me^2 k_{\rm F}}{\epsilon_0 \hbar^2 \pi^2 q^2} \quad \text{(SI units)}, \tag{4.81}$$

and from (4.75)

$$\chi_{\rm e}^{\rm TF} = -\left(\frac{\pi^2}{k_{\rm F}} + v_{\rm C}(q)\right)^{-1} \equiv -\left(\frac{\hbar^2\pi^2}{mk_{\rm F}} + \frac{e^2}{\epsilon_0 q^2}\right)^{-1} \quad \text{(SI units)}. \tag{4.82}$$

I have included here the formulae in SI units. Like many formulae in atomic units, these look dimensionally wrong, because $k_{\rm F}$ and q are both inverse lengths. In SI units the second term in (4.81) appears divided by the Bohr radius $a_0 = 4\pi\epsilon_0\hbar^2/(me^2)$ which makes the dielectric constant properly dimensionless.

The characteristic value of q below which the Thomas–Fermi dielectric constant $\epsilon_{\rm e}^{\rm TF}(q)$ exceeds 2 is given a special name, the Thomas-Fermi wavevector

$$q_{\rm TF} = 2\sqrt{\frac{k_{\rm F}}{\pi}}. \tag{4.83}$$

Its inverse is the characteristic screening length in the Thomas–Fermi approximation, which becomes exact in the limit of a high-density electron gas. Unfortunately that is denser than the electrons in materials under normal laboratory conditions, so the model is only qualitative. Thomas–Fermi screening does become an exact theory for *non-interacting* electrons, even at normal densities, in the long wavelength limit. To see this we expand formula (4.55) for the exact linear response function $\chi_{\rm s}(q)$ about $q = 0$, making use of the Taylor expansion

$$\log\left|\frac{1+q}{1-q}\right| = 2q + 2q^3/3 + \cdots \tag{4.84}$$

and we find

$$\chi_{\rm s}(q) = -\frac{k_{\rm F}}{\pi^2}\left(1 - \frac{q^2}{8k_{\rm F}^2} + O(q^3)\right), \tag{4.85}$$

which as $q \to 0$ tends to the Thomas–Fermi result.

4.9.2 The Compressibility Sum Rule

We will now see how the long wavelength limit of the response function in jellium is related to the compressibility of the uniform electron gas. Actually we work in terms of the inverse of the compressibility, namely the *bulk modulus* of the electron gas, $B^{\rm jel}$, and as a preliminary to the sum rule, we need to define this quantity. Just as for any material it is defined in terms of the pressure P^{jel} and energy E^{jel} of a piece of jellium of volume V (defined in (1.223)) by

$$B^{\text{jel}} = -V\frac{dP^{\text{jel}}}{dV} \tag{4.86}$$

where

$$P^{\text{jel}} = -\frac{dE^{\text{jel}}}{dV}, \tag{4.87}$$

so that

$$B^{\text{jel}} = V\frac{d^2E^{\text{jel}}}{dV^2}. \tag{4.88}$$

It will be useful to express this in terms of the energy per unit fixed volume, $\bar{E}^{\text{jel}}(\rho) = E^{\text{jel}}/V$. Since $\rho = N/V$, where N is the fixed number of electrons, we can express the volume derivative in terms of a density derivative:

$$\frac{d}{dV} = \frac{d\rho}{dV}\frac{d}{d\rho} = -\frac{\rho}{V}\frac{d}{d\rho} \tag{4.89}$$

the straightforward application of which leads us to the formula

$$B^{\text{jel}} = \rho^2\frac{d^2\bar{E}^{\text{jel}}}{d\rho^2}. \tag{4.90}$$

Now imagine what happens when we rumple the positive background. For simplicity let us make a sinusoidal perturbation of its density, defined by

$$\Delta\rho_Z(\mathbf{r}) = \Delta\rho_{Zq}\sin(\mathbf{q}\cdot\mathbf{r}). \tag{4.91}$$

This creates a change ΔV_{ext} and a change in the electron density (see Fig. 4.4), given by

$$\Delta V_{\text{ext}}(\mathbf{r}) = \Delta V_q\sin(\mathbf{q}\cdot\mathbf{r}) \tag{4.92}$$

and

$$\Delta\rho(\mathbf{r}) = \Delta\rho_q\sin(\mathbf{q}\cdot\mathbf{r}) \tag{4.93}$$

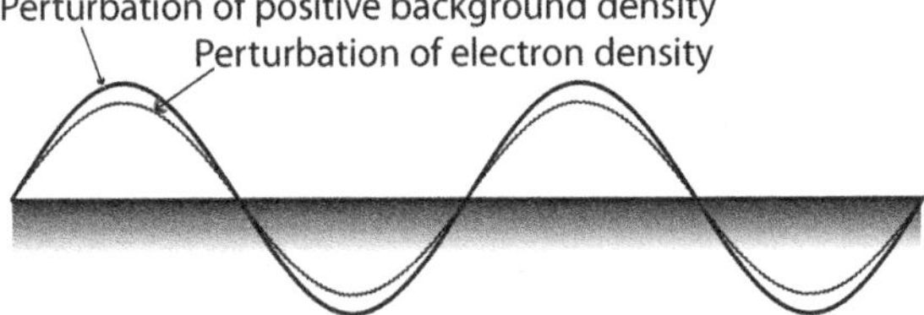

FIG. 4.4 Screening of a perturbation in jellium.

where

$$\Delta\rho_q = \chi_e(q)\Delta V_q = \chi_e(q)v_C(q)\Delta\rho_{Zq}. \tag{4.94}$$

Consider perturbations of longer and longer wavelength. As we enter the world of macroscopic physics the behaviour is dominated by electrostatic energy. As q gets smaller and smaller, $\Delta\rho_q$ will approach closer and closer to $\Delta\rho_{Zq}$ in order to minimize the electrostatic energy. In the long wavelength limit the electrons will flow so as to *exactly* neutralize the positively charged perturbation of the uniform background. At long wavelengths, before we quite reach $q = 0$, we can describe the energy in two ways, which I will formulate in turn before equating them. The first way is with the second-order term from (4.16), knowing that the first-order term vanishes because it contains equal and opposite contributions from the positive and negative regions of the perturbation. Expressed in **k**-space, the second-order energy change is easily shown to be

$$\begin{aligned}\Delta E^{\text{jel}} &= \tfrac{1}{4}\Delta\rho_q\Delta V_q + \Delta E_{ZZ} \\ &= \tfrac{1}{4}\Delta\rho_{Zq}v_C(q)\chi_e(q)v_C(q)\Delta\rho_{Zq} + \tfrac{1}{4}\Delta\rho_{Zq}v_C(q)\Delta\rho_{Zq}.\end{aligned} \tag{4.95}$$

The first term is half the electrostatic interaction between the induced charge and the inducing density, the second term is the self-energy of the inducing density. Individually these terms diverge, due to the $1/q^2$ behaviour of the Coulomb energy, but this divergence cancels when they are added. Expressing χ_e in terms of χ_s and the local field correction $\mathcal{G}$ by means of (4.20) and (4.27) we find

$$\begin{aligned}\lim_{q\to 0}(v_C^2\chi_e + v_C) &= \lim_{q\to 0} v_C\left(1 + \frac{v_C\chi_s}{1 - v_C(1-\mathcal{G})\chi_s}\right) \\ &= \lim_{q\to 0}\left(-\frac{1}{(1-\mathcal{G})\chi_s} - \frac{4\pi\mathcal{G}}{q^2(1-\mathcal{G})}\right).\end{aligned} \tag{4.96}$$

We know from (4.85) that $\lim_{q\to 0}\chi_s(q) = -k_F/\pi^2$, so the limiting value hinges upon the low q behaviour of $\mathcal{G}$. Consider an expansion of $\mathcal{G}(q)$ about $q = 0$. The first term must be quadratic:

$$\mathcal{G}(q) = \tfrac{1}{2}\mathcal{G}''q^2 + O(q^3), \tag{4.97}$$

where the second derivative

$$\mathcal{G}'' = \left.\frac{\partial^2\mathcal{G}(q)}{\partial q^2}\right|_{q=0} \tag{4.98}$$

is a function of the background density. There cannot be a constant or a linear term, otherwise the $1/q^2$ factor in the second term of (4.96) would blow it up to infinity.

The result is therefore

$$\lim_{q\to 0}(v_C^2\chi_e + v_C) = \frac{\pi^2}{k_F} - 2\pi\mathcal{G}''. \tag{4.99}$$

The second way to describe the energy at long wavelength is by integrating the local energy, which is just the energy required to compress or expand the jellium to its local density. This is $\frac{1}{2}B^{\text{jel}}(\Delta\rho_Z/\rho)^2$ per unit volume. Since $\Delta\rho_Z$ is varying sinusoidally, eq. (4.91), the average value of $\Delta\rho_Z^2$ is half its maximum value. The integral of $\Delta\rho_Z^2$ over unit volume is therefore $\frac{1}{2}\Delta\rho_{Zq}^2$, so the resulting energy per unit volume is

$$\Delta E^{\text{jel}} = \tfrac{1}{4}B^{\text{jel}}(\Delta\rho_{Zq}/\rho)^2. \tag{4.100}$$

We must now equate (4.100) to the limit of (4.95) as $q \to 0$, which gives

$$B^{\text{jel}} = \rho^2 \lim_{q\to 0}(v_C^2\chi_e + v_C) = \rho^2\left(\frac{\pi^2}{k_F} - 2\pi\mathcal{G}''\right) \tag{4.101}$$

and this is the compressibility sum rule. The importance of this equation is in pinning down the form of $\mathcal{G}$, by fixing the correct value of $\mathcal{G}''$, given that we have already determined a good approximation to B^{jel}. It is one of the remarkable results of physics that the quantum Monte Carlo calculations which have given us a good model of E^{jel} and hence B^{jel} (Ceperley and Alder, 1980) also yield direct information about the exact response function of the electron gas. For our purposes the low-q expansion that follows from (4.101) will be useful, namely:

$$\chi_e(q) = -\frac{q^2}{4\pi} + \frac{B^{\text{jel}}}{\rho^2}\left(\frac{q^2}{4\pi^2}\right)^2 + O(q^6). \tag{4.102}$$

Exercise

Show by direct differentiation that if only the kinetic energy contribution to E_{eg} is taken into account, $B^{\text{jel}} = \rho^2\pi^2/k_F$. Now include the Dirac expression for exchange energy, (1.230), into the energy of the jellium and find what value of $\mathcal{G}''$ it predicts.

4.10 Linear Response in a Perfect Crystal

For inhomogeneous electron gases we unfortunately have no analytical expressions as we do for jellium. Formula (4.17) is exact, but we have no useful formulae for the functionals $T_s[\rho]$ and $E_{xc}[\rho]$ which would serve to evaluate it, except for the LDA. This would give us at best an extended Thomas–Fermi model, which treats the kinetic as well as the exchange and correlation energies within the LDA. Such a model has been applied in solids to explain qualitative trends in cohesion, but has never been very useful for quantitative work. For this reason in practical applications it is usually more convenient to formulate the response in **k**-space, based on an evaluation of the general response function (4.46). The response equation in the form (4.51) still holds, and one can always have recourse to evaluating the integrals numerically with a computer program.

From the point of view of theory, the next best thing to jellium is a perfect crystal, which is a periodic system as described in Section 1.4.2. The periodicity, by definition, means that operators have **r**-space matrix elements satisfying

$$\begin{aligned}\langle \mathbf{r}|\hat{V}_{\text{eff}}|\mathbf{r}'\rangle &= \langle \mathbf{r}+\mathbf{R}|\hat{V}_{\text{eff}}|\mathbf{r}'+\mathbf{R}\rangle,\\ \langle \mathbf{r}|\hat{\chi}_{\text{S}}|\mathbf{r}'\rangle &= \langle \mathbf{r}+\mathbf{R}|\hat{\chi}_{\text{S}}|\mathbf{r}'+\mathbf{R}\rangle,\end{aligned} \tag{4.103}$$

etc., where as before **R** is a vector of the lattice. What does this imply about the matrix elements in **k**-space, such as $\chi_{\text{S}}(\mathbf{k},\mathbf{k}')$? Let us construct $\chi_{\text{S}}(\mathbf{k},\mathbf{k}')$ in the usual way and find out. Starting with (4.58), let us break up the domain of integration for **r** and **r**$'$, which is volume V, into N_c identical unit cells of volume V_c, as in Section (1.4.2). It then becomes

$$\begin{aligned}\chi_{\text{S}}(\mathbf{k},\mathbf{k}') &= \frac{1}{V}\sum_{\mathbf{R},\mathbf{R}'}\int_{V_c} \chi_{\text{S}}(\mathbf{R}+\mathbf{r},\mathbf{R}'+\mathbf{r}')\,\exp(\mathrm{i}(\mathbf{k}\cdot\mathbf{r}-\mathbf{k}'\cdot\mathbf{r}'))\\ &\quad\times\exp(\mathrm{i}(\mathbf{k}\cdot\mathbf{R}-\mathbf{k}'\cdot\mathbf{R}'))\\ &= \frac{1}{V}\sum_{\mathbf{R},\mathbf{R}'}\int_{V_c} \chi_{\text{S}}(\mathbf{r},\mathbf{R}'-\mathbf{R}+\mathbf{r}')\,\exp(\mathrm{i}\mathbf{k}\cdot\mathbf{r})\\ &\quad\times\exp(-\mathrm{i}\mathbf{k}'\cdot(\mathbf{R}'-\mathbf{R}+\mathbf{r}'))\exp(\mathrm{i}(\mathbf{k}-\mathbf{k}')\cdot\mathbf{R}).\end{aligned} \tag{4.104}$$

The integrations over **r** and **r**$'$ now just extend over the volumes V_c, and the sums over **R** and **R**$'$ extend over the set $\mathcal{R}$ which corresponds to all the unit cells within V. Next we have to make the very reasonable assumption that $\chi_{\text{S}}(\mathbf{r},\mathbf{r}')$ is not, for all practical purposes, infinitely long-ranged; there is a distance, let us call it d, beyond which the separation of **r** and **r**$'$ is great enough for $\chi_{\text{S}}(\mathbf{r},\mathbf{r}')$ to be negligible. The volume V is arbitrarily large, so d is tiny compared to the diameter of V. We are now going to replace the sum over **R**$'$ by the sum over the relative position $\mathbf{R}''=\mathbf{R}'-\mathbf{R}$, and allow $\mathbf{R}''$ to range over $\mathcal{R}$, the same set of vectors as **R**. What error do we make by counting $\mathbf{R}''$ over $\mathcal{R}$ rather than over its exact domain, which depends where **R** is since **R**$'$ must also be in $\mathcal{R}$? Well for most values of **R** it makes no difference, since there is no contribution from terms where $|\mathbf{R}''|>d$. Only when **R** lies within a skin of thickness d inside the surface of V are we making an error, because we are counting terms for which **R**$'$ is within a skin of thickness d on the *outside* surface of V. Thus the fractional error we make is of the order of the ratio of the number of **R** vectors within this skin to the total number of **R** vectors.

This fraction is of order $d/V^{1/3}$ and vanishes in the limit of large V. All this was a necessary preamble to writing

$$\chi_s(\mathbf{k},\mathbf{k}') = \frac{1}{V}\sum_{\mathbf{R},\mathbf{R}''}\int_{V_c}\chi_s(\mathbf{r},\mathbf{R}''+\mathbf{r}')\exp(\mathrm{i}\mathbf{k}\cdot\mathbf{r})\exp(-\mathrm{i}\mathbf{k}'\cdot(\mathbf{R}''+\mathbf{r}'))$$
$$\times\exp(\mathrm{i}(\mathbf{k}-\mathbf{k}')\cdot\mathbf{R}), \qquad (4.105)$$

by means of which we have decoupled the two summations. From the result in Section (1.4.6) we see that the $\mathbf{R}$ summation vanishes unless $\mathbf{k}-\mathbf{k}'=\mathbf{g}$, a reciprocal lattice vector, in which case it equals V/V_c. We can therefore express the response function as the following double Fourier transform:

$$\chi_s(\mathbf{k},\mathbf{g}+\mathbf{k}) = \frac{1}{V_c}\sum_{\mathbf{R}}\int_{V_c}\chi_s(\mathbf{r},\mathbf{R}+\mathbf{r}')\exp(\mathrm{i}\mathbf{k}\cdot(\mathbf{r}-\mathbf{r}'-\mathbf{R}))\exp(\mathrm{i}\mathbf{g}\cdot\mathbf{r}')). \qquad (4.106)$$

The form of χ_s (which would be exactly the same for χ_e or ϵ_e) tells us that a periodic perturbation induces a charge density not only with the same period, as is the case in jellium, but also with all the periodicities of the lattice. If we work with the Fourier transformed potential and density response, the equation in a periodic system corresponding to (4.65) in jellium is

$$\Delta\rho(\mathbf{k}) = \sum_{\mathbf{g}}\chi_s(\mathbf{k},\mathbf{k}+\mathbf{g})\Delta V_{\mathrm{eff}}(\mathbf{k}+\mathbf{g}). \qquad (4.107)$$

4.11 Non-local Potentials

4.11.1 General Ideas

In practical work there is a considerable saving in computer time for many kinds of calculations if the true external potential, the sum of atom centred Coulomb potentials, is replaced by a *pseudopotential*. In a nutshell the idea is that the tightly bound electrons, or *core* electrons in an atom really don't participate in the chemical bonding of atoms, only the outer, or *valence* electrons are important for the chemistry and other properties of a molecule, solid or liquid. Therefore it is really only the valence electrons which need to be explicitly described by Kohn–Sham equations or some approximation to them. The only role of the core electrons is to provide an effective external potential or pseudopotential in which the outer electrons move. I will return to this idea in later chapters. For the moment it is only important that there is a price to be paid for deciding to leave core electrons out of the quantum mechanical problem. Although the number of electrons per atom we have to worry about is less, it is still a major task to define the new V_{ext} which represents the core electrons as well as the nucleus. To be reliable and transferable it turns out that the pseudopotential V_{ext} must be non-local. Fortunately the results of the Hohenberg–Kohn–Sham theory continue to hold with apparently only minor modification when non-local external potentials are used. I say 'apparently', because although the formulae don't look very different, there is a large increase in complexity in their practical implementation. In this section, I present some of the key formulae for linear response purposes. More technical details can be found in the review by Baroni *et al.* (2001), the paper by Quong and Klein (1992), and references therein. We need to see how linear response theory works out with non-local potentials for two reasons. The first is in order to derive formulae for the effective pairwise interactions in simple metals, which I will be describing in detail in Chapter 6. The second is to be able to calculate the force on an atom when the position of another atom is perturbed. This is the basis of lattice dynamics, which I discuss further in Section 5.4.

Consider the expansion of the total energy to second order eq. (3.18). Let us write the energy change to second order as:

$$\Delta E^{(2)} = \text{Tr}\rho\Delta V_{\text{ext}} + \tfrac{1}{2}\text{Tr}\Delta\rho\Delta V_{\text{ext}} + \Delta E_{ZZ}. \tag{4.108}$$

The equation is unchanged when the external potential is non-local, but now we have to include the off-diagonal elements of ρ in $\mathbf{r}$-space, so that in $\mathbf{r}$-space the equation looks like:

$$\begin{aligned}\Delta E^{(2)} &= \int \rho(\mathbf{r},\mathbf{r}')\Delta V_{\text{ext}}(\mathbf{r}',\mathbf{r})\,\mathbf{dr}\,\mathbf{dr}' \\ &\quad + \frac{1}{2}\int \Delta\rho(\mathbf{r},\mathbf{r}')\Delta V_{\text{ext}}(\mathbf{r}',\mathbf{r})\,\mathbf{dr}\,\mathbf{dr}' + \Delta E_{ZZ}.\end{aligned} \tag{4.109}$$

Reviewing the derivation of the density response, (4.46), we see that it is just the same for off-diagonal elements of $\Delta\rho$. Hence we have in the eigenstate representation

$$\langle n|\Delta\hat{\rho}|n'\rangle = \Delta\rho(n,n') = 2\frac{f_{\text{F}}(\epsilon_n) - f_{\text{F}}(\epsilon_{n'})}{\epsilon_n - \epsilon_{n'}}\langle n|\Delta\hat{V}_{\text{eff}}|n'\rangle. \tag{4.110}$$

The effective potential ΔV_{eff} now has a non-local part from ΔV_{ext} together with a local part from $\Delta\rho(\mathbf{r})$, namely $\Delta V_{\text{H}} + \Delta V_{\text{xc}}$. The off-diagonal elements of $\Delta\rho$ are not relevant for $\Delta V_{\text{H}} + \Delta V_{\text{xc}}$, which is given by

$$\Delta V_{\text{H}}(\mathbf{r}) + \Delta V_{\text{xc}}(\mathbf{r}) = \int C(\mathbf{r},\mathbf{r}')\Delta\rho(\mathbf{r}')\,\mathbf{dr}', \tag{4.111}$$

where the kernel C is defined in eqs. (3.28), (4.17) or (4.20).

Hence we obtain an implicit equation for the elements of the change in the density matrix, both diagonal and off-diagonal, which in the **r**-represention looks like

$$\Delta\rho(\mathbf{r},\mathbf{r}') = \sum_{n,n'} 2\frac{f_{\mathrm{F}}(\epsilon_n) - f_{\mathrm{F}}(\epsilon_{n'})}{\epsilon_n - \epsilon_{n'}} \langle \mathbf{r}|n\rangle \{\langle n|\Delta\hat{V}_{\mathrm{ext}}|n'\rangle + \int \langle n|\mathbf{r}'''\rangle C(\mathbf{r}''',\mathbf{r}'')\Delta\rho(\mathbf{r}'')\langle \mathbf{r}'''|n'\rangle \,\mathrm{d}\mathbf{r}''\,\mathrm{d}\mathbf{r}'''\}\langle n'|\mathbf{r}'\rangle. \quad (4.112)$$

In order to calculate ΔV_{eff} we only need the diagonal elements $\Delta\rho(\mathbf{r})$, given by

$$\Delta\rho(\mathbf{r}) = \sum_{n,n'} 2\frac{f_{\mathrm{F}}(\epsilon_n) - f_{\mathrm{F}}(\epsilon_{n'})}{\epsilon_n - \epsilon_{n'}} \langle \mathbf{r}|n\rangle \{\langle n|\Delta\hat{V}_{\mathrm{ext}}|n'\rangle + \int \langle n|\mathbf{r}'\rangle C(\mathbf{r}',\mathbf{r}'')\Delta\rho(\mathbf{r}'')\langle \mathbf{r}'|n'\rangle \,\mathrm{d}\mathbf{r}''\,\mathrm{d}\mathbf{r}'\}\langle n'|\mathbf{r}\rangle. \quad (4.113)$$

The second order change in the energy becomes, in the eigenstate representation:

$$\Delta E^{(2)} = 2\sum_n f_{\mathrm{F}}(\epsilon_n)\langle n|\Delta V_{\mathrm{ext}}|n\rangle + \Delta E_{ZZ} + \sum_{n,n'} \frac{f_{\mathrm{F}}(\epsilon_n) - f_{\mathrm{F}}(\epsilon_{n'})}{\epsilon_n - \epsilon_{n'}} \langle n|\Delta\hat{V}_{\mathrm{eff}}|n'\rangle\langle n'|\Delta V_{\mathrm{ext}}|n\rangle, \quad (4.114)$$

where

$$\langle n|\Delta\hat{V}_{\mathrm{eff}}|n'\rangle = \langle n|\Delta\hat{V}_{\mathrm{ext}}|n'\rangle + \int \langle n|\mathbf{r}'\rangle C(\mathbf{r}',\mathbf{r}'')\Delta\rho(\mathbf{r}'')\langle \mathbf{r}'|n'\rangle \,\mathrm{d}\mathbf{r}''\,\mathrm{d}\mathbf{r}'. \quad (4.115)$$

Although we can no longer write down the formulae in terms of a response function or its inverse, these formulae can still be used for practical calculations. They must be solved iteratively, or the kernel of (4.113) must be inverted to get the non-local analogue of an inverse dielectric function. This is not a trivial enterprise, and various tricks are involved, notably to avoid the summation over the unoccupied states, which in principle includes all the eigenstates up to infinite energy.

Personally I find the appearance of **r**-space in eq. (4.115) is almost a shame, it is an inelegance of the very practical nature of DFT that it makes its home in **r**-space. Namely, DFT works with only the diagonal elements of the density operator in **r**-space, and not with the whole operator. This confers a special status on **r**-space, also apparent in the results of the following section.

4.11.2 Non-local Perturbations in Jellium

In this section, I describe the linear response of a uniform electron gas to a non-local potential, because we shall need it in deriving pairwise potentials in simple metals.

The perturbed charge density and energy in jellium are much easier to formulate than they are in the general case, because the unperturbed eigenstates in jellium are $|\mathbf{k}\rangle$, which in $\mathbf{r}$-space are plane waves.

It is only the change in the external potential, $\Delta V_{\text{ext}}(\mathbf{r}, \mathbf{r}')$ which is non-local, so let us first examine the local part of the change in the effective potential, namely the change in the Hartree and exchange-correlation potentials. These potentials are always local in $\mathbf{r}$-space within DFT, and it should be emphasized that this is an exact property, and has nothing to do with the local density approximation, which we do not need to make in the present analysis. The changes $\Delta V_{\text{H}}(\mathbf{r})$ and $\Delta V_{\text{xc}}(\mathbf{r})$ are given from eqs. (3.34) and (4.27) by

$$\Delta V_{\text{H}}(\mathbf{r}) + \Delta V_{\text{xc}}(\mathbf{r}) = \int \langle \mathbf{r}|v_{\text{C}}(1 - \mathcal{G})|\mathbf{r}'\rangle \Delta\rho(\mathbf{r}')\, d\mathbf{r}'. \tag{4.116}$$

The electron–electron interaction, corrected for the local field, which appears in the integrand is diagonal in $\mathbf{k}$-space by virtue of the translational invariance of jellium, so we can use the isotropic representation, by analogy with eqs. (4.60)–(4.71):

$$\langle \mathbf{k}|v_{\text{C}}(1 - \mathcal{G})|\mathbf{k}'\rangle = \frac{(2\pi)^3}{V} v_{\text{C}}(k)(1 - \mathcal{G}(k))\delta(\mathbf{k} - \mathbf{k}'). \tag{4.117}$$

This gives us a simple form for the matrix element in $\mathbf{r}$-space:

$$\langle \mathbf{r}|v_{\text{C}}(1 - \mathcal{G})|\mathbf{r}'\rangle = \sum_{\mathbf{k},\mathbf{k}'} \langle \mathbf{r}|\mathbf{k}\rangle \langle \mathbf{k}|v_{\text{C}}(1 - \mathcal{G})|\mathbf{k}'\rangle \langle \mathbf{k}'|\mathbf{r}'\rangle \tag{4.118}$$

which, when we insert the explicit plane waves, becomes

$$\langle \mathbf{r}|v_{\text{C}}(1 - \mathcal{G})|\mathbf{r}'\rangle = \frac{1}{(2\pi)^3} \int v_{\text{C}}(k)(1 - \mathcal{G}(k)) \exp(i\mathbf{k}'' \cdot (\mathbf{r} - \mathbf{r}'))\, d\mathbf{k}''. \tag{4.119}$$

I have introduced the dummy argument $\mathbf{k}''$ so that now we can substitute (4.119) into (4.116) and form the $\mathbf{k}, \mathbf{k}'$ matrix element, which is easily obtained:

$$\begin{aligned}\Delta V_{\text{H}}(\mathbf{k}, \mathbf{k}') + \Delta V_{\text{xc}}(\mathbf{k}, \mathbf{k}') = \frac{1}{(2\pi)^3} \frac{1}{V} \int & \exp(i(\mathbf{k}' - \mathbf{k} + \mathbf{k}'') \cdot \mathbf{r}) \exp(-i\mathbf{k}'' \cdot \mathbf{r}') \\ & \times v_{\text{C}}(k'')(1 - \mathcal{G}(k''))\Delta\rho(\mathbf{r}')\, d\mathbf{k}''\, d\mathbf{r}\, d\mathbf{r}'. \end{aligned} \tag{4.120}$$

Performing first the $\mathbf{r}$ integration produces a delta function that picks out $\mathbf{k}'' = \mathbf{k} - \mathbf{k}'$. Let us denote this $\mathbf{k}$-vector by $-\mathbf{q}$, so that we can make the substitution

$\mathbf{k}' = \mathbf{k} + \mathbf{q}$. The result is

$$\Delta V_{\mathrm{H}}(\mathbf{k}, \mathbf{k}+\mathbf{q}) + \Delta V_{\mathrm{xc}}(\mathbf{k}, \mathbf{k}+\mathbf{q}) = v_{\mathrm{C}}(q)(1 - \mathcal{G}(q))\Delta\rho(\mathbf{q}), \tag{4.121}$$

where we have introduced the Fourier transformed $\Delta\rho$ with our standard normalization:

$$\Delta\rho(\mathbf{q}) = \frac{1}{V}\int \exp(\mathrm{i}\mathbf{q}\cdot\mathbf{r})\Delta\rho(\mathbf{r})\,\mathrm{d}\mathbf{r}. \tag{4.122}$$

Notice that this looks a little different to the transform which defines the isotropic represention of the quantities which are translationally invariant in jellium. Unlike $\mathcal{G}(\mathbf{r}-\mathbf{r}')$ or $v_{\mathrm{C}}(\mathbf{r}-\mathbf{r}')$, e.g. $\Delta\rho(\mathbf{r}, \mathbf{r}')$ has no such special symmetry. A second point to note is that this symmetry implies that for operators which are local in $\mathbf{r}$-space, the $\mathbf{k}, \mathbf{k}'$ matrix elements are functions only of the difference $\mathbf{k} - \mathbf{k}'$.

Now we turn to the total change in the effective potential, given by (4.115). We might want to make a separation of the external potential into local and non-local parts, denoted by superscripts L and NL, in which case we can now write (4.115) in the form

$$\Delta V_{\mathrm{eff}}(\mathbf{k}, \mathbf{k}+\mathbf{q}) = \Delta V_{\mathrm{ext}}^{NL}(\mathbf{k}, \mathbf{k}+\mathbf{q}) + V_{\mathrm{ext}}^{L}(\mathbf{q}) + v_{\mathrm{C}}(q)(1 - \mathcal{G}(q))\Delta\rho(\mathbf{q}). \tag{4.123}$$

The density change $\Delta\rho$ is obtained from the linear response expression (4.47). If we take its Fourier transform in the same way as we did to arrive at (4.122) we find

$$\Delta\rho(\mathbf{q}) = \frac{2}{V}\sum_{\mathbf{k}} \frac{(f_{\mathrm{F}}(\epsilon_{\mathbf{k}}) - f_{\mathrm{F}}(\epsilon_{\mathbf{k}+\mathbf{q}}))}{(\epsilon_{\mathbf{k}} - \epsilon_{\mathbf{k}+\mathbf{q}})}\Delta V_{\mathrm{eff}}(\mathbf{k}, \mathbf{k}+\mathbf{q}). \tag{4.124}$$

Notice that if ΔV_{eff} is local, $V_{\mathrm{eff}}(\mathbf{k}, \mathbf{k}+\mathbf{q})$ becomes $V_{\mathrm{eff}}(\mathbf{q})$, which factors outside the summation. In this case the $\mathbf{k}$ summation can be done exactly, and making use of (4.55) the result is the same as given in eq. (4.65):

$$\Delta\rho(\mathbf{q}) = \chi_{\mathrm{s}}(q)\Delta V_{\mathrm{eff}}(\mathbf{q}). \tag{4.125}$$

Furthermore we can define a local external potential, let us call it $\bar{V}_{\mathrm{ext}}$, which in linear response creates the same perturbation of charge density as the non-local V_{ext}. The appropriate formula is

$$\Delta\bar{V}_{\mathrm{ext}}(\mathbf{q}) = \chi_{\mathrm{s}}^{-1}(q)\frac{2}{V}\sum_{\mathbf{k}} \frac{(f_{\mathrm{F}}(\epsilon_{\mathbf{k}}) - f_{\mathrm{F}}(\epsilon_{\mathbf{k}+\mathbf{q}}))}{(\epsilon_{\mathbf{k}} - \epsilon_{\mathbf{k}+\mathbf{q}})}\Delta V_{\mathrm{ext}}(\mathbf{k}, \mathbf{k}+\mathbf{q}). \tag{4.126}$$

To see this, substitute (4.123) into (4.124), and rearrange it to make $\Delta\rho(\mathbf{q})$ the subject. With the definition (4.126) the result is:

$$\Delta\rho(\mathbf{q}) = \frac{\chi_{\mathrm{s}}(q)}{1 - v_{\mathrm{C}}(q)(1 - \mathcal{G}(q))}\Delta\bar{V}_{\mathrm{ext}}(\mathbf{q}) = \frac{\chi_{\mathrm{s}}(q)}{\epsilon_{\mathrm{e}}(q)}\Delta\bar{V}_{\mathrm{ext}}(\mathbf{q}) = \chi_{\mathrm{e}}(q)\Delta\bar{V}_{\mathrm{ext}}(\mathbf{q}). \tag{4.127}$$

These equivalent forms use the relations (4.74) and (4.75) for the response functions derived previously. In the limit of zero non-locality, that is if $\Delta V_{\text{ext}}^{NL} = 0$, we recover $\Delta\bar{V}_{\text{ext}}(\mathbf{q}) \rightarrow \Delta V_{\text{ext}}(\mathbf{q})$.

4.11.3 Second-order Energy in Jellium

We can derive a fairly simple expression for the second-order energy change in jellium even if it is due to a non-local potential. Historically, this was a basis for self-consistent total energy calculations in the nineteen-sixties and seventies using pseudopotentials, as described for example by Harrison (1966) or Heine and Weaire (1970). With the jellium eigenstates eq. (4.114) becomes

$$\Delta E^{(2)} = 2\sum_{\mathbf{k}} f_{\text{F}}(\epsilon_{\mathbf{k}})\Delta V_{\text{ext}}(\mathbf{k},\mathbf{k}) + \Delta E_{ZZ} + \sum_{\mathbf{k},\mathbf{q}} \frac{(f_{\text{F}}(\epsilon_{\mathbf{k}}) - f_{\text{F}}(\epsilon_{\mathbf{k}+\mathbf{q}}))}{(\epsilon_{\mathbf{k}} - \epsilon_{\mathbf{k}+\mathbf{q}})} \langle\mathbf{k}|\Delta\hat{V}_{\text{eff}}|\mathbf{k}+\mathbf{q}\rangle\langle\mathbf{k}+\mathbf{q}|\Delta\hat{V}_{\text{ext}}|\mathbf{k}\rangle. \quad (4.128)$$

Care must be taken with $\mathbf{k} = 0$ and $\mathbf{k}' = 0$ terms in these summations. A $\mathbf{k} = 0$ term would describe a constant shift in the potential, and we cannot allow our perturbation to have a constant component, because that would immediately produce a diverging electrostatic energy.

If the perturbation is a local potential eq. (4.128) reduces immediately to

$$\Delta E^{(2)} = 2\sum_{\mathbf{k}} f_{\text{F}}(\epsilon_{\mathbf{k}})\Delta V_{\text{ext}}(\mathbf{k},\mathbf{k}) + \frac{V}{2}\sum_{\mathbf{q}} \chi_{\text{e}}(q)\,|\Delta V_{\text{ext}}(\mathbf{q})|^2 + \Delta E_{ZZ}. \quad (4.129)$$

We have made use of the relations $V_{\text{eff}} = V_{\text{ext}}/\epsilon_{\text{e}}$ and $\chi_{\text{e}} = \chi_{\text{s}}/\epsilon_{\text{e}}$ derived previously in Section 4.4. With a non-local potential we can arrive at something rather similar, as follows. We write the effective potential in terms of its external and response parts. With a little rearrangement, the product of potentials in (4.128) can be written as

$$\begin{aligned}\langle\mathbf{k}|\Delta\hat{V}_{\text{eff}}|\mathbf{k}+\mathbf{q}\rangle\langle\mathbf{k}+\mathbf{q}|\Delta\hat{V}_{\text{ext}}|\mathbf{k}\rangle = {} & (\Delta V_{\text{ext}}(\mathbf{k},\mathbf{k}+\mathbf{q}) - \Delta\bar{V}_{\text{ext}}(q) + \Delta\bar{V}_{\text{ext}}(q) \\ & + v_{\text{C}}(q)(1-\mathcal{G}(q))\chi_{\text{e}}(q)\Delta\bar{V}_{\text{ext}}(q)) \\ & \times \left(\Delta\bar{V}_{\text{ext}}^{*}(q) + \Delta V_{\text{ext}}(\mathbf{k}+\mathbf{q},\mathbf{k}) - \Delta\bar{V}_{\text{ext}}^{*}(q)\right).\end{aligned} \quad (4.130)$$

The local potential $\Delta\bar{V}_{\text{ext}}$ has been added and subtracted within each factor, in order to create the factor $\Delta V_{\text{ext}}(\mathbf{k},\mathbf{k}+\mathbf{q}) - \Delta\bar{V}_{\text{ext}}(q)$ which contains all the complications of non-locality. Any terms which do not depend on $\mathbf{k}$ will be untouched by the summation, which simply multiplies them by $(V/2)\chi_s(q)$. We can also use the results of Sections 4.4 and 4.8 to replace the factor $1+v_C(q)(1-\mathcal{G}(q))$ by $\epsilon_e(q)^{-1}$. We now divide up the terms in (4.130) as follows:

$$\langle \mathbf{k}|\Delta\hat{V}_{\text{eff}}|\mathbf{k}+\mathbf{q}\rangle\langle \mathbf{k}+\mathbf{q}|\Delta\hat{V}_{\text{ext}}|\mathbf{k}\rangle$$
$$= \Delta\bar{V}_{\text{ext}}(q))\epsilon_e^{-1}(q)\Delta\bar{V}^*_{\text{ext}}(q) + \left|\Delta V_{\text{ext}}(\mathbf{k},\mathbf{k}+\mathbf{q}) - \Delta\bar{V}_{\text{ext}}(q)\right|^2 + \text{rem}. \tag{4.131}$$

It is easy to verify that the terms included as 'rem' will sum to zero in the final expression for $\Delta E^{(2)}$, which is:

$$\Delta E^{(2)} = 2\sum_{\mathbf{k}} f_F(\epsilon_{\mathbf{k}})\Delta V_{\text{ext}}(\mathbf{k},\mathbf{k}) + \frac{V}{2}\sum_{\mathbf{q}} \chi_e(q)\left|\Delta\bar{V}_{\text{ext}}(\mathbf{q})\right|^2 + \Delta E_{ZZ}$$
$$+ \sum_{\mathbf{k},\mathbf{q}} \frac{(f_F(\epsilon_{\mathbf{k}}) - f_F(\epsilon_{\mathbf{k}+\mathbf{q}}))}{(\epsilon_{\mathbf{k}} - \epsilon_{\mathbf{k}+\mathbf{q}})}\left|\Delta V_{\text{ext}}(\mathbf{k},\mathbf{k}+\mathbf{q}) - \Delta\bar{V}_{\text{ext}}(q)\right|^2. \tag{4.132}$$

The result of the foregoing manipulations is an expression which is as close as possible to the simple result for local potentials, eq. (4.129). Similar expressions were derived by Rasolt and Taylor (1975) and others as a basis for generating interatomic potentials. This will be done later in the book. The physical content of the above equation should be clear however. The first term on the right of eq. (4.132) is the first-order term in perturbation theory, namely the interaction of the perturbing potential with the uniform electron gas. The second term represents the interaction between the screening charge induced by the external perturbation $\Delta\bar{V}_{\text{ext}}$ and the perturbing potential itself. The third term, which is the change in self-energy of the external potential, may be represented in **k**- or **r**-space as we shall see. The final term is the explicit correction for non-locality of the perturbing potential, bearing in mind that some of the non-locality has been folded into the definition of $\Delta\bar{V}_{\text{ext}}(\mathbf{q})$.

PART II

MODELLING ATOMS WITHIN SOLIDS

5

TESTING AN INTERATOMIC FORCE MODEL

All models have to jump through a number of hoops in order to be evaluated. They have to be tried out on some simple physical properties before you use them to calculate something more interesting. In the case of models with a semi-empirical construction, this is also part of the process of fitting their parameters. The fitting of parameters is often difficult and sometimes controversial. It is said that Enrico Fermi once remarked: 'Give me two free parameters and I will fit an elephant; give me three, and I will wag its tail . . .'. However, there is absolutely nothing wrong with fitting parameters if it is done in a transparent way, because there is no doubt that some of the most empirical models can produce interesting new science if they are used intelligently. One only has to think of the earliest molecular dynamics simulations which studied atomic collisions in metals induced by irradiation, demonstrating the possibility of focussed collision sequences (Gibson *et al.*, 1960). Or the more recent studies of fracture, in which the the strong effect of lattice trapping has been demonstrated, and more recently the possibility of dislocations travelling faster than the speed of sound (Gumbsch and Gao, 1999). Some of the most important results in the molecular dynamics of liquids, the oscillating tails of temporal and spacial correlation functions, came from the model of hard spheres. There are many other examples.

The main virtue which is sought in any model is *transferability*. That is to say, it is not usually good enough for it to work well in one structure for one property, and it is of no value at all if it only correctly reproduces properties to which it was fitted. A truly transferable model would correctly reproduce all experimental data, but that can never be achieved. However, if you accept that scientific progress is measured less by the accuracy of numerical agreement with experimental data than by the quality of scientific insight, you will find that poorly transferable models can and do yield qualitative insights about mechanisms and processes in materials. At the same time a model which is poorly transferable will always lead to some wrong predictions, and may generate complete nonsense. So one had better be aware of the limitations of the model one uses. This means understanding its physical basis and making extensive tests of its quantitative accuracy.

Tests such as described in this section involve the calculation of static, zero temperature properties. The comparison is then with either better quality calculations (e.g. the best first principles calculations) or with experimental data extrapolated to $T = 0\,\mathrm{K}$. I will describe here some of the commoner tests for testing or fitting models, pointing out the pitfalls for the unwary. These tests constitute a minimum set which a model would be expected to meet. However, if a model is to be used for simulations under extreme conditions of temperature and/or pressure, or for large plastic deformations, they may be supplemented by other tests which more closely approach the configurations of interest.

5.1 The Cohesive Energy and Crystal Structures

The cohesive energy of a particular crystal structure is defined for our purposes as the difference in energy between separated free atoms and the same atoms in the perfect crystal, divided by the number of atoms.

$$E^{\mathrm{coh}} = \frac{1}{N_{\mathrm{a}}} \left(E^{\mathrm{tot}}(\text{free atoms}) - E^{\mathrm{tot}}(\text{crystal}) \right). \tag{5.1}$$

This is an obvious target for model calculations. However, it is asking a lot of any model to predict free atomic energies and the energies of condensed phases to the same accuracy. Even first-principles calculations have to go beyond the LDA to achieve that, since the energies of free atoms are especially sensitive to unpaired spins. It is more reasonable, and useful for applications, if E^{coh} as a function of crystal structure and density is correctly reproduced. One calculates the total energy of a crystal as a function of volume per atom for several competing crystal structures. If the model is good, it will predict the lowest energy in the experimentally observed crystal structure at the experimentally observed volume. Other structures which may or may not be observed in nature will have higher energies. This way of proceeding is an approximation to the truth because the experimental equilibrium structures are never available at $0\,\mathrm{K}$. The process of equilibration slows down so much at low temperature that one can never be sure if the lowest temperature crystal structures observed are truly in equilibrium at $0\,\mathrm{K}$. For example even in simple metals like lithium, sodium and potassium there has been controversy over whether the lowest temperature structures are pure $9R$ or something else (Ashcroft, 1989). Furthermore, our simple description of the total energy so far is ignoring zero-point energy and its possible structure dependence. For practical purposes however we are not usually concerned with the very low temperature limit, and calculations which ignore temperature and zero-point motion will provide an adequate test of the model if the energy differences are not too small.

The curvature of the energy versus volume at the minimum gives the bulk modulus. This is discussed in Section 5.3, where I also describe how the curvature of energy versus other strains gives the other elastic constants.

5.2 The Structural Energy Difference Theorem

Pettifor's structural energy difference theorem (SEDT) was a a breakthrough in the comparison of the energies of different crystal structures within tight-binding models, since it offers a very simple procedure for understanding *total* energy differences ΔE^{tot} between two crystal structures based purely on the difference in the sums of their one-electron eigenvalues, or bond energies (Pettifor, 1986, 1995). It is essentially the same as the theorem also referred to as Pettifor's force theorem, which serves the same purpose for *ab initio* calculations in close-packed crystal structures. Since it is such a general result, it is worth devoting a section of this chapter to it, although it presupposes that you already know a little about tight-binding. It is discussed and proved within the tight-binding context by Pettifor (1985, p. 81).

The difference in eigenvalue sums had already been used for many years as a surrogate for the total energy, but crystal structures had been compared at the same bond length or atomic volume. It was of course always known that this is not the only term in the total energy; there had to be a repulsive term as well, in order to counterbalance the attraction of the eigenvalue sum. The repulsive term was simply assumed not to make such an important contribution to the energy differences ΔE^{tot}. The SEDT says that the difference between bond energies is indeed a legitimate substitute for the difference between the total energies, but only if it is calculated *at the same repulsive energy*. Normally different crystal structures at the same atomic volume will have different repulsive energies, likewise even if the bond lengths are the same, if the coordination (the number of nearest neighbours) is different then so is the repulsive energy. The SEDT tells us we have to adjust the volumes to make the repulsive energy the same before we compare the bond energies. The following proof shows the generality of the theorem, of which the above statement is a specific example.

Let us suppose that the total energy E^{tot} of a particular crystal structure $s1$ is a function of some parameter x which we can think of for simplicity as the volume, but it could be any strain or some other continuous parameter. We write $E^{\mathrm{tot}}(s1, x)$. The important thing is that x can also be used to specify something about a second crystal structure $s2$, so there is a second function $E^{\mathrm{tot}}(s2, x)$. Thus $s1$ and $s2$ are discrete labels of two different energy functions, not continuous arguments of the same function. Suppose the equilibrium value of the parameter x depends on the crystal structure, so that $s1$ is in equilibrium when $x = x_1$, and similarly $s2$ is in equilibrium at x_2. This means:

$$\left.\frac{\partial E^{\mathrm{tot}}(s1, x)}{\partial x}\right|_{x=x_1} = 0$$

$$\left.\frac{\partial E^{\mathrm{tot}}(s2, x)}{\partial x}\right|_{x=x_2} = 0. \tag{5.2}$$

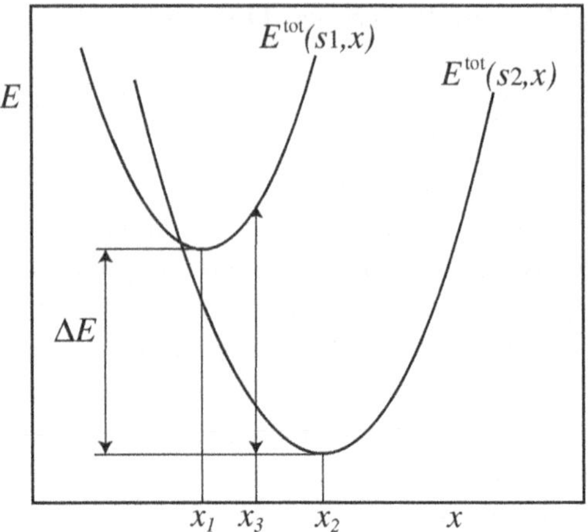

FIG. 5.1 Two energies compared to first-order in the difference in x.

We wish to calculate the total energy difference between $s1$ and $s2$ in equilibrium, that is,

$$\Delta E^{\text{tot}} = E^{\text{tot}}(s1, x_1) - E^{\text{tot}}(s2, x_2). \tag{5.3}$$

We can express this in terms of the total energy difference at other values of x by expanding to second-order about x_1 and x_2. Let us express the difference when structure $s1$ is at x_3 and structure $s2$ is at x_4:

$$\Delta E^{\text{tot}} = E^{\text{tot}}(s1, x_3) - E^{\text{tot}}(s2, x_4) + O((x_3 - x_1)^2) - O((x_4 - x_2)^2). \tag{5.4}$$

The first-order terms vanish by (5.2). The first conclusion to be drawn is that to first-order in the errors in x we can compare the total energies at any values of x. This is another example of the variational principle. See Fig. 5.1 for a hypothetical example which exaggerates the variation with x.

Now suppose that the energy can be divided into two parts, E_{A} and E_{B}. In the usual SEDT we think of E_{A} as the bond energy and E_{B} as the repulsive energy, but they could be the other way round or we could make a quite different division of the energy. We have:

$$\begin{aligned}\Delta E^{\text{tot}} = {} & E_{\text{A}}(s1, x_3) - E_{\text{A}}(s2, x_4) + E_{\text{B}}(s1, x_3) - E_{\text{B}}(s2, x_4) \\ & + O((x_3 - x_1)^2) - O((x_4 - x_2)^2).\end{aligned} \tag{5.5}$$

It is obvious that if we choose x_3 and x_4 such as to make $E_{\text{B}}(s1, x_3) = E_{\text{B}}(s2, x_4)$, then to first-order ΔE^{tot} is given just by $E_{\text{A}}(s1, x_3) - E_{\text{A}}(s2, x_4)$ and vice versa. That is the SEDT. In applications of the SEDT x_3 or equivalently x_4 is chosen to be the same as one of the equilibrium structures x_1 or x_2.

5.2.1 Examples of the SEDT

As an example consider a total energy with the form:

$$E^{\text{tot}}(x) = -C_b\beta(x) + C_r\beta(x)^{\lambda}. \tag{5.6}$$

The simplest tight-binding model looks exactly like this, in which β is the hopping integral which has some inverse power law behaviour as a function of bond length x. The range of interaction of the Hamiltonian is restricted here to nearest neighbours, so that β factors out of the bond energy. C_b and C_r are constants depending only on the crystal structure and the number of electrons per atom, not on the volume. The simplest approximation (Pettifor, 1995, p. 65) would be to set $\lambda = 2$. This can be justified if the main contribution to the repulsive term is from the overlap repulsion. Pettifor (1995, chapter 4). studied a model of this form to discuss the relative stability of four structures of a four-atom molecule, namely the linear chain(l) the square(s) the rhombus(r) and the tetrahedron(t). He studied the dependence on λ and the accuracy of the SEDT versus an exact solution of the model. The close packed t structure always had the lowest energy and the l structure the highest, while the s and r structures were close together in energy between the other two. The SEDT predicted the differences very well, in the worst case the error was only 11%. Furthermore it predicted the crossover in energy as the r became more stable than the s with increasing 'hardness' λ.

We can also see how the Hellmann–Feynman theorem falls out of the SEDT. All we have to do is to let x be any parameter describing the electron density distribution, and define $E_{\text{A}} = E_{eZ} + E_{ZZ}$. Comparing two structures at fixed x means considering a different arrangement of nuclei within the same electron distribution, including the case in which just one nucleus is slightly displaced in structure $s2$ compared to structure $s1$. In this case if x is fixed so is E_{B}, since all the dependence on the nuclear positions is within E_{A}. The SEDT now says that in this situation the energy difference between $s2$ and $s1$ is given by the change in E_{A}, and this is precisely the same as the Hellmann–Feynman theorem.

5.3 Elastic Constants

The chances are if you are reading this book you already know something about elastic constants. If not, you had better start with the basics by looking for example at the book by Nye (1985), which covers the continuum theory. The theory in crystalline materials and the relationship of elastic constants to interatomic potentials is covered in the book by Wallace (1972). Both books are outstandingly good. Given that you want to apply some model for the total energy E^{tot}, somebody should have calculated its single-crystal elastic constants, but if the model is a new one, or the crystal structure is unusual, perhaps this job has fallen to you. In any case it is a good preliminary test of the model and the program. As textbooks explain clearly, the elastic constants are proportional to the second derivatives of the total energy with respect to strain. However the standard textbooks do not give you such a clear strategy for calculating them from a given model. I will summarize the main points of the theory here so that you will be in a position to calculate any elastic constant for any crystal structure without too much of a headache.

The fundamental measure of a homogeneous strain is the matrix

$$\epsilon = \begin{pmatrix} \epsilon_{11} & \epsilon_{12} & \epsilon_{13} \\ \epsilon_{12} & \epsilon_{22} & \epsilon_{23} \\ \epsilon_{13} & \epsilon_{23} & \epsilon_{33} \end{pmatrix}. \tag{5.7}$$

The elements of ϵ are labelled by the Cartesian axes, and are defined as follows. If the strain carries the point $\mathbf{r} = (x_1, x_2, x_3)$ to a new position $\mathbf{r} + \mathbf{u}$, then the elements of ϵ are defined by

$$\epsilon_{\alpha\beta} = \frac{1}{2}\left(\frac{\partial u_\alpha}{\partial x_\beta} + \frac{\partial u_\beta}{\partial x_\alpha}\right). \tag{5.8}$$

The point of symmetrizing like this is to eliminate rotations of the body, which are associated with a purely *antisymmetric* matrix. This is not the only possible measure of strain, although it is the one we will use for practical applications. Another measure, to which I will return later, is the *Lagrangian strain* (Wallace, 1972, p. 19), defined by the symmetric matrix

$$\eta_{\alpha\beta} = \frac{1}{2}\left(\frac{\partial u_\alpha}{\partial x_\beta} + \frac{\partial u_\beta}{\partial x_\alpha} + \sum_\gamma \frac{\partial u_\gamma}{\partial x_\alpha}\frac{\partial u_\gamma}{\partial x_\beta}\right). \tag{5.9}$$

The Lagrangian strain describes the change in length of a vector which is strained from $\mathbf{R}^0$ to $\mathbf{R}$ as

$$|\mathbf{R}|^2 - |\mathbf{R}^0|^2 = 2\sum_{\alpha,\beta} \eta_{\alpha\beta} R^0_\alpha R^0_\beta. \tag{5.10}$$

The applied stress is likewise expressed in the form of a symmetric matrix with elements $\sigma_{\alpha\beta}$. The stress is related to the strain in linear elasticity theory by

$$\sigma_{\alpha\beta} = \sum_{\gamma\delta} C_{\alpha\beta\gamma\delta}\epsilon_{\gamma\delta}, \tag{5.11}$$

an equation which defines 81 elastic constants (or stiffness constants) $C_{\alpha\beta\gamma\delta}$.

Fortunately not all these 81 quantities are independent, and there are two wonderful simplifications which symmetry gives us. The first of these allows a change of notation to what is generally called Voigt notation, after one of the founders of the subject W. Voigt. In Voigt notation we write the stresses and strains as single index quantities according to the correspondences:

$$\begin{pmatrix} \epsilon_1 & \epsilon_6 & \epsilon_5 \\ \epsilon_6 & \epsilon_2 & \epsilon_4 \\ \epsilon_5 & \epsilon_4 & \epsilon_3 \end{pmatrix} \equiv \begin{pmatrix} \epsilon_{11} & 2\epsilon_{12} & 2\epsilon_{13} \\ 2\epsilon_{12} & \epsilon_{22} & 2\epsilon_{23} \\ 2\epsilon_{13} & 2\epsilon_{23} & \epsilon_{33} \end{pmatrix}$$

and

$$\begin{pmatrix} \sigma_1 & \sigma_6 & \sigma_5 \\ \sigma_6 & \sigma_2 & \sigma_4 \\ \sigma_5 & \sigma_4 & \sigma_3 \end{pmatrix} \equiv \begin{pmatrix} \sigma_{11} & \sigma_{12} & \sigma_{13} \\ \sigma_{12} & \sigma_{22} & \sigma_{23} \\ \sigma_{13} & \sigma_{23} & \sigma_{33} \end{pmatrix}. \tag{5.12}$$

The great simplification is that the elastic constant matrix can be written as a 6×6 symmetric matrix with elements C_{ij} such that the stress–strain relationships are

$$\sigma_i = \sum_j C_{ij} \epsilon_j. \tag{5.13}$$

Furthermore, the elastic energy per unit volume stored in the homogeneously strained body is given by

$$E^{\text{elas}} = \frac{1}{2} \sum_{ij} C_{ij} \epsilon_i \epsilon_j. \tag{5.14}$$

The C_{ij} are the things we want to calculate for our model by constructing the second derivative of the energy with respect to strain from a set of total energy calculations. At this stage there are still 21 independent elements in the elastic constant matrix (not 36, because the matrix is symmetric). But except in crystals of lowest symmetric (triclinic) this number can be reduced further. In cubic crystals there are only three independent elastic constants C_{11}, C_{12} and C_{44}. All the others either vanish or are equal to one of these three. Altogether there are just ten distinct types of crystals from the point of view to the symmetry of their elastic constants, including the isotropic solid which has complete rotational symmetry. They are identified together with the number of independent C_{ij} in Table 5.1. Complete information about the relationships between the elastic constants is to be found in Table 9 of Nye's book.

From the information in Nye's Table 9 we can make the energy expression (5.14) explicit for any system. We can then design a particular strain so as to pick out the elastic constant we want to calculate. As an example of this procedure I will describe the simplest and most common case, the cubic crystal.

TABLE 5.1 The distinct crystal types for elastic constants

System	No. of independent C_{ij}
Triclinic	21
Monoclinic	13
Orthorhombic	9
Cubic	3
Tetragonal, Classes 4, $\bar{4}$, $4/m$	7
Tetragonal, Classes $4mm$, $\bar{4}2m$, 422, $4/mmm$	6
Trigonal Classes 3, $\bar{3}$	7
Trigonal, Classes 32, $\bar{3}m$, $3m$	6
Hexagonal	5
Isotropic	2

5.3.1 Cubic Crystals

For cubic crystals the elastic constant matrix looks like

$$C = \begin{pmatrix} C_{11} & C_{12} & C_{12} & 0 & 0 & 0 \\ C_{12} & C_{11} & C_{12} & 0 & 0 & 0 \\ C_{12} & C_{12} & C_{11} & 0 & 0 & 0 \\ 0 & 0 & 0 & C_{44} & 0 & 0 \\ 0 & 0 & 0 & 0 & C_{44} & 0 \\ 0 & 0 & 0 & 0 & 0 & C_{44} \end{pmatrix}. \tag{5.15}$$

The elastic energy per unit volume, writing out eq. (5.14), is

$$E^{\mathrm{elas}} = \tfrac{1}{2}C_{11}(\epsilon_1^2 + \epsilon_2^2 + \epsilon_3^2) + C_{12}(\epsilon_1\epsilon_2 + \epsilon_2\epsilon_3 + \epsilon_3\epsilon_1) + C_{44}(\epsilon_4^2 + \epsilon_5^2 + e_6^2). \tag{5.16}$$

Now we have to decide what strain to apply. Suppose we want to calculate C_{11}. We can apply the strain ϵ_1 to our unit cell and construct the curve $E^{\mathrm{tot}}(\epsilon_1)$. This is done by making a number of calculations of total energy at equally spaced values of ϵ_1 ranging from say -0.02 to $+0.02$ in steps of say 0.002. A low order polynomial, typically of order 5, is fitted through the calculated energies. The curvature at $\epsilon_1 = 0$ then gives the estimated value of C_{11}, not forgetting to divide the total energy by the volume of the unit cell V_{c}. You might ask, since we only want the second derivative, why not calculate energies at just two values of ϵ_1? By symmetry $E(\epsilon_1) = E(-\epsilon_1)$, and two energies, $E(0)$ and $E(\epsilon_1)$, would therefore be enough to define a parabola, with second derivative $2(E(\epsilon_1) - E(0))/\epsilon_1^2$. This would not be a very good idea except for a crude estimate, because we would not see the error introduced by the *anharmonicity* of the crystal. Terms in the energy of third and higher order in ϵ_1 do affect the energy significantly at rather small strains, so the polynomial fitting procedure is a more reliable way to extract the second-order term. Some researchers prefer to use functions other than a simple polynomial, but the result is the same. Another reason for calculating several points on the energy versus strain curve is to monitor any random numerical error in the numbers.

Let us look in more detail at how this procedure can be carried out in practice. Applying the strain ϵ_1 to a cubic unit cell simply means extending the side $\mathbf{a}_1$ by a factor $1 + \epsilon_1$. In the case of a general strain, we will need to transform all the basis vectors $\mathbf{a}_i$. For this purpose it is convenient to separate the *symmetry* of a strain from a scalar measure of its magnitude and sign, which we will call γ. We introduce a symmetric transformation matrix T such that

$$\epsilon_{\alpha\beta} = \gamma T_{\alpha\beta}. \tag{5.17}$$

The strain is achieved in practice by applying this transformation γT to the basis vectors, taking $\{\mathbf{a}_1, \mathbf{a}_2, \mathbf{a}_3\}$ to $\{\mathbf{a}'_1, \mathbf{a}'_2, \mathbf{a}'_3\}$, where

$$a'_{i\alpha} = a_{i\alpha} + \gamma \sum_\beta T_{\alpha\beta} a_{i\beta}, \tag{5.18}$$

in which $a_{i\alpha}$ is the Cartesian component along the x_α axis of basis vector $\mathbf{a}_i$. Now to recap our example above, for the calculation of C_{11} we would choose $T_{xx} = 1$ and $T_{\alpha\beta} = 0$ for the other elements. The strain parameter γ is then varied in steps and the points are fitted to produce a curve of total energy E^{tot} versus γ, based on a unit cell of volume V_{c}. At each value of γ there must be no forces on any atoms within the unit cell, so in all but the simplest crystal structures it would be necessary to relax the atomic positions. The coefficient of $\gamma^2/2$ in the fitted polynomial gives us directly $\partial^2 E^{\text{tot}}/\partial\gamma^2$ and

$$C_{11} = \frac{1}{V_{\text{c}}} \left. \frac{\partial^2 E^{\text{tot}}}{\partial \gamma^2} \right|_{\gamma=0}. \tag{5.19}$$

You can calculate C_{44} in just the same way, as we discuss a little later. However, C_{12} is different. You can see from (5.16) that there is no strain that picks out C_{12} and only C_{12}, it always appears in combination with C_{11}. However, we can obtain any linear combination of C_{11} and C_{12} we like by applying an appropriate combination of, say, ϵ_1 and ϵ_2. Usually the combination $\frac{1}{2}(C_{11} - C_{12})$ is calculated, for reasons below. In this way, all three elastic constants can be extracted from three curves of E^{tot} versus γ.

Usually the *bulk modulus* is the first thing to be calculated; it is defined by

$$B = V_{\text{c}} \frac{\partial^2 E^{\text{tot}}}{\partial V_{\text{c}}^2}, \tag{5.20}$$

evaluated at zero stress and strain. To get it we simply have to vary the volume of the crystal, for which the transformation matrix T is the unit matrix. B is a linear combination of C_{11} and C_{12} which is exposed by rearranging (5.14), using (5.12) and (5.17), to give

$$E^{\text{elas}} = \tfrac{1}{6}(C_{11} + 2C_{12})(T_{xx} + T_{yy} + T_{zz})^2\gamma^2 + \tfrac{1}{3}(C_{11} - C_{12})(T_{xx}^2 + T_{yy}^2 + T_{zz}^2 - T_{yy}T_{zz} - T_{zz}T_{xx} - T_{xx}T_{yy})\gamma^2 + 2C_{44}(T_{yz}^2 + T_{zx}^2 + T_{xy}^2)\gamma^2. \quad (5.21)$$

Consider now the case where T is the unit matrix. Returning to the unit cell transformation (5.18), you should be able to see that a strain parameter γ, takes $V_c(0)$ to $V_c(\gamma)$, which is given by

$$V_c(\gamma) = (1+\gamma)^3 V_c(0). \quad (5.22)$$

Now we make the transformation of derivatives:

$$\frac{\partial}{\partial V_c} = \frac{1+\gamma}{3V_c}\frac{\partial}{\partial \gamma} \quad (5.23)$$

and

$$\begin{aligned} V_c\frac{\partial^2}{\partial V_c^2} &= V_c\frac{\partial}{\partial V_c}\left[\left(\frac{\partial V_c}{\partial \gamma}\right)^{-1}\frac{\partial}{\partial \gamma}\right] \\ &= V_c\frac{\partial}{\partial V_c}\left(\frac{1+\gamma}{3V_c(\gamma)}\frac{\partial}{\partial \gamma}\right) \\ &= -\frac{2}{3}\frac{\partial}{\partial V_c} + \frac{1}{9V_c}\frac{\partial^2}{\partial \gamma^2}, \end{aligned} \quad (5.24)$$

where we have set $\gamma = 0$ only *after* taking the second derivative. The pressure is given by

$$P = -\frac{\partial E^{\text{tot}}}{\partial V_c}, \quad (5.25)$$

so when the crystal is in equilibrium at zero pressure.

$$V_c\frac{\partial^2 E^{\text{tot}}}{\partial V_c^2} = \frac{1}{9V_c}\frac{\partial^2 E^{\text{tot}}}{\partial \gamma^2}. \quad (5.26)$$

With the unit T matrix, only the first term in (5.21) is not zero, and the total energy is a constant plus the elastic energy:

$$E^{\text{elas}} = \tfrac{1}{6}(C_{11} + 2C_{12})9\gamma^2. \quad (5.27)$$

By combining (5.20), (5.26) and (5.27) we can extract B directly from the curve of energy versus γ:

$$B = \frac{1}{9V_c}\frac{\partial^2 E^{\text{tot}}}{\partial \gamma^2}. \quad (5.28)$$

Remember that a curve of energy versus volume is also needed in the first place, before calculating any elastic constants, in order to obtain the cohesive energy and equilibrium lattice parameter, at which $\partial E^{\text{tot}}/\partial\gamma$ vanishes. So we obtain three physical parameters for the price of one from this curve.

The quantity $\frac{1}{2}(C_{11} - C_{12})$, like C_{44}, is referred to as a *shear modulus*. Shear moduli correspond to volume conserving strains, for which the trace of the strain matrix (or equivalently of T) is zero. Usually the quantity $\frac{1}{2}(C_{11} - C_{12})$ is given the special symbol C'. C' describes the elastic energy of volume conserving strains with no off-diagonal components in an analogous way that C_{44} describes the elastic energy of volume conserving strains with no diagonal components. To calculate C' we could set

$$T = T^t = \begin{pmatrix} 1.0 & 0 & 0 \\ 0 & -0.5 & 0 \\ 0 & 0 & -0.5 \end{pmatrix} \tag{5.29}$$

in which case, as seen by inspecting (5.21), we would extract

$$C' = \frac{1}{3V_c}\frac{\partial^2 E^{\text{tot}}}{\partial\gamma^2}. \tag{5.30}$$

This corresponds to a tetragonal distortion, a change in the c/a ratio of the unit cell. Alternatively we could shear the cell on a (110) plane, as described by

$$T = T^{110} = \begin{pmatrix} 1.0 & 0 & 0 \\ 0 & -1.0 & 0 \\ 0 & 0 & 0 \end{pmatrix}, \tag{5.31}$$

in which case

$$C' = \frac{1}{4V_c}\frac{\partial^2 E^{\text{tot}}}{\partial\gamma^2}. \tag{5.32}$$

In a similar way we might calculate C_{44} by applying a *rhombohedral shear*, which consists of stretching the cube along a [111] direction, while maintaining the volume constant. An appropriate transformation is

$$T = T^r = \begin{pmatrix} 0 & 0.5 & 0.5 \\ 0.5 & 0 & 0.5 \\ 0.5 & 0.5 & 0 \end{pmatrix}, \tag{5.33}$$

giving

$$C_{44} = \frac{1}{3V_c}\frac{\partial^2 E^{\text{tot}}}{\partial\gamma^2}. \tag{5.34}$$

Alternatively, we could shear on a (100) plane, for example in the y direction, with

$$T = T^{100} = \begin{pmatrix} 0 & 1 & 0 \\ 1 & 0 & 0 \\ 0 & 0 & 0 \end{pmatrix}, \tag{5.35}$$

in which case

$$C_{44} = \frac{1}{4V_c}\frac{\partial^2 E^{\text{tot}}}{\partial \gamma^2}. \tag{5.36}$$

The similar form of C' and C_{44} is no coincidence of course. In an isotropic material these elastic constants are identical. That's why the factor of $\frac{1}{2}$ appears in the definition of $C' = \frac{1}{2}(C_{11} - C_{12})$.

We have rearranged the expression for elastic energy from three terms proportional to C_{11}, C_{12} and C_{44} to three terms proportional to B, C' and C_{44}, each of which can be calculated by applying separate strains. This is an example of a general procedure which you could apply to any of the other eight classes of crystal. It has the additional property that a necessary condition for the crystal to be mechanically stable is that each of the independent terms must be positive. In the cubic case this means the three elastic constants B, C' and C_{44} must be positive. Note that there is no requirement on C_{12} to be positive.

At a deeper level, what we are doing here is to sort the set of strain parameters $\{\epsilon_{\alpha\beta}\}$ into irreducible representations of the point symmetry class of the crystal. Group theory says that each irreducible representation has one and only one second-order invariant. For the cubic crystal discussed here in detail, the three representations A (the identity), E and T correspond to the elastic constants B, C' and C_{44}. They are also familiar in electronic structure as the symmetry of an s orbital and the E_g and T_{2g} symmetries of d orbitals in a cubic environment. I have tossed in these group theory concepts without proper explanation just to whet your appetite, because they would certainly be helpful if you have to understand the elastic constants of any non-cubic crystals.

I will not discuss non-cubic crystals further except to make one point. Even the bulk modulus is not trivial to calculate. The reason is that the volume change is imagined to be due to an applied pressure (positive or negative), and unless the crystal has cubic symmetry there is no reason why it should preserve its shape as the volume changes. So you do not know what matrix T to use. Be careful.

The stability of a crystal is not guaranteed by the fact that all the strains you could possibly apply lead to a positive elastic energy, although that is a good start. In fact we can never guarantee to have found the crystal structure of lowest energy with a given model. There might always be large distortions, homogeneous or inhomogeneous, which lead to a lowering of the energy. Besides simply enumerating and calculating as many of these as you have the time and patience to look at, there is a way to check that the energy of *all* the inhomogeneous distortions is positive, but only if the distortions are small. That is to calculate the frequencies of vibration, a matter we describe in Section 5.4.

5.3.2 Some Subtleties: Pair Potentials and Cauchy Pressure

When the total energy is made up of more than one kind of term, you might ask the question: what is the contribution of each term to the elastic constants? In particular, there are models of the total energy which consist of a sum of pairwise interactions plus something else. It is a little worrying to discover in the literature that the contribution of each term to an elastic constant appears not to be unique! Different authors give different expressions which it turns out are equally valid when properly interpreted, but it is easy to make mistakes. The ambiguity arises because as mentioned above there are alternative measures of the strain, which however agree to first-order. If the internal energy is expanded to second-order in the strain, there will be a first-order term which only vanishes when the crystal is in equilibrium at zero strain. But if we change the definition of strain say to the Lagrangian strain $\eta_{\alpha\beta}$, this first-order term includes a second-order term which is picked up in the corresponding alternative definition of the elastic constant. We have seen this happen in the case of the bulk modulus discussed above. To make this idea clear we will consider the simplest example which is a purely volume dependent energy, $E^{\mathrm{vol}}(V_{\mathrm{c}})$. Another perfectly natural measure of the volume strain is

$$\eta = \tfrac{1}{3}(V_{\mathrm{c}}/V_{\mathrm{c}}(0) - 1) \tag{5.37}$$

instead of the one we used:

$$\gamma = \left(\frac{V_{\mathrm{c}}}{V_{\mathrm{c}}(0)}\right)^{1/3} - 1 \tag{5.38}$$

where to first-order $\eta = \gamma$. Whereas we have adopted the convention that the bulk modulus is given in terms of the second derivative with respect to γ according to eq. (5.28):

$$B = \frac{1}{3}(C_{11} + 2C_{12}) = \frac{1}{9V_{\mathrm{c}}}\frac{\partial^2 E^{\mathrm{tot}}}{\partial \gamma^2}, \tag{5.39}$$

the original definition is in terms of the second derivative with respect to volume or equivalently with respect to η:

$$B = V_{\mathrm{c}}\frac{\partial^2 E^{\mathrm{tot}}}{\partial V_{\mathrm{c}}^2} \equiv \frac{1}{9V_{\mathrm{c}}}\frac{\partial^2 E^{\mathrm{tot}}}{\partial \eta^2}. \tag{5.40}$$

As we have seen, these expressions are only the same when the pressure P is zero. However, when only *part* of the total energy is given by E^{vol}, the pressure P^{vol} given by

$$P^{\mathrm{vol}} = -\frac{\partial E^{\mathrm{vol}}}{\partial V_{\mathrm{c}}} \equiv -\frac{1}{3V_{\mathrm{c}}(0)}\frac{\partial E^{\mathrm{vol}}}{\partial \gamma} \equiv -\frac{1}{3V_{\mathrm{c}}(0)}\frac{\partial E^{\mathrm{vol}}}{\partial \eta} \tag{5.41}$$

will *not* vanish when the crystal is in equilibrium, it will be balanced by the pressure from the other terms in the energy. As a consequence when we calculate B^{vol} from the strain γ we obtain from (5.24)

$$B^{\mathrm{vol}} = \frac{1}{9V_{\mathrm{c}}}\frac{\partial^2 E^{\mathrm{vol}}}{\partial \gamma^2} \equiv V_{\mathrm{c}}\frac{\partial^2 E^{\mathrm{vol}}}{\partial V_{\mathrm{c}}^2} - \frac{2}{3}P^{\mathrm{vol}}. \tag{5.42}$$

The pressure term that has appeared will also enter the shear moduli as follows.

To obtain general expressions we need the first and second derivatives of the volume, since the contribution of E^{vol} to the elastic constants is obtained from

$$\frac{\partial^2 E^{\mathrm{vol}}}{\partial \gamma^2} = \frac{\partial}{\partial \gamma}\left(\frac{\partial V}{\partial \gamma}\frac{\partial E^{\mathrm{vol}}}{\partial V}\right) = \frac{\partial^2 V}{\partial \gamma^2}\frac{\partial E^{\mathrm{vol}}}{\partial V} + \left(\frac{\partial V}{\partial \gamma}\right)^2 \frac{\partial^2 E^{\mathrm{vol}}}{\partial V^2}. \tag{5.43}$$

If the strain γ did not change the volume, this term would vanish. However, even the tetragonal and rhombohedral shears described by (5.29) and (5.33), although volume conserving to *first* order in γ, do cause *second*-order volume changes. To be precise, we start from the formula for the ratio of final to initial volumes produced by a strain γT, which is given by the determinant

$$\frac{V_{\mathrm{c}}(\gamma)}{V_{\mathrm{c}}(0)} = \begin{vmatrix} 1+\gamma T_{xx} & \gamma T_{xy} & \gamma T_{xz} \\ \gamma T_{yx} & 1+\gamma T_{yy} & \gamma T_{yz} \\ \gamma T_{xz} & \gamma T_{zy} & 1+\gamma T_{zz} \end{vmatrix}. \tag{5.44}$$

Differentiating (5.44) we find

$$\left.\frac{1}{V_{\mathrm{c}}(0)}\frac{\partial V_{\mathrm{c}}}{\partial \gamma}\right|_{\gamma=0} = \operatorname{Tr} T \tag{5.45}$$

and

$$\left.\frac{1}{V_{\mathrm{c}}(0)}\frac{\partial^2 V_{\mathrm{c}}}{\partial \gamma^2}\right|_{\gamma=0} = 2(T_{xx}T_{yy} + T_{yy}T_{zz} + T_{zz}T_{xx} - T_{xy}^2 - T_{yz}^2 - T_{zx}^2). \tag{5.46}$$

Inserting these formulae into (5.43) and using the matrices T^t and T^r we find the following contributions of E^{vol} to the shear moduli:

$$C'^{\mathrm{vol}} = \tfrac{1}{2}P^{\mathrm{vol}}$$
$$C_{44}^{\mathrm{vol}} = \tfrac{1}{2}P^{\mathrm{vol}}. \tag{5.47}$$

It seems paradoxical that an energy which depends only on volume should contribute to these shear moduli, and there are two points to make here. The first is that the paradox is a result of using the simplest possible definition of strains and their associated elastic constants, such that even a shear strain involves a second-order volume change, although the first order volume change described by $\operatorname{Tr} T$ vanishes. A more sophisticated definition of strain could be made which would shear the crystal at strictly constant volume.

The second point is that there is no ambiguity in any measurable elastic constant at zero total pressure. The rest of the energy provides a pressure which balances P^{vol}. Indeed we can choose to associate all the contributions from P^{vol} with the rest of the energy.

To understand these points better let us study a model crystal in which the total energy is described by E^{vol} together with a sum of pairwise potentials between atoms:

$$E^{\text{tot}} = E^{\text{vol}} + E^{\text{pair}} \tag{5.48}$$

where

$$E^{\text{pair}} = \frac{1}{2}\sum_{I,J} V_{IJ}(|\mathbf{R}_I - \mathbf{R}_J|). \tag{5.49}$$

The summation over I runs over all atoms within the unit cell of volume V_c, while the summation over J runs over all atoms within range, and the factor $1/2$ is the usual double counting correction. The second derivatives of E^{pair} with respect to the strains, $\epsilon_{\alpha\beta}$ can be worked out explicitly. They will clearly involve first and second derivatives of the interatomic potentials $V_{IJ}(R)$ evaluated at the interatomic distances $|\mathbf{R}_I - \mathbf{R}_J|$. Let us denote these derivatives by V'_{IJ} and V''_{IJ}. Furthermore let us introduce the shorter notation $\mathbf{R}_{IJ}$ and R_{IJ} for $\mathbf{R}_I - \mathbf{R}_J$ and $|\mathbf{R}_I - \mathbf{R}_J|$. From now on we restrict our discussion for simplicity to the class of crystals of sufficiently high symmetry that homogeneous strains will apply the same transformation to each interatomic vector $\mathbf{R}_{IJ} = (R_{IJx}, R_{IJy}, R_{IJz})$.

Let us now derive expressions for B^{pair}, C'^{pair} and C_{44}^{pair}, the contribution of E^{pair} to the elastic constants in a cubic crystal. These will involve combination of first and second derivatives of V_{IJ}. First, let us derive general expressions for $\partial^2 E^{\text{vol}}/\partial\gamma^2$ for a general symmetric T matrix. We shall need the derivatives of V_{IJ} with respect to the strain parameter γ, which have the form:

$$\begin{aligned}
\frac{\partial V_{IJ}}{\partial \gamma} &= V'_{IJ}\frac{\partial R_{IJ}}{\partial \gamma}, \\
\frac{\partial^2 V_{IJ}}{\partial \gamma^2} &= V'_{IJ}\frac{\partial^2 R_{IJ}}{\partial \gamma^2} + V''_{IJ}\left(\frac{\partial R_{IJ}}{\partial \gamma}\right)^2,
\end{aligned} \tag{5.50}$$

where $\mathbf{R}_{IJ}$ is related to $\mathbf{R}^0_{IJ}$ through the transformation matrix and strain parameter:

$$\mathbf{R}_{IJ} = (1 + \gamma T)\cdot \mathbf{R}^0_{IJ}. \tag{5.51}$$

It is convenient to work with the squared length of $\mathbf{R}_{IJ}$, which is

$$R^2_{IJ} = \mathbf{R}^0_{IJ}\cdot(1+\gamma T)(1+\gamma T)\cdot\mathbf{R}^0_{IJ}, \tag{5.52}$$

Differentiating (5.52) twice before setting $\gamma = 0$ gives the completely general results:

$$\frac{\partial R_{IJ}}{\partial \gamma} = \frac{1}{R_{IJ}} \mathbf{R}_{IJ} \cdot T \cdot \mathbf{R}_{IJ}$$

$$\frac{\partial^2 R_{IJ}}{\partial \gamma^2} = \frac{1}{R_{IJ}} \mathbf{R}_{IJ} \cdot T^2 \cdot \mathbf{R}_{IJ} - \frac{1}{R_{IJ}^3} (\mathbf{R}_{IJ} \cdot T \cdot \mathbf{R}_{IJ})^2. \tag{5.53}$$

The pressure from the pair potential follows as

$$P^{\text{pair}} = -\frac{1}{2} \frac{\partial}{\partial V_{\text{c}}} \sum V_{IJ} = -\frac{1}{2V_{\text{c}}} \sum \tfrac{1}{3} V'_{IJ} R_{IJ}. \tag{5.54}$$

Assembling the above ingredients for the second derivative gives us

$$\frac{1}{V_{\text{c}}} \frac{\partial^2 E^{\text{pair}}}{\partial \gamma^2} = \frac{1}{2V_{\text{c}}} \sum_{I,J} \left\{ V'_{IJ} \left(\frac{1}{R_{IJ}} \mathbf{R}_{IJ} \cdot T^2 \cdot \mathbf{R}_{IJ} - \frac{1}{R_{IJ}^3} (\mathbf{R}_{IJ} \cdot T \cdot \mathbf{R}_{IJ})^2 \right) \right.$$
$$\left. + V''_{IJ} \frac{1}{R_{IJ}^2} (\mathbf{R}_{IJ} \cdot T \cdot \mathbf{R}_{IJ})^2 \right\}, \tag{5.55}$$

which can be used in the expressions for the elastic constants derived previously. We now apply it to cubic crystals.

For the bulk modulus, T is the unit matrix, so the coefficient of V'_{IJ} vanishes and we have

$$B^{\text{pair}} = \frac{1}{9V_{\text{c}}} \frac{\partial^2 E^{\text{pair}}}{\partial \gamma^2} = \frac{1}{18V_{\text{c}}} \sum_{I,J} R_{IJ}^2 V''_{IJ}. \tag{5.56}$$

For the shear moduli we find:

$$C'^{\text{pair}} = \frac{1}{24V_{\text{c}}} \sum_{I,J} \left\{ 3R_{IJ} V'_{IJ} \left[1 - \frac{1}{R_{IJ}^4} \left(R_{IJx}^4 + R_{IJy}^4 + R_{IJz}^4 \right) \right] \right.$$
$$\left. + R_{IJ}^2 V''_{IJ} \left[-1 + \frac{3}{R_{IJ}^4} \left(R_{IJx}^4 + R_{IJy}^4 + R_{IJz}^4 \right) \right] \right\}, \tag{5.57}$$

and

$$C_{44}^{\text{pair}} = \frac{1}{12V_{\text{c}}} \sum_{I,J} \left\{ R_{IJ} V'_{IJ} \left[1 - \frac{2}{R_{IJ}^4} (R_{IJx}R_{IJy} + R_{IJy}R_{IJz} + R_{IJz}R_{IJx})^2 \right] \right.$$
$$\left. + R_{IJ}^2 V''_{IJ} \frac{2}{R_{IJ}^4} (R_{IJx}R_{IJy} + R_{IJy}R_{IJz} + R_{IJz}R_{IJx})^2 \right\}. \tag{5.58}$$

In addition, we can construct

$$C_{12}^{\text{pair}} = B^{\text{pair}} - \tfrac{2}{3}C'^{\text{pair}}$$

$$= \frac{1}{12V_c}\sum_{I,J}\left\{R_{IJ}V'_{IJ}\left[-1+\frac{1}{R_{IJ}^4}\left(R_{IJx}^4+R_{IJy}^4+R_{IJz}^4\right)\right]\right.$$

$$\left. + R_{IJ}^2 V''_{IJ}\left[1-\frac{1}{R_{IJ}^4}\left(R_{IJx}^4+R_{IJy}^4+R_{IJz}^4\right)\right]\right\}. \tag{5.59}$$

A definition of shear strain which exactly conserves volume would modify these expressions in an obvious way. To the shear constants from E^{vol} we would have to subtract $\frac{1}{2}P^{\text{vol}}$, making them identically zero. Since the equilibrium condition $P^{\text{vol}} + P^{\text{pair}} = 0$ holds, the term $\frac{1}{2}P^{\text{pair}}$ must also be subtracted from C'^{pair} and C_{44}^{pair}.

Cauchy relations are relationships between the elastic constants which arise when the total energy is described by E^{pair} alone. They are a result of the constraint $P^{\text{pair}} = 0$. With some algebra, which includes exploiting the cubic symmetry, you can show from the above equations that

$$C_{12}^{\text{pair}} - C_{44}^{\text{pair}} = \tfrac{1}{2}P^{\text{pair}}. \tag{5.60}$$

Within a purely pair potential model, in cubic crystals the zero pressure condition for equilibrium implies the Cauchy relation

$$C_{12} - C_{44} = 0. \tag{5.61}$$

Other relationships apply in other symmetries and there are analogous relationships between higher order elastic constants. A full derivation is given by Wallace (1972).

The quantity $C_{12} - C_{44}$ is sometimes called the *Cauchy pressure*. In real cubic materials is usually found to be non-zero, which proves that the total energy of that material cannot be described by pair potentials alone.

Before leaving the subject of subtleties, the importance of *sublattice displacements*, sometimes described as *internal strains*, should be emphasized. I already mentioned that when applying a strain to the unit cell, the atoms within it should be allowed to relax. In the above expressions only the simple class of structures was considered in which the same strain applies to all the relative atomic positions in the unit cell. It includes the simple metallic structures and compounds such as NaCl and CsCl, but it obviously does not include molecular crystals, in which the bonds within an individual molecule will certainly not follow the strains of the unit cell. Although they are relatively simple crystals, even fluorite and zincblende fall outside this class. This effect is sometimes overlooked when it occurs in rather simple structures. Consider for example the fluorite (CaF_2) structure, which is illustrated

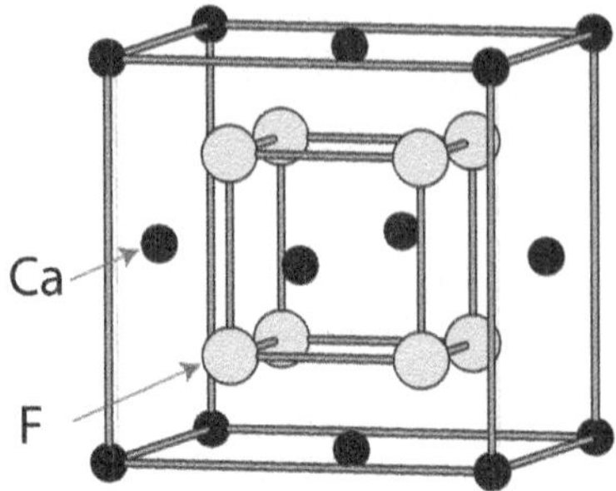

FIG. 5.2 The crystal structure of fluorite.

in Fig. 5.2. The structure can be thought of as a face-centred cubic lattice of Ca with a F atom in each of the tetrahedral interstices. If now the cell is stretched along the z-axis, usually referred to as [0 0 1], the tetrahedron of 4 Ca atoms surrounding a F will be stretched but its centre remains a point of special symmetry. There is therefore no relaxation of the F sublattice relative to the unit cell. If on the other hand the cell is stretched along [1 1 1], one of the Ca neighbours of each F atom becomes different to the others; the bond from F to the Ca atom along [1 1 1] is stretched differently to the bonds to its other three neighbours. Under such a strain the centre of the distorted tetrahedron is no longer definable by any special symmetry, and the Ca sublattice will shift along [1 1 1] so as to minimize the energy. A related structure is that of GaAs, in which however only half the tetrahedral sites are occupied, such that both types of atom form a FCC lattice. Pure diamond, Si and Ge structures are the same. All display internal strains when stretched along [1 1 1], which is the mode of strain associated with C_{44}. First principles calculations of Nielsen and Martin (1985) showed that internal strains in the latter three materials lower C_{44} by about 20% compared to what it would be if all the vectors between atoms were strained by the same amount.

Exercises

1. Prove eq. (5.60). *Hint:* cubic symmetry requires that for each set of equidistant neighbours J of atom I:

$$\sum_J (R_{IJx}+R_{IJy}+R_{IJz})(R_{IJx}R_{IJy}+R_{IJy}R_{IJz}+R_{IJz}R_{IJx}) = 0. \qquad (5.62)$$

2. Derive expressions for the elastic constants of a face centred cubic crystal whose energy is described by a central potential acting between nearest neighbours together with a volume dependent energy.

5.4 Phonons

The concept of normal modes of vibration and their frequencies is entirely independent of quantum mechanical notions. They are determined in principle by solving the classical equations of motion of the ions when they undergo small displacements from their equilibrium sites. The concept of *phonons* on the other hand is purely quantum mechanical. But the quantum mechanics is exactly as in the treatment of a simple harmonic oscillator. Namely, when the classical frequency ω of a normal mode of vibration has been found, then the actual energy in that mode is only allowed to be an integer multiple of $\hbar\omega$, plus the zero-point energy $\frac{1}{2}\hbar\omega$. These packets of energy are the phonons. Remember that these are angular frequencies, or $\omega = 2\pi\nu$ if ν is the frequency in cycles per second.

Good data on the phonon frequencies $\omega(\mathbf{k})$ of a crystal as a function of wavevector $\mathbf{k}$ along high symmetry directions is available for many materials and provides a useful database for testing or fitting models. The practical calculation of phonon frequencies assumes that the Born–Oppenheimer approximation is valid, so that the potential energy when the ions are not in their equilibrium positions is a function only of the positions of the ions and does not depend on their velocities or history.

5.4.1 Lattice Dynamics in the Harmonic Approximation

There are many good books on the theory of lattice vibrations in the harmonic approximation, still one of the best being the classic textbook of Born and Huang (1954), so it is not necessary for me to revise the whole theory here. There are however a few standard methods and pitfalls which are worth describing. These days one can also make useful calculations of quantities such as vibrational entropy and coefficient of thermal expansion going just beyond the harmonic approximation. I refer the reader to the literature for a discussion of how to do such calculations. Again a good starting point for the theory is Wallace (1972).

Just as the theory of electronic states in crystals, the theory of lattice vibrations is set up within the framework of periodic boundary conditions within a large (tending to infinity) volume V, within which there are smaller repeating unit cells of volume V_c. And just as for electronic states, Bloch's theorem applies to the normal modes of vibration, which are analogous to the eigenstates of the Hamiltonian for electrons. The main difference is that the displacements of atoms are the variables rather than the wavefunction amplitudes, and these are defined on discrete sites rather than being continuous functions of position.

In order to proceed we need to set up some notation. Let us label a unit cell by l and denote its position by $\mathbf{R}(l)$. The first unit cell $l = 0$ has its corner at the origin. The actual atoms within the unit cell, sometimes called the *basis*, will be labelled κ. Each atom in the crystal is specified by its static equilibrium position

$\mathbf{R}(l\kappa)$ and the mass M_κ of its nucleus. The small displacements of atoms from their static equilibrium positions (or mean positions) will be denoted $\mathbf{U}(l\kappa)$. The Cartesian components of $\mathbf{U}(l\kappa)$ will be denoted $U_\alpha(l\kappa)$ as for other vectors.

Within the harmonic approximation the potential energy is expanded to second order as

$$E^{\text{tot}} = E_0^{\text{tot}} + \frac{1}{2} \sum_{l\kappa\alpha\, l'\kappa'\beta} U_\alpha(l\kappa)\Phi_{\alpha\beta}(l\kappa, l'\kappa')U_\beta(l'\kappa'), \tag{5.63}$$

where the coefficients Φ form a matrix that is symmetric under interchange of the indices $\alpha l\kappa$ with $\beta l'\kappa'$. If we introduce the 3×3 matrices $\Phi(l\kappa, l'\kappa')$ we can write (5.63) in the more compact form

$$E^{\text{tot}} = E_0^{\text{tot}} + \frac{1}{2} \sum_{l\kappa\, l'\kappa'} \mathbf{U}(l\kappa)\Phi(l\kappa, l'\kappa')\mathbf{U}(l'\kappa'). \tag{5.64}$$

The classical equation of motion of atom $l\kappa$ is therefore

$$M_\kappa \ddot{\mathbf{U}}(l\kappa) = -\sum_{l'\kappa'} \Phi(l\kappa, l'\kappa')\mathbf{U}(l'\kappa'). \tag{5.65}$$

The elements of Φ (or its negative) are sometimes called *force constants*. They represent the force on atom $l\kappa$ when atom $l'\kappa'$ is slightly displaced, all other atoms being held fixed. These are the basic quantities we need to calculate with our interatomic force models.

Bloch's theorem now suggests that the normal mode displacements, by analogy with eq. (1.97), should take the form:

$$\mathbf{U}(l\kappa) = \mathbf{u}_{kn}(\kappa) \exp\left[\mathrm{i}(\mathbf{k} \cdot \mathbf{R}(l\kappa) - \omega_{kn}t)\right]. \tag{5.66}$$

I have used the notation $\mathbf{u}_{kn}(\kappa)$ here for the pattern of displacements which has the periodicity of the lattice or superlattice because it is exactly analogous to the $u_{kn}(\mathbf{r})$ of eq. (1.97) which has the periodicity of the lattice in electron theory. $\mathbf{U}(l\kappa)$ represents a travelling wave with wavevector $\mathbf{k}$, which modulates a periodic pattern of displacements $\mathbf{u}_{kn}$, just like the Bloch functions $\psi_{kn}(\mathbf{r})$. In this case it is understood that actual displacements are given by the real part of $\mathbf{U}$. By analogy with the electron case, n labels the normal mode. By substituting this form into the equation of motion we obtain an eigenvalue equation for the normal mode frequencies and their eigenvectors. To ensure that the eigenvalues are real and make the solution easier we want a Hermitian matrix, which is achieved by replacing the vectors $\mathbf{u}$ by vectors $\mathbf{w}$ defined by the scaling:

$$\mathbf{w}_{kn}(\kappa) = \sqrt{M_\kappa}\,\mathbf{u}_{kn}(\kappa). \tag{5.67}$$

After a little algebra, replacement of **u** by **w** in the equation of motion gives us the standard form of eigenvalue equation

$$\sum_{\kappa'} \mathbf{D}_k(\kappa, \kappa')\mathbf{w}(\kappa') = \omega^2 \mathbf{w}(\kappa), \tag{5.68}$$

in which **D** is called the *dynamical matrix*, defined by

$$\mathbf{D}_k(\kappa, \kappa') = \frac{\exp\left[\mathrm{i} \cdot \left(\mathbf{R}(0, \kappa') - \mathbf{R}(0, \kappa)\right)\right]}{\sqrt{M_\kappa M_{\kappa'}}} \sum_{l'} \Phi(0\kappa, l'\kappa') \exp\left(\mathrm{i}\mathbf{k} \cdot \mathbf{R}(l')\right). \tag{5.69}$$

Our introduction of the vectors **w** has ensured the Hermiticity of **D**:

$$D_{k\alpha\beta}(\kappa, \kappa') = D^*_{k\beta\alpha}(\kappa', \kappa), \tag{5.70}$$

which means the eigenvalues ω^2_{kn} are real and the eigenvectors $\mathbf{w}_{kn}$ are orthogonal. We choose their normalization such that

$$\sum_{\kappa} \mathbf{w}_{kn}(\kappa)^* \cdot \mathbf{w}_{kn'}(\kappa) = \delta_{nn'}. \tag{5.71}$$

Remember that the index κ runs over the N_c atoms in a supercell (which may or may not be the smallest, or primitive, unit cell), so the dimensions of the dynamical matrix are $3N_c \times 3N_c$, and there are $3N_c$ eigenvalues and eigenvectors at each point in **k**-space. In the simplest possible case of just one atom per unit cell, the three eigenvectors are called *acoustic modes*, and they are referred to as having different *polarizations*. Along high symmetry directions in **k**-space the polarisation, which is defined by the direction of **w**, may be parallel or perpendicular to **k**. Such modes are referred to as *longitudinal* and *transverse*, respectively. The characteristic of acoustic modes is that their frequency tends to zero at the gamma point ($\mathbf{k} = 0$), close to which they become the macroscopic sound waves, which are described by the dynamical theory of elasticity. With more than one atom per unit cell there are still always, in all directions in **k**-space, just three of the eigenvalues that tend to zero at the gamma point, corresponding to three acoustic modes.

The choice of cell or supercell does not change the normal modes themselves of course but it does make a difference to the way they are described. The dynamical matrix of a large supercell has correspondingly more eigenstates at each **k**-point. What is happening here is the 'folding back' of **k**-space into the first Brillouin zone (BZ). The larger the supercell, the smaller its first BZ. However, this BZ still has to accommodate the same total number of modes at a density $V/(2\pi)^3$, where V you remember is the large (tending to infinity) volume which is divided into supercells and to which periodic boundary conditions apply. These modes are accommodated by increasing the number of bands as the BZ gets smaller. Thus for example a mode which is at the BZ boundary in one supercell finds itself at the gamma point if we double the linear dimensions of the supercell. It works the same way for electron states. The *kn* label is not absolute.

In a simple crystal structure, quite a lot can be learned from the $\omega(\mathbf{k})$ curves, called dispersion curves, along high symmetry directions. Think of the line in $\mathbf{k}$-space along $\langle 1\,0\,0 \rangle$ in a FCC crystal of an element such as copper. At each $\mathbf{k}$-value there is a longitudinal mode, in which all the movements are parallel to the $\mathbf{k}$-vector, and two degenerate transverse modes, in which the movements are in the y, z plane. perpendicular to $\mathbf{k}$. The problem maps onto the dynamics of a one-dimensional chain, the force constants of which are *interplanar* force constants of the three-dimensional system, which couple the motion of $(1\,0\,0)$ planes. These interplanar force constants can therefore be uniquely extracted from experimental data, which is not the case for interatomic force constants. In fact they are Fourier transforms of the dispersion curves. Each interplanar force constant is a linear combination of all the force constants linking atoms between the two planes. The interplanar force constants therefore give a direct indication of the range of interatomic force constants. For example, if a dispersion curve displays wiggles or kinks, which require many Fourier components to represent them, it is evidence that the force constants are of correspondingly long range.

Exercise

If we denote by $\mathbf{R}$ the vector $\mathbf{R}(l\kappa) - \mathbf{R}(l'\kappa')$, show that for a central potential $V(R)$ the force constant matrix has the form:

$$\Phi(l\kappa, l'\kappa')$$
$$= \begin{pmatrix} \frac{1}{R}V'\left(\frac{R_x^2}{R^2}-1\right) - V''\frac{R_x^2}{R^2}, & \frac{1}{R}V'\frac{R_xR_y}{R^2} - V''\frac{R_xR_y}{R^2}, & \frac{1}{R}V'\frac{R_xR_z}{R^2} - V''\frac{R_xR_z}{R^2} \\ \frac{1}{R}V'\frac{R_yR_x}{R^2} - V''\frac{R_yR_x}{R^2}, & \frac{1}{R}V'\left(\frac{R_y^2}{R^2}-1\right) - V''\frac{R_y^2}{R^2}, & \frac{1}{R}V'\frac{R_yR_z}{R^2} - V''\frac{R_yR_z}{R^2} \\ \frac{1}{R}V'\frac{R_zR_x}{R^2} - V''\frac{R_zR_x}{R^2}, & \frac{1}{R}V'\frac{R_zR_y}{R^2} - V''\frac{R_zR_y}{R^2}, & \frac{1}{R}V'\left(\frac{R_z^2}{R^2}-1\right) - V''\frac{R_z^2}{R^2} \end{pmatrix} \tag{5.72}$$

Notice that the first derivative of the potential only enters the force constant when the movement has a component tangential to the bond ('bond bending'), whereas the second derivative enters both in bond stretching and bond bending modes.

5.4.2 Calculating Force Constants

The main difficulty in calculating phonon frequencies is in the calculation of the force constants, which may be long ranged. An expression for the force induced on atom I in direction α when atom J moves slightly in direction β is straightforward to write down based on the Hellmann–Feynman theorem and linear response theory, and it was derived from first principles by Pick *et al.* (1970):

$$\Phi_{\alpha\beta}(I, J) = \frac{\partial}{\partial R_{I\alpha}\partial R_{J\beta}} \left(\int \epsilon^{-1}(\mathbf{R}_I, \mathbf{r}) \frac{Z_I Z_J}{|\mathbf{r} - \mathbf{R}_J|} \, d\mathbf{r} \right). \tag{5.73}$$

This is the second derivative of the energy, as in (5.63), which is strictly speaking the negative of the force defined above. The site labels I and J now appear in place of the $l\kappa$ just to emphasize that the definition of Φ need not be restricted to periodic systems. The inverse dielectric function ϵ^{-1} was defined in eq. (4.30), and it describes the change in the total potential as seen by a test charge for a given change in the external potential. A small displacement $u_{J\beta}$ of nucleus J induces a change in the charge density $\Delta\rho$, which together with ΔV_{ext} due to the movement of the nucleus itself gives rise to a change in the electrostatic potential $\epsilon^{-1}\Delta V_{\text{ext}}$. More explicitly this potential at site I is given by

$$\int \epsilon^{-1}(\mathbf{R}_I, \mathbf{r}) \Delta V_{\text{ext}}(\mathbf{r}) \, d\mathbf{r} = \int \epsilon^{-1}(\mathbf{R}_I, \mathbf{r}) \frac{\partial}{\partial R_{J\beta}} \frac{-Z_J}{|\mathbf{r} - \mathbf{R}_J|} u_{J\beta} \, d\mathbf{r}. \tag{5.74}$$

Remember our sign and units conventions are such that the electrostatic potential has the same sign and magnitude as the potential energy of a particle with an electronic charge. Equation (5.73) for the force constant follows simply from the Hellmann–Feynman theorem when we differentiate the potential at site I by $R_{I\alpha}$. It is feasible to calculate ϵ^{-1} and hence force constants within a specific model for the total energy, even with the full HKS formalism, but it is a complicated and computationally difficult task. For a simple orthogonal tight-binding model of transition metals, force constants were evaluated by Finnis *et al.* (1984a), see also Finnis and Pettifor (1985), who calculated phonon spectra with them. Force constants out to more than six shells of neighbours, the lattice vector $(2\,0\,0)a$ in BCC crystals, were obtained. Their magnitude depends strongly on the band structure and the value of the Fermi energy, and does not fall off rapidly or even monotonically with distance, which is unfortunate from the point of view of simple interatomic potential models.

As usual, periodic systems offer the possibility of formulating the equations in **k**-space. Unless you want to learn explicitly about the interatomic forces in **r**-space, or the system is an isolated molecule, it is much easier in the case of a periodic system to evaluate the dynamical matrix directly in **k**-space. I shall not embark on a derivation here, but note one particular advantage of the **k**-space approach. If you look at the form of the response function in **k**-space, as derived in Section 4.10, you can see how by transforming (5.74) into **k**-space, that is by writing the response function and the Coulomb potential as their Fourier transforms, the derivatives $\partial/\partial R_{I\alpha}$ simply become multiplicative factors like $\mathrm{i}q_\alpha$. Even then, the expression for the inverse dielectric function in **k**-space is not trivial to evaluate in general.

It has been more common to calculate phonon frequencies by ‘brute force’. In a ‘brute force’ method the total energy changes are evaluated by actually moving a selected atom or atoms in the supercell and recalculating the total energy and possibly at the same time the Hellmann–Feynman forces. In simple crystals the eigenvector of the phonon of interest may be completely determined by the symmetry; think of the zone boundary phonons in a FCC crystal for example. In such cases the frequency of the phonon can be calculated directly by subtracting the total energy of the perfect lattice from its total energy when the atoms are given small static displacements corresponding to the eigenvector. This is therefore called the *frozen phonon* method. In effect, to do this one must be able to diagonalise the dynamical matrix by hand. Otherwise, by a number of brute force calculations one can build up the elements of the dynamical matrix. From the dynamical matrix so calculated, all those phonon frequencies can be obtained which have wavevectors commensurate with the supercell, those are the frequencies at the gamma point of its Brillouin zone. The larger the supercell, the smaller its Brillouin zone and the larger the dimensions of the dynamical matrix and the number of frequencies that can therefore be obtained. This restriction on the **k**-vectors of the phonons that can be calculated has to be weighed against the advantage of the method, namely that it can be carried through without the need to calculate dielectric functions.

Exercise
If a FCC crystal is represented by its primitive unit cell, containing 1 atom, no frequencies can be obtained by the 'brute force' methods. How many non-zero frequencies can be obtained if it is represented by a cubic supercell of side a, containing 4 atoms? The answer should be 9, but they only take two distinct values: there are three symmetrically equivalent longitudinal modes and six symmetrically equivalent transverse modes.

5.5 Point Defects

The subject of point defects covers vacancies and interstitial atoms, which destroy the perfect crystal periodicity of an element or a compound. Interstitial or substitutional impurites are also features which require similar techniques for calculating their properties. The properties we can calculate and compare with experiment are mainly the energies of formation (or heats of solution in the case of impurities), and in a few cases the positions of the relaxed atoms around the defect. In addition migration energies of defects may be of interest.

The practical difficulties in doing calculations for point defects are of several kinds. First, there is the technical problem that a point defect may perturb the positions of atoms out to quite a long range, in principle to infinity. The energy of formation of a single isolated defect may depend quite sensitively on these long range relaxations of atomic positions, which in an infinite crystal are described far from the defect by an elastic strain field. However, a calculation is normally carried out in a supercell with periodic boundary conditions. This procedure not only ignores the long ranged strain field, it introduces the interactions of a periodic lattice of point defects, which is a spurious contribution when what we want is the energy of an isolated defect. Calculations by whatever means and with whatever models must therefore be set up and carried out with an eye to their convergence with respect to the size of the supercell. If the calculation is not done with periodic boundary conditions then similar effort goes into making sure there is an adequate match between a region within which atomic positions are explicitly relaxed and an outer region which may be treated as an elastic medium. In any case, in the worst cases accurate calculations can require thousands of atoms.

A second difficulty, which is common to calculations of surface energy and comparisons of structural energy, is the need to achieve high numerical accuracy. This is simply because the energy of interest is of the order of one or two eV, which is obtained as a difference between two energies which may be tens of thousands of eV. This is really only a problem for the kind of electronic structure methods that involve calculations of eigenvalues at sample points in **k**-space. Then wherever possible the calculation should be set up so that there is a tendency for errors in **k**-space sampling to cancel. This can be achieved for example by making sure that when two large total energies are subtracted they have both been calculated with the same supercell and the same set of **k**-points.

A third difficulty concerns charged defects. In semiconductors and insulators a vacancy may trap one or more electrons and if it is charged it may be screened by a long ranged compensating cloud of carrier density, whose nature and extent is critically dependent on levels of doping in the material.

Finally, there is the practical difficulty of knowing whether what one is calculating is actually the quantity that is being measured experimentally! The following section describes how to formulate a calculation of the vacancy formation energy, which is probably the most favourable case as regards the interpretation of experimental data. Nevertheless, what the experimentalists measure, in a sample at equilibrium, is a *concentration* of vacancies as a function of temperature, obtained from positron lifetimes or from the simultaneous determination of lattice parameter and volume per atom of a crystalline sample (the 'Simmons and Balluffi' method). When the vacancies do not interact with each other significantly their equilibrium concentration per atom follows an Arrhenius law

$$c_{\mathrm{v}} = c_0 \exp\left(-\frac{\Delta H_{\mathrm{f}}^{\mathrm{v}}}{k_{\mathrm{B}}T}\right), \tag{5.75}$$

where

$$c_0 = \exp(\Delta S_{\mathrm{f}}^{\mathrm{v}}/k_{\mathrm{B}}) \tag{5.76}$$

and

$$\Delta H_{\mathrm{f}}^{\mathrm{v}} = \Delta E_{\mathrm{f}}^{\mathrm{v}} + P\Delta V_{\mathrm{f}}^{\mathrm{v}}. \tag{5.77}$$

ΔE_f^v, ΔV_f^v, ΔS_f^v, and ΔH_f^v are the formation energy, formation volume, formation entropy and formation enthalpy of a vacancy. This at least is the case for elements; for multi-component systems the situation is trickier. For ordered compounds the appropriate theory is described in the paper of Hagen and Finnis (1998).

In experiments ΔH_f^v is obtained from the slope of log c_v versus $1/T$ (an *Arrhenius plot*), which should be a straight line. Whether the pressure is one atmosphere or zero is of no significance since ΔV_f^v is typically less that one atomic volume. Thus for an atomic volume of say $0.3\,\mathrm{nm}^3$, $P\Delta V_f^v$ contributes only 0.2 meV to ΔH_f^v. The main difficulty is that at high temperature, where the number of vacancies is reasonably large (although their concentration is always less than about 10^{-4} per atom), the Arrhenius plot tends to deviate from a straight line, for example due to the presence of divacancies or other defects. Finally, there is bound to be some part of ΔH_f^v which is not captured by the $T = 0$ calculations of the formation energy ΔE_f^v which are normally made. If there is a linear temperature dependence, it simply joins the formation entropy within the preexponential factor, and does not influence the slope, but if not, e.g. the zero-point energy contribution, an error is automatically introduced.

Having said all that, it is still worth doing the calculations when data are available for comparison, because it is well established that first principles LDA calculations in metals do give the right answer to within about 10%, and they also provide useful benchmark data for testing models. Interstitial formation energies ΔE_f^i can be calculated in the same way as vacancies, but since these are larger numbers, the equilibrium concentration of interstitials is very small, so the experimental data are few and obtained indirectly.

5.5.1 Definition of Vacancy Formation Energies

The microscopic theory of the vacancy concentration leading to (5.75) supposes that vacancies are created by thermal excitations of the crystal. For example a random fluctuation in its kinetic energy causes an atom to jump from a site *in* the surface to a site *on* the surface, leaving a surface vacancy. This vacancy, again by random fluctuations, is filled by a neighbouring atom below the surface and so migrates into the bulk, statistically maintaining the population given by (5.75). In a similar way, vacancies may be generated at a dislocation core or at a grain boundary rather than at a surface. We can therefore think of ΔE_f^v as the energy required to take an atom from the bulk and put it on the surface, which is the conventional picture, but it seems as if alternative processes are equally valid, and surely at least we need to know to what particular site on a surface the atom goes?

Fortunately there is no need to worry at all about *where* on the surface it goes since we are dealing with a statistical average. One way of seeing this is the following. Consider a macroscopic piece of material containing 10^{30} atoms. Some 10^{20} of them, give or take a power of ten, will be in surface sites. If we now create a small concentration of vacancies, say $c_v = 10^{-6}$, by removing atoms from the bulk and putting them on the surface, 99.99% of all the 10^{24} atoms we remove will be buried again. So to calculate ΔE_f^v for statistical purposes we do not need to put the atom we remove on the surface at all, we have to bury it in the bulk. This leads to the formula

$$\Delta E_f^v = \lim_{N_a \to \infty} \{E(N_a, 1) - E(N_a - 1, 0)\}, \tag{5.78}$$

where the first argument of E denotes the number of lattice sites and the second the number of vacancies. Thus $E(N_a, 1)$ denotes the energy of $N_a - 1$ atoms on N_a lattice sites with 1 vacancy and with the same notation $E(N_a - 1, 0)$ is the energy of $N_a - 1$ atoms on $N_a - 1$ perfect lattice sites with no vacancies. All the atoms should be relaxed to their equilibrium positions. The analysis here is based on the usual concept that the energy of a supercell is calculated, the supercell having been set up to contain the specified number of sites.

As an amusing aside, it is still possible to think in terms of putting the atom we have removed onto a surface site if it is a special surface site that leaves the surface structure self-similar. In particular, we should add the atom to a surface ledge at a kink site (see Fig. 5.3), because this process does not change the surface structure. In equilibrium, the atoms in the neighbourhood of the kink site are relaxed to some extent from their perfect lattice positions, and the whole pattern of relaxation shifts with the kink site as it is moved along by adding an atom. The energy change when an atom is moved from isolation in vacuum to such a surface site must therefore be $-E^{coh}$, a quantity we normally calculate much more easily just by comparing the energies of bulk material with free atoms! There are many such special kink sites, depending on the crystallography of the chosen surface, yet the energy of binding a free atom to each of them is the same. One can imagine building a bulk crystal line-by-line and plane-by-plane by laying individual atoms down one at a time at such a kink site. We disregard of course the faces of this bulk crystal, which are defined where we start and finish and at the places where the line of atoms turns around like a row of knitting. The surfaces contribute a negligible part of the total

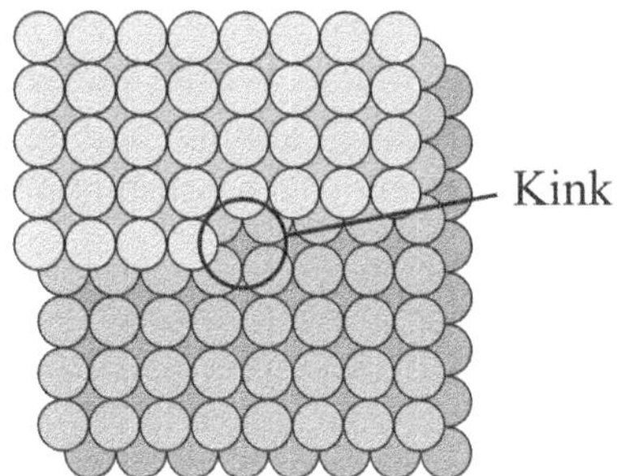

FIG. 5.3 A kink site.

energy for this purpose. Another way to think of the vacancy formation energy is therefore as the energy E_{rem} required to pluck an atom out of the bulk and remove it to infinity, less E^{coh}, the energy released when it is returned to the surface.

$$\Delta E_{\text{f}}^{\text{v}} = E_{\text{rem}} - E^{\text{coh}}. \tag{5.79}$$

It is noteworthy that in a simple central potential model, neglecting relaxation if we remove atom I:

$$E_{\text{rem}} = -\sum_{J} V_{IJ}(\mathbf{R}_{IJ}), \tag{5.80}$$

and if all the atoms are of the same species, remembering the factor of a half for double counting:

$$E^{\text{coh}} = -\frac{1}{2}\sum_{J} V_{IJ}(\mathbf{R}_{IJ}). \tag{5.81}$$

So in this artificial model the vacancy formation energy is the same as the cohesive energy. Plucking the atom out breaks all its bonds, of which on average half are remade when the atom is returned to the surface. In reality vacancy formation energies in metals are typically only one quarter to one third of the cohesive energy in magnitude, which underlines the inadequacy of a pairwise bond-breaking model of cohesion.

In practice a good estimate of $\Delta E_{\text{f}}^{\text{v}}$ can be had with N_{a} in the range from 16 to several thousand, depending on the material and on the accuracy sought. The convergence with N_{a} must be tested in each individual case, and an extrapolation to $N_{\text{a}} = \infty$ can be made. It is not geometrically possible to construct a supercell which has $N_{\text{a}} - 1$ sites given that we already have one with N_{a} sites. Therefore (5.78) is re-expressed as

$$\Delta E_{\text{f}}^{\text{v}} = \lim_{N_{\text{a}}\to\infty} \left\{ E(N_{\text{a}}, 1) - \frac{N_{\text{a}} - 1}{N_{\text{a}}} E(N_{\text{a}}, 0) \right\}. \tag{5.82}$$

It is not necessary to relax the volume of the supercell when calculating $E(N_{\text{a}}, 1)$ as long as the atomic positions within it are relaxed. To see this let us add the volume as an argument of these energies, so $E(N_{\text{a}}, 1, V)$ denotes the energy of the supercell containing a vacancy at volume V after relaxation of atomic positions. Let V_0 be the equilibrium perfect lattice volume and V_{v} the equilibrium volume with the vacancy present. If the system is at a constant pressure P we have

$$\left.\frac{\mathrm{d}E(N_{\text{a}}, 0, V)}{\mathrm{d}V}\right|_{V_0} = -P \tag{5.83}$$

and

$$\left.\frac{\mathrm{d}E(N_{\text{a}}, 1, V)}{\mathrm{d}V}\right|_{V_{\text{v}}} = -P. \tag{5.84}$$

The best estimate of ΔE_f^v with N_a lattice sites would be given by

$$\Delta E_f^v(N_a) = E(N_a, 1, V_v) - \frac{N_a - 1}{N_a} E(N_a, 0, V_0). \tag{5.85}$$

We now make a Taylor expansion of the energies to second-order in the volume:

$$\begin{aligned} E(N_a, 1, V) &= E(N_a, 1, V_v) - P(V - V_v) + \frac{1}{2} B_v \frac{(V - V_v)^2}{V_v} \cdots \\ E(N_a, 0, V) &= E(N_a, 0, V_0) - P(V - V_0) + \frac{1}{2} B_0 \frac{(V - V_0)^2}{V_0} \cdots, \end{aligned} \tag{5.86}$$

where the B_v and B_0 would be bulk moduli in a cubic system, otherwise they are simply some appropriate elastic moduli. In any case they only differ because of the presence of the vacancy, and so

$$B_v - B_0 = O(N_a^{-1}) B_0. \tag{5.87}$$

Finally, since V, V_c and V_0 only differ by amounts which are of the order of an atomic volume $\Omega_0 = V_0/N_a$, we find

$$\begin{aligned} \Delta E_f^v(N_a) &= E(N_a, 1, V) - \frac{N_a - 1}{N_a} E(N_a, 0, V) \\ &\quad - P \Delta V_f^v(N_a) + P \Omega_0 O(N_a^{-1}) + B_0 \Omega_0 O(N_a^{-1}). \end{aligned} \tag{5.88}$$

The last two terms are of the order of a few electron volts scaled by N_a^{-1}. To this order therefore the calculation of $\Delta E_f^v(N_a)$ can be done at constant volume if the pressure, and hence the formation volume term, is zero. Furthermore, if we do the calculation at a number of different volumes, and evaluate the pressure derivative of ΔE_f^v, we obtain a calculated value of the formation volume ΔV_f^v. This is completely analogous to the way ΔV_f^v is obtained experimentally.

6

PAIRWISE POTENTIALS IN SIMPLE METALS

6.1 Introduction

In this chapter, I show how pairwise potentials for describing interatomic forces in simple metals can be derived from first principles. The derivation includes an unusually detailed treatment of the volume dependent terms in the total energy, which in many cases are essential if such potentials are to be used for simulations, and the formalism includes alloys. I also take explicit account of non-local pseudopotentials. The central result of this chapter is an equation for the total energy in terms of volume dependent pairwise potentials and other volume dependent terms, namely eq. (6.51). In proposing this as a useful model, I am straying from the beaten track, where effective medium or embedded atom models became more fashionable in the nineteen eighties. However, I think there is a case for reviving pairwise potentials of this old fashioned kind for simulating some systems. Their relative neglect probably stems from uncertainty about how to deal with the volume dependent terms, so I finish the chapter with a suggestion as to how this might be done.

By 'simple' we mean metals whose conduction electrons are in bands formed by the overlap of atomic s and p orbitals. This distinguishes them from the transition metals, in which electrons in bands formed by overlapping d-orbitals play a dominant part in the conduction bands, and hence in the nature of the bonding and the form of the interatomic forces. The noble metals (Cu, Ag and Au) are usually singled out of the transition series for special treatment, since although their d-bands are full, they still participate in the bonding and interatomic forces, so their d-electrons cannot be regarded on the same footing as tightly bound core electrons. The simple metals are the nearest approximation nature gives us to jellium, and are where the free electron description, the Sommerfeld model one learns about near the beginning of textbooks on solid state physics, comes closest to reality. Historically they were also where the 'nearly free electron' approximation was successful for describing the band structure. The Fermi surfaces of simple metals proved to be nearly spherical, and the deviations from sphericity could be described by second-order perturbation theory. In fact, before the advent of the theory of pseudopotentials in the nineteen fifties and sixties it was something of a paradox that a metal made of atoms in every core of which there is a singular, Coulomb potential, could display any nearly free electron like properties at all.

The replacement of the core of an atom by a pseudopotential not only made the electronic structure understandable, it soon led to the application of second-order perturbation theory for the total energy, in which the starting point is jellium and the perturbation is the bare ion pseudopotential. The developments are described in the book by Harrison (1966). Second-order perturbation theory led to a description of the total energy that divided it into two terms. The first depends only on the total volume V, the second depends on the arrangement of the ions within that volume, and is the sum of a pairwise interatomic potential, which in its most general form looks like $V_{IJ}(|\mathbf{R}_I - \mathbf{R}_J|, \rho)$. This pair potential depends on the species of the atoms which are at $\mathbf{R}_I$ and $\mathbf{R}_J$, which I will identify by labels $\tau(I)$ and $\tau(J)$ respectively. It also depends implicitly on the density ρ of the jellium in which it is derived. The density or volume dependence is of critical importance, as a number of authors have discussed (Finnis, 1974; Rosenfeld and Stott, 1987). It implies that the pair potential describes the *rearrangement energy* of the ions at constant total volume, but not the changes in energy associated with changes in volume. With this restriction it has been used successful in explaining trends in many metallic structures, including quasicrystals. These applications are well summarized in the book by Hafner (1987) and papers by Hafner and Heine (1986).

Although the pair potential description is physically appealing and leads to useful insights, the fact that the pair potential only describes the rearrangement energy is a serious limitation which has led to some misapplications and to a well-known paradox. The paradox usually refers to a discrepancy which appears between the bulk modulus as calculated from the second volume derivative of the total energy and as calculated from the slopes in the long wavelength limit of the phonon dispersion curves; that is from the elastic constants obtained from the low $\mathbf{q}$ limit of the dynamical matrix. The first approach, sometimes called the method of *homogeneous deformation*, involves changing the total volume of the sample whereas the second does not, yet the theory of lattice dyamics tells us they must give the same answer. They do not, because the lattice dynamics calculation only involves the pair potential, and the homogeneous deformation method includes all the other volume dependent terms in the total energy which it turns out do not sum to zero. An even starker form of the long-wave paradox as I shall call it can be constructed which does not depend on phonon dispersion curves. Here is a thought experiment. Consider a volume V containing a large number of atoms N_{a} roughly uniformly distributed throughout the volume. The volume is large enough that the atoms are much farther apart than in the equilibrium metal, say a volume more typical of a vapour. Now *formally* the total energy can be derived to second-order, and the resulting pair potential refers to this peculiarly low density jellium. Keeping the volume constant, now collapse the atoms together to form a lump having normal metallic density, surrounded by vacuum, all within the same constant volume V (see Fig. 6.1). The energy change per atom we know physically *should* just be the cohesive energy. Furthermore, since we have made a constant volume change in

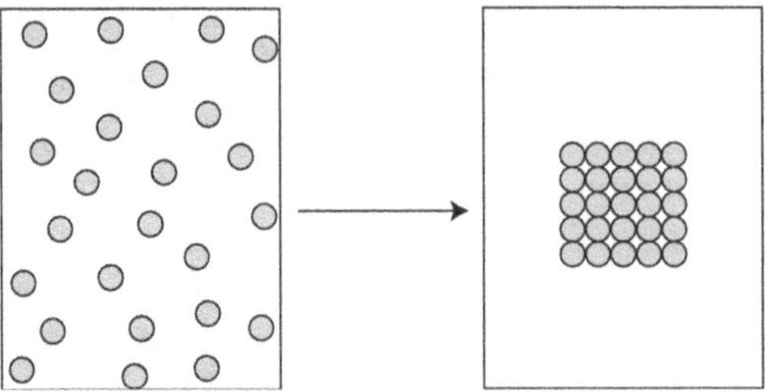

FIG. 6.1 'Condensation' at constant volume.

atomic positions, it should be described by V_{IJ}. We can ignore a contribution from the surface energy which for large N_a is $O(N_a^{-1/3})$ per atom and so negligible compared to the bulk energy per atom. Although geometrically we have just rearranged the atoms at constant volume, V_{IJ} certainly does not describe the cohesive energy. In fact the sum of V_{IJ} contributes a relatively small fraction of the cohesive energy.

The long-wave paradox was resolved formally in 1969 by Brovman and coworkers (Brovman and Kagan, 1969; Brovman *et al.*, 1969). It turned out to be a discrepancy which disappears only if the perturbation theory is extended to fourth order, something which is very complicated to implement in practice, and physically rather difficult to understand. Later explanations of the paradox by Wallace (1972), Finnis (1974) and Rosenfeld and Stott (1987) further clarified the situation. The latter is an excellent review and source for anyone wishing to dig deeper into this aspect of pairwise potentials in simple metals.

In this chapter, I derive expressions for the total energy in both **k**- and **r**-space, making clear the form of the pair potential, and showing how the above paradox can be resolved. The derivation here differs from what you will find in most of the literature in several respects. For one thing, the presence of atoms of more than one species requires such a trivial extension of the theory that it seems reasonable to include it from the start. The role of the non-locality of the pseudopotentials is more complex, but by including non-locality from the start I also show how the resulting equations reduce to nearly the same form as for simple local pseudopotentials. Finally, although not mathematically rigorous, the derivation here should be detailed enough for you to follow all the steps yourself, with the help of the preceeding chapters. In particular, before we arrive at the form of the interatomic potentials there are some tricky details about how the divergent electrostatic terms involved are made to cancel. These problems are usually skated over in the literature, but can cause great difficulties to anyone trying to understand fully what the potentials mean. As usual, the reader not concerned with such details can skip to the later sections.

The use of pair potentials in simple metals became unfashionable in around the mid nineteen eighties. This was due partly to the advent of first-principles total energy calculations, and partly to the introduction of effective medium potentials which enabled rapid large scale simulations to be carried out avoiding the difficulties referred to above. Nevertheless, I believe the time is overdue for a modest revival of this kind of model, and I will suggest how it might be adapted and put to good use in future atomistic simulations.

6.2 The Energy in Terms of Pseudopotentials

6.2.1 Structure Factors

A good starting point to derive the form of pair potentials is provided by the second-order change in total energy, eq. (4.132). The perturbation has the form of a sum of the pseudopotentials of the individual ions, less the potential of the positive jellium background which has to be removed in step with switching on the ions:

$$\Delta V_{\text{ext}}(\mathbf{r}, \mathbf{r}') = \sum_I v_\tau(\mathbf{r} - \mathbf{R}_I, \mathbf{r}' - \mathbf{R}_I) - V^{\text{jel}}\delta(\mathbf{r} - \mathbf{r}'). \tag{6.1}$$

The pseudopotential $v_{\tau(I)}$ describes an ion of type τ. The I dependence of τ is included to make it clear that that there may be several elements in the system, different sites I being occupied by different elements. In $\mathbf{k}$-space the perturbing potential becomes

$$\Delta V_{\text{ext}}(\mathbf{k}, \mathbf{k}') = \sum_I \frac{1}{V} \int \exp(\mathrm{i}(\mathbf{k} \cdot \mathbf{r} - \mathbf{k}' \cdot \mathbf{r}'))v_{\tau(I)}(\mathbf{r} - \mathbf{R}_I, \mathbf{r}' - \mathbf{R}_I)\, \mathrm{d}\mathbf{r}\, \mathrm{d}\mathbf{r}' - V^{\text{jel}}(|\mathbf{k} - \mathbf{k}'|). \tag{6.2}$$

The integrand in expression (6.2) can be factored into a part that depends only on the arrangement of the ions and a part that depends only on the pseudopotentials. Such a factorisation is familiar in the theory of diffraction. It is done by changing the variables temporarily from $\mathbf{r}$ and $\mathbf{r}'$ to $\mathbf{u}$ and $\mathbf{u}'$, where

$$\mathbf{u} = \mathbf{r} - \mathbf{R}_I \quad \mathbf{u}' = \mathbf{r}' - \mathbf{R}_I. \tag{6.3}$$

We have simply shifted the origin to ion I. The $\mathbf{R}_I$ dependence can then be factored outside the integral in (6.2) which becomes

$$\Delta V_{\text{ext}}(\mathbf{k}, \mathbf{k}') = \sum_I \exp(\mathrm{i}(\mathbf{k} - \mathbf{k}') \cdot \mathbf{R}_I)\frac{1}{V} \times \int \exp(\mathrm{i}(\mathbf{k} \cdot \mathbf{u} - \mathbf{k}' \cdot \mathbf{u}'))v_{\tau(I)}(\mathbf{u}, \mathbf{u}')\, \mathrm{d}\mathbf{u}\, \mathrm{d}\mathbf{u}' - V^{\text{jel}}(|\mathbf{k} - \mathbf{k}'|). \tag{6.4}$$

The result can now be expressed as a sum of contributions from each type of ion. First however note that in **k**-space the pseudopotential is as in eq. (1.247):

$$V_\tau(\mathbf{k}, \mathbf{k}') = \frac{1}{\Omega} \int \exp(\mathrm{i}(\mathbf{k} \cdot \mathbf{u} - \mathbf{k}' \cdot \mathbf{u}')) v_\tau(\mathbf{u}, \mathbf{u}') \, \mathrm{d}\mathbf{u} \, \mathrm{d}\mathbf{u}'. \tag{6.5}$$

We can therefore express (6.4) as

$$\Delta V_{\text{ext}}(\mathbf{k}, \mathbf{k} + \mathbf{q}) = \sum_\tau S_\tau(\mathbf{q}) V_\tau(\mathbf{k}, \mathbf{k} + \mathbf{q}) - V^{\text{jel}}(q), \tag{6.6}$$

in which we have introduced as previously $\mathbf{q} = \mathbf{k}' - \mathbf{k}$. The *structure factors* are defined by

$$S_\tau(\mathbf{q}) = \frac{1}{N_{\text{a}}} \sum_{I_\tau} \exp(-\mathrm{i}\mathbf{q} \cdot \mathbf{R}_I), \tag{6.7}$$

and as before

$$V_\tau(\mathbf{k}, \mathbf{k}') = N_{\text{a}} v_\tau(\mathbf{k}, \mathbf{k}'). \tag{6.8}$$

The summation runs only over those sites I which are occupied by atoms of type τ.

Equation (6.6) is the desired factorization of the potential into structure dependent and structure independent parts. It is easy to see that this kind of factorisation can be made for any potential which is the summation of atom-centred contributions.

6.2.2 The $q = 0$ Problem

The jellium background potential is the only tricky aspect to the mathematics which follow. Its one and only role is to cancel out the divergences at $\mathbf{q} = 0$ and it must vanish for q infinitesimally greater than zero. Formally, it is helpful to write it as a contribution from each ion core as if that ion core had been 'smeared out' over the whole volume V. This will enable us to write the jellium potential as a sum over atomic quantities like the pseudopotential, whereby we will be able to make the $q = 0$ terms cancel explicitly. To fix ideas, suppose we represent the jellium background by a sum of atom-centred Gaussian charge distributions of the form:

$$\rho_\tau^{\text{G}}(\mathbf{r}, q_0) = \frac{Z_\tau q_0^3}{\pi^{3/2}} \exp(-q_0^2 r^2), \tag{6.9}$$

which will give the correct uniform density as these distributions spread out when we let the parameter q_0 tend to zero. The normalisation of the Gaussian ensures that each spherical density contains a total charge Z_τ. The Fourier transform of one of these Gaussian charge distributions is, using the convention of (4.122):

$$\rho_\tau^{\mathrm{G}}(q, q_0) = \frac{Z_\tau}{V} \exp\left(-\frac{q^2}{4q_0^2}\right), \tag{6.10}$$

and the **k**-space representation of its potential is

$$v_\tau^{\mathrm{G}}(q, q_0) = -\frac{4\pi Z_\tau}{Vq^2} \exp\left(-\frac{q^2}{4q_0^2}\right). \tag{6.11}$$

For practical purposes we can also define a scaled version which remains finite as $V \to \infty$, just as we did for the pseudopotential:

$$V_\tau^{\mathrm{G}}(q, q_0) = -\frac{4\pi Z_\tau}{\Omega q^2} \exp\left(-\frac{q^2}{4q_0^2}\right). \tag{6.12}$$

The sign is negative here because we are dealing with a positive charge density, and our convention is that the potential always refers to the potential as felt by an electron. Adding the atomic contributions gives by analogy with (6.6):

$$V^{\mathrm{jel}}(q) = -\lim_{q_0 \to 0} \sum_\tau S_\tau(\mathbf{q}) V_\tau^{\mathrm{G}}(q, q_0). \tag{6.13}$$

As expected, when we eventually take the limit $q_0 \to 0$, only the divergent $q = 0$ part remains. Notice that if we were to take the opposite limit, $q_0 \to \infty$, we would recover the electrostic energy of a system of point charges, and we will also make use of this shortly.

The energy change to second-order, eq. (4.132), now takes the form:

$$\begin{aligned}
\Delta E^{(2)} &= 2\sum_{\mathbf{k}} f_{\mathrm{F}}(\epsilon_{\mathbf{k}}) \lim_{q_0 \to 0} \sum_I \lim_{q \to 0} \left\{ v_{\tau(I)}(\mathbf{k}, \mathbf{k}+\mathbf{q}) - v_{\tau(I)}^{\mathrm{G}}(q, q_0) \right\} \\
&+ \lim_{q_0 \to 0} \frac{V}{2} \sum_{\mathbf{q}} \chi_e(q) \left| \sum_\tau S_\tau(\mathbf{q}) \left(\bar{V}_\tau(\mathbf{q}) + \frac{4\pi Z_\tau}{\Omega q^2} \exp\left(-\frac{q^2}{4q_0^2}\right) \right) \right|^2 \\
&+ \sum_{\mathbf{k},\mathbf{q}} \frac{(f_{\mathrm{F}}(\epsilon_{\mathbf{k}}) - f_{\mathrm{F}}(\epsilon_{\mathbf{k}+\mathbf{q}}))}{(\epsilon_{\mathbf{k}} - \epsilon_{\mathbf{k}+\mathbf{q}})} \left| \sum_\tau S_\tau(\mathbf{q}) \left(V_\tau(\mathbf{k}, \mathbf{k}+\mathbf{q}) - \bar{V}_\tau(\mathbf{q}) \right) \right|^2 + \Delta E_{ZZ}.
\end{aligned} \tag{6.14}$$

The quantities $\bar{V}_\tau(\mathbf{q})$ are defined from $V_\tau(\mathbf{k}, \mathbf{k}+\mathbf{q})$ as equivalent local potentials by the same operation as we used to define $\bar{V}_{\mathrm{ext}}(\mathbf{q})$ from $V_{\mathrm{ext}}(\mathbf{k}, \mathbf{k}+\mathbf{q})$, namely:

$$\bar{V}_\tau(q) = \frac{2}{V} \sum_{\mathbf{k}} \frac{(f_{\mathrm{F}}(\epsilon_{\mathbf{k}}) - f_{\mathrm{F}}(\epsilon_{\mathbf{k}+\mathbf{q}}))}{(\epsilon_{\mathbf{k}} - \epsilon_{\mathbf{k}+\mathbf{q}})} \frac{V_\tau(\mathbf{k}, \mathbf{k}+\mathbf{q})}{\chi_{\mathrm{s}}(q)}. \tag{6.15}$$

Notice that the first summation over I in eq. (6.14) is merely to count atoms of each type, and it does not involve the atomic positions.

The low-q limits of the pseudopotentials which appear here need to be treated with respect, to avoid the trap of assuming that infinity minus infinity is zero. So let us take a closer look. We can write any non-local pseudopotential as the sum of a part $v_\tau^{\rm core}(\mathbf{r},\mathbf{r}')$, with $r, r' < R_\tau$, for a suitably chosen value of R_τ, plus the corresponding Ashcroft potential.

$$v_\tau(\mathbf{r},\mathbf{r}') = v_\tau^{\rm core}(\mathbf{r},\mathbf{r}')\Theta(R_\tau - r)\Theta(R_\tau - r') - \frac{Z_\tau}{r}\delta(\mathbf{r}-\mathbf{r}')\Theta(r - R_\tau). \tag{6.16}$$

In $\mathbf{k}$-space, using the result (1.250) for the Ashcroft potential derived previously this becomes

$$V_\tau(\mathbf{k},\mathbf{k}+\mathbf{q}) = V_\tau^{\rm core}(\mathbf{k},\mathbf{k}+\mathbf{q}) - \frac{4\pi Z_\tau}{\Omega q^2}\cos(qR_\tau). \tag{6.17}$$

The Fourier transform of a function that is bounded to a region of space around the origin tends to a constant independent of q. In this case, the contribution of the core to $V_\tau(\mathbf{k},\mathbf{k}+\mathbf{q})$ is given from (1.247) by

$$V_\tau^{\rm core}(\mathbf{k},\mathbf{k}+\mathbf{q}) = \frac{1}{\Omega}\int_{r,r'\le R_\tau} \exp(\mathrm{i}(\mathbf{k}\cdot(\mathbf{r}-\mathbf{r}')-\mathbf{q}\cdot\mathbf{r}'))v_\tau^{\rm core}(\mathbf{r},\mathbf{r}')\,\mathrm{d}\mathbf{r}\,\mathrm{d}\mathbf{r}', \tag{6.18}$$

and as $q \to 0$:

$$\lim_{q\to 0} V_\tau^{\rm core}(\mathbf{k},\mathbf{k}+\mathbf{q}) = V_\tau^{\rm core}(\mathbf{k},\mathbf{k})$$

which is bounded for any reasonably well-behaved function, and we do not need to consider pseudopotentials which are less well behaved than this. The Ashcroft potential in $\mathbf{k}$-space obviously has the same low-q limit as the Coulomb potential of Z_τ. The low q behaviour of V_τ is therefore also given by

$$V_\tau(\mathbf{k},\mathbf{k}+\mathbf{q}) \approx -\frac{4\pi Z_\tau}{\Omega q^2}, \tag{6.19}$$

or more precisely

$$\lim_{q\to 0}\left\{V_\tau(\mathbf{k},\mathbf{k}+\mathbf{q}) + \frac{4\pi Z_\tau}{\Omega q^2}\right\} = V_\tau^{\rm core}(\mathbf{k},\mathbf{k}) - \frac{2\pi Z_\tau R_\tau^2}{\Omega}. \tag{6.20}$$

These limits are very plausible, because at low q the Fourier transforms are dominated by the long ranged Coulomb interaction. In the first term of (6.14) we therefore have the limit:

$$
\begin{aligned}
&2\sum_{\mathbf{k}} f_{\mathrm{F}}(\epsilon_{\mathbf{k}}) \lim_{q_0\to 0} \sum_{I} \lim_{q\to 0} \left\{ v_I(\mathbf{k}, \mathbf{k}+\mathbf{q}) - v_I^{\mathrm{G}}(q, q_0) \right\} \\
&= 2\sum_{\mathbf{k}} f_{\mathrm{F}}(\epsilon_{\mathbf{k}}) \frac{1}{N_{\mathrm{a}}} \lim_{q_0\to 0} \sum_{I} \lim_{q\to 0} \left\{ V_I(\mathbf{k}, \mathbf{k}+\mathbf{q}) - V_I^{\mathrm{G}}(q, q_0) \right\} \\
&= \sum_{I} \frac{\Omega}{4\pi^3} \int f_{\mathrm{F}}(\epsilon_{\mathbf{k}}) \left\{ V_I^{\mathrm{core}}(\mathbf{k}, \mathbf{k}) - \frac{2\pi Z_I R_{\tau(I)}^2}{\Omega} \right\} \mathrm{d}\mathbf{k}. \qquad (6.21)
\end{aligned}
$$

Where it is unambiguous I have for the sake of simpler notation used the site label I rather then the atom type label $\tau(I)$ when the latter is site-dependent. Soon we shall also need the long wavelength limit of $\bar{V}_\tau(q)$, eq. (6.15), which turns out to be almost the same as that of the non-local potential whose induced charge density it was designed to reproduce. The derivation runs as follows. Notice in (6.15) the non-locality is expressed by the $\mathbf{k}$-dependence of $V_\tau(\mathbf{k}, \mathbf{k}+\mathbf{q})$, without which we could factor V_τ out of the summation. The summation over Fermi factors would then reproduce $\chi_{\mathrm{s}}(q)$, which would cancel the $\chi_{\mathrm{s}}(q)$ in the denominator. Luckily, a similar thing happens as $q \to 0$, because the only contributions from $f_{\mathrm{F}}(\epsilon_{\mathbf{k}}) - f_{\mathrm{F}}(\epsilon_{\mathbf{k}+\mathbf{q}})$ are from an ever-thinner shell of $\mathbf{k}$-vectors which survives as $\mathbf{k}$ approaches the Fermi surface $|\mathbf{k}| = k_{\mathrm{F}}$. We now make use of eq. (6.20), expanding it about the nearest point to $\mathbf{k}$ on the Fermi sphere, so that $|\mathbf{k} - \mathbf{k}_{\mathrm{F}}| \leq q$. For small q we have:

$$
\begin{aligned}
\bar{V}_\tau(q) = &-\frac{4\pi Z_\tau}{\Omega q^2} + V_\tau^{\mathrm{core}}(\mathbf{k}_{\mathrm{F}}, \mathbf{k}_{\mathrm{F}}) - \frac{2\pi Z_\tau R_\tau^2}{\Omega} + O(q) \\
&+ \frac{2}{V} \sum_{\mathbf{k}} \frac{(f_{\mathrm{F}}(\epsilon_{\mathbf{k}}) - f_{\mathrm{F}}(\epsilon_{\mathbf{k}+\mathbf{q}}))}{(\epsilon_{\mathbf{k}} - \epsilon_{\mathbf{k}+\mathbf{q}})} \frac{O(q)}{\chi_{\mathrm{s}}(q)}. \qquad (6.22)
\end{aligned}
$$

The last term also goes to zero with q, and we are left with the same limit as for the non-local potential, except the core potential is evaluated at $|\mathbf{k}| = k_{\mathrm{F}}$, namely

$$
\lim_{q\to 0} \left\{ \bar{V}_\tau(q) + \frac{4\pi Z_\tau}{\Omega q^2} \right\} = V_\tau^{\mathrm{core}}(\mathbf{k}_{\mathrm{F}}, \mathbf{k}_{\mathrm{F}}) - \frac{2\pi Z_\tau R_\tau^2}{\Omega}. \qquad (6.23)
$$

Notice that by virtue of its spherical symmetry, $V_\tau^{\mathrm{core}}(\mathbf{k}, \mathbf{k})$ is a function of just the modulus $|\mathbf{k}|$.

6.2.3 The Madelung Energy ΔE_{ZZ}

An important part of the pairwise potential is in the term ΔE_{ZZ} which we now discuss. If we were to evaluate the Coulomb interactions between the ions on their own, the result would of course be divergent, in the sense that for a single ion the result would be infinite. However, in ΔE_{ZZ} the divergent part is cancelled by subtraction of the self-energy of the jellium background. The non-divergent, structure dependent term which is left in ΔE_{ZZ} is best evaluated by a technique which divides it into real and reciprocal space parts, the *Ewald–Fuchs method*, as described for example by Wallace (1972, pp. 265–271), Ziman (1972, pp. 37–42) or Hafner (1987, pp. 331–339). I will not describe this method in detail again here. We need to deal with ΔE_{ZZ}, however, because it includes the ion-ion Coulomb interaction which is an important contribution to the pairwise potential we are deriving. Since ΔE_{ZZ} is the difference of two divergent parts, our strategy will be to pair each part off with divergent terms in the electronic energy so that none of the terms in our final expression is divergent. Before proceeding it is worth mentioning an important and remarkable identity. Imagine that our ions are immersed in a uniform background of compensating negative charge, having the same density as the electron gas we started with. The total electrostatic energy of this system per ion, in which the ions are treated as point charges, is known as the *Madelung energy* and given the symbol E_{M}. The remarkable thing is that this equals the energy ΔE_{ZZ}. It is perhaps not so remarkable if you think about it. Since both include the ion-ion interaction, the only difference is in the $q = 0$ interactions. Whereas ΔE_{ZZ} includes the background-background interaction with a *negative* sign, E_{M} includes it with a *positive* sign but also includes the ion-background interaction which is negative and at $q = 0$ exactly equal to minus twice the self-energy of the background. So these $q = 0$ terms amount to the same thing in ΔE_{ZZ} and E_{M}. Now let us formulate ΔE_{ZZ} so as to expose these cancelling divergences.

To remain compatible with the way we have dealt with the electronic energy, let us replace the point ions by small Gaussian charge distributions with a width parameter Q_0, which we will allow to go to infinity in order to recover point charges, but only when there is no resulting divergence in the potential:

$$\lim_{Q_0 \to \infty} \rho_\tau^{\mathrm{G}}(\mathbf{r}, Q_0) = Z_\tau \delta(\mathbf{r}), \tag{6.24}$$

where the function ρ_τ^{G} is defined by (6.9). An expression for the electrostatic energy of superimposed Gaussian charge distributions is straightforward to obtain in $\mathbf{k}$-space. We have rehearsed the necessary manipulations as we derived the linear response expression for $E^{(2)}$ in jellium. The result is:

$$E^{\mathrm{G}}(Q_0) = \frac{\Omega}{2}\sum_{\mathbf{q}}\sum_{IJ}\exp(-\mathrm{i}\mathbf{q}\cdot\mathbf{R}_{IJ})V_I^{\mathrm{G}}(q,Q_0)\rho_J^{\mathrm{G}}(q,Q_0)$$

$$= \frac{1}{2V}\sum_{\mathbf{q}}\sum_{IJ}\frac{4\pi}{q^2}\exp(-\mathrm{i}\mathbf{q}\cdot\mathbf{R}_{IJ})Z_I Z_J\exp\left(-\frac{q^2}{2Q_0^2}\right). \tag{6.25}$$

We shall also have to subtract the self-energy of the point charges, which would be a divergent contribution if we kept it, and which like the internal energy of the nuclei is not part of what we call the total energy. It is the 'on site' part of (6.25):

$$E^{\mathrm{GSE}}(Q_0) = \frac{1}{2V}\sum_{\mathbf{q}}\sum_{I}\frac{4\pi Z_I^2}{q^2}\exp\left(-\frac{q^2}{2Q_0^2}\right). \tag{6.26}$$

Standing alone (6.25) is still an infinite quantity because of the $q = 0$ divergence. One way to deal with it is just to omit the $q = 0$ term in the summation, bravely confident that it has no role to play in the final total energy. However, you would justifiably be nervous at doing this, since we always go to the limit $V \to \infty$ when the $\mathbf{q}$-vectors become continuously distributed. A more satisfactory procedure is to pair up the divergent terms from the start. Thus we can write ΔE_{ZZ} as the limit in which the $q = 0$ divergences explicitly cancel:

$$\Delta E_{ZZ} = \lim_{\substack{q_0\to 0\\ Q_0\to\infty}}\left\{E^{\mathrm{G}}(Q_0) - E^{\mathrm{GSE}}(Q_0) - E^{\mathrm{G}}(q_0)\right\}. \tag{6.27}$$

The first term here is the total electrostatic energy of the Gaussian charges, the second term subtracts the self-energy of each Gaussian and the third term subtracts the self-energy of the background. Notice how (6.27) naturally splits up into 'on-site' terms and the sum of pairwise contributions of the form:

$$\lim_{\substack{q_0\to 0\\ Q_0\to\infty}}\sum_{\mathbf{q}} Z_\tau Z_{\tau'}\frac{4\pi}{Vq^2}\exp\left(\mathrm{i}\mathbf{q}\cdot\mathbf{R}_{IJ}\right)\left\{\exp\left(-\frac{q^2}{2Q_0^2}\right) - \exp\left(-\frac{q^2}{2q_0^2}\right)\right\}. \tag{6.28}$$

The inverse Fourier transform of ΔE_{ZZ} recovers the pairwise Coulomb interaction as follows. The sum over q can be replaced by an integral, and the first term of (6.28) then tends to the Coulomb interaction between a pair of ions:

$$\frac{Z_\tau Z_{\tau'}}{(2\pi)^3}\int\frac{4\pi}{q^2}\exp\left(-\mathrm{i}\mathbf{q}\cdot\mathbf{R}_{IJ}\right)\mathrm{d}\mathbf{q} \equiv \frac{Z_\tau Z_{\tau'}}{R_{IJ}}. \tag{6.29}$$

We shall see later what happens to the other terms. An analogous expression can be written down for E_{M}, namely

$$E_{\mathrm{M}} = \lim_{\substack{q_0 \to 0 \\ Q_0 \to \infty}} \left\{ E^{\mathrm{G}}(Q_0) - E^{\mathrm{GSE}}(Q_0) + E^{\mathrm{G}}(q_0) - E^{\mathrm{G}}(Q_0, q_0) \right\}, \tag{6.30}$$

where, similarly to (6.25):

$$\begin{aligned} E^{\mathrm{G}}(Q_0, q_0) &= \Omega \sum_{\mathbf{q}} \sum_{IJ} \exp(-i\mathbf{q} \cdot \mathbf{R}_{IJ}) V_I^{\mathrm{G}}(q, Q_0) \rho_J^{\mathrm{G}}(q, q_0) \\ &= \frac{1}{V} \sum_{\mathbf{q}} \sum_{IJ} \frac{4\pi}{q^2} \exp(-i\mathbf{q} \cdot \mathbf{R}_{IJ}) Z_I Z_J \exp\left\{ -\frac{q^2}{4} \left(\frac{1}{Q_0^2} + \frac{1}{q_0^2} \right) \right\}. \end{aligned} \tag{6.31}$$

This term is included to describe the interaction of the point charges with the background when the limits are taken.

The difference is thus

$$E_{\mathrm{M}} - \Delta E_{ZZ} = \lim_{\substack{q_0 \to 0 \\ Q_0 \to \infty}} \left\{ 2E^{\mathrm{G}}(q_0) - E^{\mathrm{G}}(Q_0, q_0) \right\} = 0. \tag{6.32}$$

The Madelung energy has been calculated to high accuracy using Ewald–Fuchs methods for a number of crystal structures; some useful tables of results are given by Hafner (1987, pp. 331–338) who also describes the method. For elements, it is commonly represented as an energy *per atom* and written in the form

$$E_{\mathrm{M}} = \alpha_{\mathrm{M}} \frac{Z^2}{2R_{\mathrm{a}}}, \tag{6.33}$$

where Z is the charge on the ion, R_a is the atomic radius, sometimes called the *Wigner–Seitz radius*, and it is defined in terms of the volume per atom by

$$\frac{4\pi R_{\mathrm{a}}^3}{3} = \Omega. \tag{6.34}$$

The coefficient α_{M}, known as the Madelung constant, depends on the crystal structure but not the volume, and it is usually the quantity found in tables. It is perhaps surprising how little α_{M} varies with structure. Some common values are given in Table 6.1.

TABLE 6.1 Madelung constants for some simple structures of elements

Structure	α_M
FCC	−1.791747
BCC	−1.791858
HCP	−1.791676
Diamond	−1.670856
Simple cubic	−1.760119

The following approximation to E_M accounts for the fact that the values are close to −1.8. Imagine the crystal structure as an assembly of identical, space-filling polygons centred on the atoms. These are the *Wigner–Seitz cells*. In face-centred cubic (FCC) and hexagonal-close-packed (HCP) crystals they are regular dodecahedra, in the body-centred cubic (BCC) structure they are tetrakaidecahedra (14 faces) and in simple cubic they are simple cubes. Let us label them $I, J, \ldots$ like the atomic sites. We can now divide up E_M in the following way:

$$E_M = \{E(Z \leftrightarrow e) + E(e \leftrightarrow e)\}_I + \frac{1}{2}\sum_J \{E(Z \leftrightarrow Z) + E(Z \leftrightarrow e) + E(e \leftrightarrow e)\}_{IJ} . \tag{6.35}$$

The term $E(e \leftrightarrow e)$ means the electrostatic energy of a uniform polyhedron of charge, which in the first parenthesis is interacting with itself and in the second parenthesis interacting with another Wigner–Seitz cell. Similarly the term $E(Z \leftrightarrow Z)$ is the interaction between positive ions, and $E(Z \leftrightarrow e)$ is the interaction between a positive charge and a polyhedron of negative charge. If we had simple expressions for the electrostatic interactions between polyhedra of charge this would be a very rapidly convergent way to calculate E_M. The polyhedra are overall neutral and almost spherical in the close packed structures, so the second term is a rather small interaction between multipoles. If we approximate the polyhedra as spheres, the result is trivial, because the second term vanishes. The first term is obtained by elementary electrostatics and gives $-0.9Z^2/R_a$, equivalent to $\alpha_M = -1.8$.

6.2.4 The Total Energy and the Energy-wavenumber Characteristic

To obtain an expression for the total energy, we now have to add the energy of the unperturbed jellium E^{jel}. In addition, let us factor out the structure factors in (6.14) and insert (6.21). Our total energy becomes

$$E^{\text{tot}} = E^{\text{jel}} + \sum_{I} \frac{\Omega}{4\pi^3} \int f_{\text{F}}(\epsilon_{\mathbf{k}}) \left\{ V^{\text{core}}_{\tau(I)}(\mathbf{k},\mathbf{k}) - \frac{2\pi Z_\tau R_\tau^2}{\Omega} \right\} \mathbf{dk}$$
$$+ N_{\text{a}} \lim_{q_0 \to 0} \sum_{\mathbf{q}\tau\tau'} \left\{ F_{\tau\tau'}(q) + \frac{Z_\tau Z_{\tau'}}{2\Omega} \chi_{\text{e}}(q) v_{\text{C}}(q)^2 \exp\left(-\frac{q^2}{2q_0^2}\right) \right.$$
$$\left. + Z_{\tau'} \chi_{\text{e}}(q) v_{\text{C}}(q) \bar{V}_\tau(q) \exp\left(-\frac{q^2}{4q_0^2}\right) \right\} S_\tau(\mathbf{q}) S_{\tau'}(-\mathbf{q}) + \Delta E_{ZZ}, \quad (6.36)$$

where the quantity $F_{\tau\tau'}$ is called the *energy-wavenumber characteristic*, and is defined by:

$$F_{\tau\tau'}(q) = \frac{\Omega}{2} \chi_{\text{e}}(q) \bar{V}_\tau(q) \bar{V}_{\tau'}(q)$$
$$+ \frac{1}{N_{\text{a}}} \sum_{\mathbf{k}} \frac{(f_{\text{F}}(\epsilon_{\mathbf{k}}) - f_{\text{F}}(\epsilon_{\mathbf{k}+\mathbf{q}}))}{(\epsilon_{\mathbf{k}} - \epsilon_{\mathbf{k}+\mathbf{q}})} (V_\tau(\mathbf{k},\mathbf{k}+\mathbf{q})$$
$$- \bar{V}_\tau(\mathbf{q})) \left(V_\tau(\mathbf{k},\mathbf{k}+\mathbf{q}) - \bar{V}_\tau(\mathbf{q}) \right)^*. \quad (6.37)$$

Thanks to the spherical symmetry of individual pseudopotentials, it is a function of the modulus of $\mathbf{q}$ but not of its direction. At low q, the asymptotic forms of χ_{s} (4.85), χ_{e} (4.102), V_τ (6.20) and $\bar{V}_{\tau'}$ (6.23) can be inserted and we see that the leading term in (6.37) is the first one. After a little straightforward algebra we find to second-order in q:

$$F_{\tau\tau'}(q) = -\frac{Z_\tau Z_{\tau'}}{2\Omega} \frac{4\pi}{q^2} + \frac{Z_\tau Z_{\tau'}}{2\Omega} \frac{B_{eg}}{\rho^2} + \frac{1}{2} \left(Z_\tau V^{\text{core}}_{\tau'}(\mathbf{k}_{\text{F}},\mathbf{k}_{\text{F}}) \right.$$
$$\left. + Z_{\tau'} V^{\text{core}}_{\tau}(\mathbf{k}_{\text{F}},\mathbf{k}_{\text{F}}) \right) + O(q^2). \quad (6.38)$$

The other terms in the braces can likewise be expanded about $q = 0$ and we find that all the $q = 0$ terms cancel, as expected. Hence we can write (6.36) as

$$E^{\text{tot}} = E^{\text{jel}} + \sum_{I} \frac{\Omega}{4\pi^3} \int f_{\text{F}}(\epsilon_{\mathbf{k}}) \left\{ V^{\text{core}}_{\tau(I)}(\mathbf{k},\mathbf{k}) - \frac{2\pi Z_\tau R_\tau^2}{\Omega} \right\} \mathbf{dk}$$
$$+ N_{\text{a}} {\sum_{\mathbf{q}\tau\tau'}}' F_{\tau\tau'}(q) S_\tau(\mathbf{q}) S_{\tau'}(-\mathbf{q}) + \Delta E_{ZZ}. \quad (6.39)$$

The prime on the summation is a convention that indicates $\mathbf{q} = 0$ terms are excluded. Notice how essential it was to take the $q \to 0$ limit before taking the $q_0 \to 0$ limit. If we had done otherwise, $q = 0$ divergent terms would have remained.

Equation (6.39) deserves some discussion before we move on to derive an expression for the total energy in terms of pair potentials, because it has been used for many total energy calculations in crystals. The summation over $F_{\tau\tau'}(q)$ is referred to as the *band structure energy*, because it describes the energy change induced by the scattering of the free electrons by the ions, which is the origin of band structure. Band structure is normally something we associate with perfect crystals, and indeed the above expression is very convenient in this case because the summation over $\mathbf{q}$ becomes restricted to reciprocal lattice vectors as shown in Section 6.3. However, the above derivation makes no assumptions about crystallinity. The point about this form of the total energy is that all the structure dependence is contained in the factors $S_\tau(q)$, together with the classical Coulomb interaction between the positive ions, which is described by ΔE_{ZZ} as discussed previously. The energy-wavenumber characteristics are structure independent and given the density ρ they can be calculated once and for all for any two elements which are amenable to the present treatment by second-order perturbation theory.

6.3 Periodic Boundary Conditions

If our system has a crystalline structure, periodic boundary conditions on a supercell of volume $V_c \ll V$ lead to the selection of reciprocal lattice vectors from the sums over $\mathbf{q}$. We have seen the same effect for wavefunctions in periodic structures, described in Section 1.4.6. It is worth digressing to consider this important situation here. To obtain the appropriate formulae we break the sum over all I into a double summation. Each supercell has a label l and its origin is at $\mathbf{R}(l)$. Following the notation used in Section 5.4.1, within each supercell the atoms are labelled κ and have positions $\mathbf{R}(l\kappa)$. We further define the positions $\mathbf{R}(\kappa)$ relative to the origin of a supercell, so that

$$\mathbf{R}_I \equiv \mathbf{R}(l\kappa) \equiv \mathbf{R}(l) + \mathbf{R}(\kappa). \tag{6.40}$$

As before we make use of the result

$$\sum_l \exp(\mathrm{i}\mathbf{q} \cdot \mathbf{R}(l)) = \frac{V}{V_c}\delta_{\mathbf{qg}} \tag{6.41}$$

where $\mathbf{g}$ is a reciprocal lattice vector. The structure factors (6.7) are now only non-vanishing at reciprocal lattice vectors and become

$$S_\tau(\mathbf{g}) = \frac{1}{N_\mathrm{c}} \sum_{\kappa_\tau} \exp(-\mathrm{i}\mathbf{g} \cdot \mathbf{R}(\kappa)), \tag{6.42}$$

where N_c is the total number of atoms in the supercell.

The formula for the second-order energy is as before, but the structure factors mean that the summations over $\mathbf{q}$ become summations over reciprocal lattice vectors $\mathbf{g}$. Total energies in perfect crystals are most naturally calculated using these formulae, inserted into (6.39), rather than making the transformation to **r**-space and pairwise potentials which I will describe in Section 6.4. The sum over q, the energy-wavenumber characteristic, in which the $q = 0$ term is excluded, becomes a sum over reciprocal lattice vectors **g**, which must be truncated at some finite range $g_{\max}$. For practical purposes of comparing total energies of different crystal structures, as described in Heine and Weaire (1970) or Pettifor (1995), experience has shown that values of $g_{\max} \approx 10\, k_{\mathrm{F}}$ or less could be taken, while the Madelung energy could also be obtained by rapidly convergent Ewald summations.

6.4 The Effective Pairwise Interaction

We have seen from the form of (6.36) how all the structure dependence of the band structure energy is described by factors

$$S_\tau(\mathbf{q})S_{\tau'}(-\mathbf{q}) = \sum_{I(\tau),J(\tau')} \frac{1}{N_{\mathrm{a}}^2} \exp(-\mathrm{i}\mathbf{q}\cdot\mathbf{R}_{IJ}), \tag{6.43}$$

which are multiplied by a function of **q**, the energy-wavenumber characteristic. Pairwise interactions emerge as the IJ term in this triple summation over I, J and **q**, just as they do in the Madelung energy ΔE_{ZZ} as described by eq. (6.29). As in other places, we have assumed without paying due attention to the demands of mathematical rigour, that the order of these summations over **q** and $\mathbf{R}_{IJ}$ can be interchanged. This is because we do all our summation within a large but finite volume V, and then let $V \to \infty$ as and when required to convert sums in **k**-space to integrals.

The term in E^{tot} that depends on $\mathbf{R}_{IJ}$ is not merely pairwise, it is *isotropic*, that is to say, it does not depend on the direction of $\mathbf{R}_{IJ}$. This is not obvious from formula (6.14), but it is physically obvious that it must be so. One way to see this is as follows. First, we note that each IJ term is independent of the others. The pairwise term would be the same if atoms I and J were the only two real atoms in the jellium. In that case it is obvious that the interaction between them cannot depend on the direction of the vector connecting them, because jellium itself is isotropic. The mathematical reason appears in the fact that the pseudopotentials of the ions and the response function are invariant to rotation.

To obtain a useful expression for the pairwise part of the total energy we need to convert the sum over **q** to an integral in the usual way. However, we cannot do this without including the $q = 0$ term; it would not be correct to write the *primed* summation of the energy-wavenumber characteristic as an integral, because $F_{\tau\tau'}(q = 0)$ diverges when it stands alone, although it causes no problem when it is part of an integrand. Mathematically speaking, $1/q^2$ is an integrable singularity in three dimensions. Physically, we need to account for all the long range Coulomb terms at the same time.

The way to proceed is to go back to (6.36) and regroup the terms as follows:

$$
\begin{aligned}
E^{\text{tot}} &= E^{\text{jel}} + \sum_I \frac{\Omega}{4\pi^3} \int f_{\text{F}}(\epsilon_{\mathbf{k}}) \left\{ V^{\text{core}}_{\tau(I)}(\mathbf{k},\mathbf{k}) - \frac{2\pi Z_\tau R_\tau^2}{\Omega} \right\} d\mathbf{k} \\
&+ \lim_{\substack{q_0\to 0 \\ Q_0\to\infty}} \left\{ N_{\text{a}} \sum_{\mathbf{q}\tau\tau'} \left\{ F_{\tau\tau'}(q) + \frac{Z_\tau Z_{\tau'}}{2\Omega}\frac{4\pi}{q^2} \exp\left(-\frac{q^2}{2Q_0^2}\right) \right.\right. \\
&- \frac{Z_\tau Z_{\tau'}}{2\Omega}\frac{4\pi}{q^2} \exp\left(-\frac{q^2}{2q_0^2}\right) + \frac{Z_\tau Z_{\tau'}}{2\Omega} \chi_{\text{e}}(q) v_{\text{C}}(q)^2 \exp\left(-\frac{q^2}{2q_0^2}\right) \\
&\left.\left. + Z_{\tau'} \chi_e(q) v_{\text{C}}(q) \bar{V}_\tau(q) \exp\left(-\frac{q^2}{4q_0^2}\right)\right\} S_\tau(\mathbf{q}) S_{\tau'}(-\mathbf{q}) - E^{\text{GSE}}(Q_0) \right\}.
\end{aligned}
\tag{6.44}
$$

We have taken the Madelung energy in the representation (6.27) into the summation with the energy-wavenumber characteristic. Now we see that the $q = 0$ divergence of $F_{\tau\tau'}(q)$ is exactly cancelled by the same divergence in $E^{\text{G}}(Q_0)$ and it is left to $E^{\text{G}}(q_0)$ to cancel the divergence in the last two terms within the braces, which it does as follows. Let us give this group of three terms a name e_0 as $q \to 0$:

$$
\begin{aligned}
e_0 = \lim_{q\to 0} &\left\{ -\frac{Z_\tau Z_{\tau'}}{2\Omega}\frac{4\pi}{q^2} \exp\left(-\frac{q^2}{2q_0^2}\right) + \frac{Z_\tau Z_{\tau'}}{2\Omega} \chi_{\text{e}}(q) v_{\text{C}}(q)^2 \exp\left(-\frac{q^2}{2q_0^2}\right) \right. \\
&\left. + Z_{\tau'} \chi_{\text{e}}(q) v_{\text{C}}(q) \bar{V}_\tau(q) \exp\left(-\frac{q^2}{4q_0^2}\right) \right\}.
\end{aligned}
\tag{6.45}
$$

We now need to insert the low-q limiting forms of $\chi_{\text{e}}(q)$, given by the compressibility sum rule (4.101), and of $\bar{V}_\tau(q)$ given by (6.23). The result after doing the straightforward cancellations is

$$
e_0 = -\frac{Z_\tau Z_{\tau'}}{2\Omega}\frac{B^{\text{jel}}}{\rho^2} - Z_{\tau'} \left\{ V^{\text{core}}_\tau(\mathbf{k}_{\text{F}}, \mathbf{k}_{\text{F}}) - \frac{2\pi Z_\tau R_\tau^2}{\Omega} \right\}. \tag{6.46}
$$

Only at $q = 0$ do these three terms in (6.45) make a contribution; at any non-zero value of q they become zero as the limit $q_0 \to 0$ is taken. When the structure factors are evaluated at $q = 0$ we can use the following identity that equates the total charge in the system on the ions to the total charge on the electrons:

$$\sum_{\tau} Z_\tau S_\tau(0) = \Omega\rho. \tag{6.47}$$

We can also use the charge conservation relation:

$$\int f_{\mathrm{F}}(\epsilon_{\mathbf{k}})\,\mathrm{d}\mathbf{k} = 4\pi^3\rho. \tag{6.48}$$

These formulae help us to incorporate e_0 into (6.36) and to write the total energy in the form:

$$\begin{aligned} E^{\mathrm{tot}} &= E^{\mathrm{jel}} - \frac{1}{2}VB^{\mathrm{jel}} + \sum_I \frac{\Omega}{4\pi^3}\int f_{\mathrm{F}}(\epsilon_{\mathbf{k}})\left\{V^{\mathrm{core}}_{\tau(I)}(\mathbf{k},\mathbf{k}) - V^{\mathrm{core}}_{\tau(I)}(\mathbf{k}_{\mathrm{F}},\mathbf{k}_{\mathrm{F}})\right\}\mathrm{d}\mathbf{k} \\ &+ \frac{1}{N_{\mathrm{a}}}\sum_{\mathbf{q}IJ}\left\{F_{IJ}(q) + \frac{Z_I Z_J}{2\Omega}\frac{4\pi}{q^2}(1-\delta_{IJ})\right\}\exp(-\mathrm{i}\mathbf{q}\cdot\mathbf{R}_{IJ}). \end{aligned} \tag{6.49}$$

The last term has the form of a pairwise summation, including 'on-site' terms for which $I = J$, for which the δ_{IJ} term, which has come from the limit of E^{GSE}, removes the self-energy of the point charges. We can convert the sum over $\mathbf{q}$ to an integral in the usual way:

$$\begin{aligned} &\frac{1}{N_{\mathrm{a}}}\sum_{\mathbf{q}IJ}\left\{F_{IJ}(q) + \frac{Z_I Z_J}{2\Omega}\frac{4\pi}{q^2}(1-\delta_{IJ})\right\}\exp(-\mathrm{i}\mathbf{q}\cdot\mathbf{R}_{IJ}) \\ &= \frac{\Omega}{(2\pi)^3}\sum_{IJ}\int\left\{F_{IJ}(q) + \frac{Z_I Z_J}{2\Omega}\frac{4\pi}{q^2}(1-\delta_{IJ})\right\}\exp(-\mathrm{i}\mathbf{q}\cdot\mathbf{R}_{IJ})\mathrm{d}\mathbf{q}. \end{aligned} \tag{6.50}$$

As before the subscripts have been written as I and J to simplify the notation a little, where it is understood that $F_{IJ} \equiv F_{\tau(I)\tau'(J)}$. From (6.29) we see that for $I \neq J$ the second term in the integrand gives precisely the direct Coulomb interaction between I and J, which is screened by the first term. We will interpret later the $I = J$ term. If we express the integral in polar coordinates, with the polar axis along $\mathbf{R}_{IJ}$, the θ integral can be done analytically and we find

$$\begin{aligned} E^{\mathrm{tot}} &= E^{\mathrm{jel}} - \frac{1}{2}VB^{\mathrm{jel}} + \sum_I \frac{\Omega}{4\pi^3}\int f_{\mathrm{F}}(\epsilon_{\mathbf{k}})\left\{V^{\mathrm{core}}_{\tau(I)}(\mathbf{k},\mathbf{k}) - V^{\mathrm{core}}_{\tau(I)}(\mathbf{k}_{\mathrm{F}},\mathbf{k}_{\mathrm{F}})\right\}\mathrm{d}\mathbf{k} \\ &+ \sum_I V_I + \frac{1}{2}\sum_{I\neq J} V_{IJ}, \end{aligned} \tag{6.51}$$

where

$$V_{IJ} = \frac{2\Omega}{(2\pi)^3} \int \left\{ F_{IJ}(q) + \frac{Z_I Z_J}{2\Omega} \frac{4\pi}{q^2} \right\} \exp(-\mathrm{i} q R_{IJ} \cos\theta) 2\pi q^2 \sin\theta \, \mathrm{d}q \, \mathrm{d}\theta$$

$$= \frac{\Omega}{\pi^2} \int_0^\infty q^2 \left\{ F_{IJ}(q) + \frac{Z_I Z_J}{2\Omega} \frac{4\pi}{q^2} \right\} \frac{\sin q R_{IJ}}{q R_{IJ}} \mathrm{d}q \quad (I \neq J). \tag{6.52}$$

We could separate explicitly the direct Coulomb repulsion:

$$V_{IJ} = \frac{\Omega}{\pi^2} \int_0^\infty q^2 F_{IJ}(q) \frac{\sin q R_{IJ}}{q R_{IJ}} \mathrm{d}q + \frac{Z_I Z_J}{R_{IJ}} \quad (I \neq J). \tag{6.53}$$

However, in practical calculations this separation is not useful. Another way in which the integral is sometimes written exposes the cancellation between the Coulomb repulsion and the band structure term. For this purpose we introduce *normalised* energy-wavenumber characteristics which tend to 1 as $q \to 0$, defined by:

$$F^N_{\tau\tau'}(q) = -\frac{\Omega q^2}{2\pi Z_\tau Z_{\tau'}} F_{\tau\tau'}(q). \tag{6.54}$$

The expression (6.52) can now be written as

$$V_{IJ} = \frac{2 Z_I Z_J}{\pi} \int_0^\infty \left\{ 1 - F^N_{IJ}(q) \right\} \frac{\sin q R_{IJ}}{q R_{IJ}} \mathrm{d}q \quad (I \neq J). \tag{6.55}$$

The on-site term is simply

$$V_I = -\frac{Z_I^2}{\pi} \int_0^\infty F^N_{II}(q) \, \mathrm{d}q. \tag{6.56}$$

Equation (6.51) for the total energy in **r**-space is the central result of this chapter. An expression equivalent to it was first derived for non-local pseudopotentials by Walker and Taylor (1990).

6.5 Example: The Ashcroft Empty-core Potential

In this section, we work out some of the properties of Ashcroft's empty-core pseudopotential, in order to get a feeling for the features of pair potentials in simple metals. The main advantages of the empty-core potential are that it is local and has a simple analytical form in **k**-space, with one parameter which is the core radius. The extra complexity we would introduce if we used non-local pseudopotentials has significant effects in many systems, but at first sight it obscures the physics, so we avoid it for the time being.

The energy-wavenumber characteristic (EWNC) is obtained from (6.37) and is

$$F_{\tau\tau'}(q) = \frac{\Omega}{2}\chi_{\mathrm{e}}(q)V_{\tau}(q)V_{\tau'}(q). \tag{6.57}$$

This form is valid for any local pseudopotential. Now we want to work out the form of the pair potential V_{IJ}, which as we have seen in Section 6.4 is a Fourier transform of the EWNC plus the direct Coulomb interaction between ions. Analytical results are not available unless we approximate the response function $\chi_{\mathrm{e}}(q)$ which in its exact form is given by (4.75). However, we can get analytic results if we make the Thomas–Fermi approximation, which we have seen is correct in the low-q limit. Inserting the Thomas–Fermi response function (4.82) we get

$$\begin{aligned} F^{\mathrm{TF}}_{\tau\tau'}(q) &= \frac{\Omega}{2}\left(\frac{\pi^2}{k_{\mathrm{F}}} + v_C(q)\right)^{-1} V_{\tau}(q)V_{\tau'}(q) \\ &= -\frac{Z_{\tau}Z_{\tau'}}{2\Omega}v_C(q)^2\left(\frac{\pi^2}{k_{\mathrm{F}}} + v_C(q)\right)^{-1} \cos(qR_{\tau})\cos(qR_{\tau'}). \end{aligned} \tag{6.58}$$

The Fourier transformation of (6.58) can now be done analytically, which would not be the case if the exact response function were used. The resulting form of the pair potential in this case was derived by Faber (1972, p. 67) and others (Hafner and Heine, 1986; Hafner, 1987). The derivation is quite instructive, so I will give it in some detail here. The first term in (6.53) is sometimes called the band-structure contribution to the pair potential. It can be written as

$$V^{\mathrm{TF}}_{IJbs}(R) = -\frac{8k_{\mathrm{F}}Z_IZ_J}{\pi^2 R}\int_0^{\infty}\frac{\sin(qR)}{q}\frac{\cos(qR_{\tau(I)})\cos(qR_{\tau'(J)})}{\left(q^2 + (4k_{\mathrm{F}}/\pi)\right)}\,\mathrm{d}q. \tag{6.59}$$

This is evaluated by contour integration as follows. If you are not familiar with contour integration, skip to the result, eq. (6.63). The integrand is an even function of q, so we can double the range of integration to $(-\infty, \infty)$. Then we express $\sin(qR)$ as an exponential, which takes us to the form:

$$V^{\mathrm{TF}}_{IJbs}(R) = -\frac{2k_{\mathrm{F}}Z_IZ_J}{\mathrm{i}\pi^2 R}\int_{-\infty}^{\infty}\frac{\exp(\mathrm{i}qR) - \exp(-\mathrm{i}qR)}{q\left(q^2 + (4k_{\mathrm{F}}/\pi)\right)}\cos(qR_{\tau(I)})\cos(qR_{\tau'(J)})\,\mathrm{d}q. \tag{6.60}$$

We now just have to close the path of integration (the *contour*) in the upper or lower half of the complex plane in the usual way, collecting $\pm 2\pi\mathrm{i}$ times the residue for each term. The radius of the large semicircles will be allowed to go to infinity when the integral along their curved perimeter thereby goes to zero. This arrangement is illustrated in Fig. 6.2. The poles at $q = 0$ have to be treated with care. We know

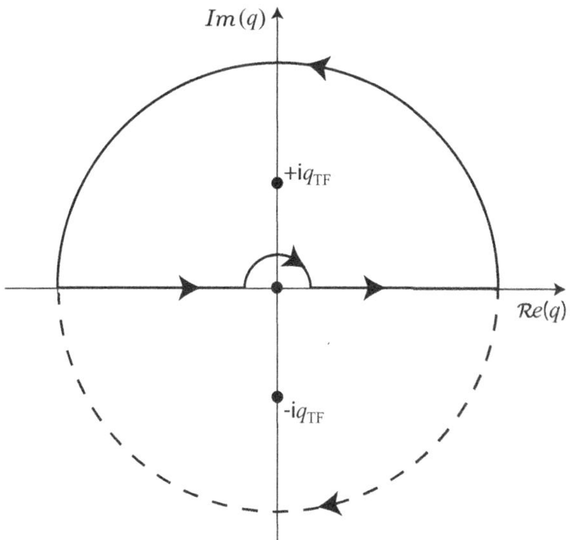

FIG. 6.2 Contour integration for the Thomas–Fermi screening function.

there is no singularity in the total integrand at $q = 0$, so we can take the contour on a semicircle around the origin either above or below the real axis. Let us take it above the real axis. If the contour is closed in the lower half plane, this gives us a factor $-\pi\mathrm{i}$ time times the residue at $q = 0$. If the contour is closed in the upper half plane, it does not encircle the origin, which therefore does not contribute. In the first term of the integrand, the coefficient of $\mathrm{i}q$ in the exponent is positive, and we close the contour in the upper half-plane, in order that the contribution from the large semicircle goes to zero as we let its radius go to infinity. This is only guaranteed if $R > R_\tau + R_{\tau'}$, because for imaginary arguments the cosines themselves become hyperbolic cosines which go to infinity as $\exp(qR_\tau)$:

$$\cos(\mathrm{i}qR_\tau) \equiv \tfrac{1}{2}(\exp(qR_\tau) + \exp(-qR_\tau)). \tag{6.61}$$

The only poles besides the origin are the ones at

$$q = \pm\mathrm{i}q_{\mathrm{TF}},$$

where q_{TF} is the Thomas–Fermi wavevector defined by (4.83).

The result after a little algebra is

$$V^{\mathrm{TF}}_{IJbs}(R) = \frac{Z_\tau Z_{\tau'}}{R}\left[\exp(-q_{\mathrm{TF}}R)\cosh(q_{\mathrm{TF}}R_\tau)\cosh(q_{\mathrm{TF}}R_{\tau'}) - 1\right]. \tag{6.62}$$

The first term comes from the poles at $q = \pm q_{\mathrm{TF}}$ and the second term comes from the pole at $q = 0$. Combining this with the direct Coulomb repulsion gives us the final result for the pair potential:

$$V_{IJ}^{\mathrm{TF}}(R) = \frac{Z_I Z_J}{R} \exp(-q_{\mathrm{TF}} R) \cosh(q_{\mathrm{TF}} R_{\tau(I)}) \cosh(q_{\mathrm{TF}} R_{\tau'(J)}). \tag{6.63}$$

It has the same screened Coulomb form as a Yukawa potential.
The low-q behaviour of the pseudopotential is critical in this derivation, namely

$$V_\tau(q) \to -\frac{4\pi}{\Omega q^2} Z_\tau. \tag{6.64}$$

The on-site terms in the total energy from (6.56) become after a little manipulation:

$$V_I^{\mathrm{TF}} = -\frac{Z_I^2}{\pi} \int_0^\infty \frac{\cos^2 q R_{\tau(I)}}{1 + q^2/q_{\mathrm{TF}}^2} dq. \tag{6.65}$$

The integral can be done analytically and we find

$$V_I^{\mathrm{TF}} = \frac{Z_I^2}{4} q_{\mathrm{TF}} \left\{1 + \exp(-2 q_{\mathrm{TF}} R_{\tau(I)})\right\}. \tag{6.66}$$

The on-site term V_I is clearly much larger in magnitude than an interatomic potential, which is already strongly screened at nearest neighbours. For example in aluminium, with $k_{\mathrm{F}} = 0.927$ au, the nearest neighbours are at $R = 5.40$ au and $q_{\mathrm{TF}} = 1.09$ au. A reasonable value of the core radius is $R_\tau = 1.11$ au (Hafner, 1987, p. 67). The bare Coulomb potential is therefore heavily screened at the nearest neighbours, by the factor $\exp(-q_{\mathrm{TF}} R) = 0.0028$. As a consequence, the pairwise contribution never amounts to more than a few percent of the total energy of simple metals within the Thomas–Fermi model.

The third term of (6.51) vanishes for all local pseudopotentials, simply because the matrix elements $V_\tau(\mathbf{k}, \mathbf{k} + \mathbf{q})$ are independent of $\mathbf{k}$ when the pseudopotential is local, corresponding to the delta function in real space; that's what being local means.

6.6 Asymptotic Forms of the Pair Potential

How does the above form of the pair potential vary if we start with another form of pseudopotential or model potential? The empty core of Ashcroft could be filled with an arbitrary function as long as it reproduces the scattering of the real core within some energy range. A wide class of local potentials can be written in the form:

$$V_\tau(q) = -\frac{4\pi}{\Omega q^2} Z_\tau M_\tau(q), \tag{6.67}$$

where $M_\tau(q)$ is an analytic function of q such that

$$\lim_{q\to 0} M_\tau(q) = 1. \tag{6.68}$$

By following through the analogous derivation to the one above where $M_\tau(q)$ had the Ashcroft form $\cos(qR_\tau)$ we find that the pair potential in the Thomas–Fermi limit is given by the following simple generalization (Hafner, 1987):

$$V_{IJ}^{\mathrm{TF}}(R) = \frac{Z_\tau Z_{\tau'}}{R} \exp(-q_{\mathrm{TF}}R) M(\mathrm{i}q_{\mathrm{TF}}R_\tau) M(\mathrm{i}q_{\mathrm{TF}}R_{\tau'}). \tag{6.69}$$

This is the repulsive form of the pair potential in the high density limit, and also at short range at normal metallic densities.

With a more realistic screening function, at long range, the pair potential becomes strongly influenced by the fact that the screening function has a singular first derivative at $q = 2k_{\mathrm{F}}$, which arises at zero temperature because of the abruptness of the Fermi sphere as described in connection with eq. (4.55). Although hardly apparent in a graph of $\chi_s(q)$ versus q (see Fig. 4.2), this singularity is sufficient to dominate the Fourier transform of the EWNC and it can be shown (Heine and Weaire, 1970, p. 276) that the asymptotic form of the pair potential at large distances becomes

$$V_{IJ}(R) \approx A \frac{\cos(2k_{\mathrm{F}}R + \phi)}{(2k_{\mathrm{F}}R)^3}, \tag{6.70}$$

where the amplitude A and the phase shift ϕ depend on the pseudopotential. A pair potential for Al calculated with the Ashcroft pseudopotential and the response function of eq. (4.75), Fig. 4.3, is illustrated in Fig. 6.3, showing the oscillations at long range.

This long ranged tail to V_{IJ} is enough to make the use of such pair potentials impractical. The summations needed to calculate an elastic constant for example must be carried out to exceedingly long range with such a potential. A careful study of the effect of truncating the potential at some value R_{max} was made by Hartmann and Milbrodt (1971), who calculated the phonon dispersion curves and related quantities, including elastic constants, in aluminium. For calculating the potential and its derivatives, integrals over the EWNC were calculated by Simpson's rule with 2400 points out to $q = 34k_{\mathrm{F}}$, a typical range in **k**-space needed for accurate evaluation of V_{IJ}. They conclude that if one calculates C_{11} from the first eight shells of force constants, the oscillations in the final value as further shells of neighbours are included in the sum amount to about 25% of the mean. High frequency phonons

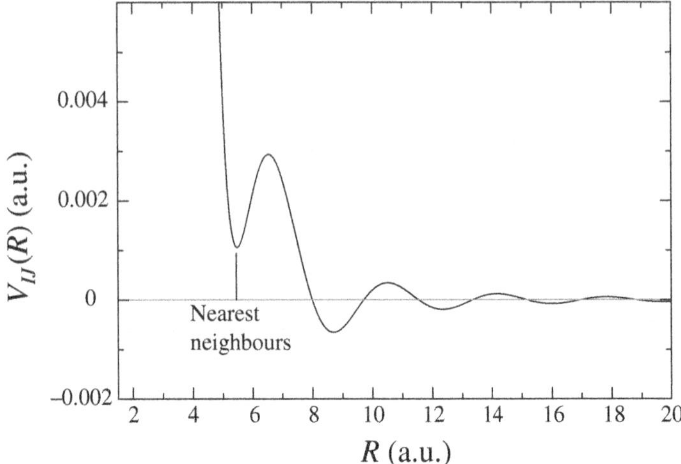

FIG. 6.3 A pair potential for aluminium, calculated with the Ashcroft empty core pseudopotential and the response function of Fig. 4.3.

on the other hand could be calculated satisfactorily with the pair potential. For elastic constants, the representation of the energy and its strain derivatives in **k**-space turned out to converge much better than the **r**-space representation.

One way to make the pairwise potential shorter in range is to take account of the finite temperature of the electrons, which means that the factors $f_F(\mathbf{k})$ no longer go abruptly from 1 to 0 as **k** goes through the Fermi surface. The effect of smearing the Fermi surface in this way is to remove the singularity in χ_e, and the resulting pairwise potential then becomes damped exponentially. Another way to damp the potential exponentially, without much changing the value of total energies calculated with it, was devised by Pettifor and Ward (1984), and is described in the book by Pettifor (1995). Their method is to smooth out the singularity in $\chi_e(q)$ by representing it with an analytic Pade approximant. However it is done, the exponential damping makes the pair potential a feasible model for atomistic simulation. Nevertheless, there is no getting round the fact that if the results are to be quantitatively accurate, the potentials even when damped exponentially have to be of rather long range. That the long range is a real physical attribute of pair potentials can be deduced directly from experimental phonon dispersion curves, as described in Section 5.4.

6.7 The Pseudoatom Picture

The concept of a *pseudoatom* helps us to visualise what these pair potentials actually mean. A pseudoatom is the entity formed by an ion core and its own screening cloud of electrons. The pseudoatom charge density in the case of a local potential for an atom of type τ is

$$\rho_\tau(\mathbf{r}) = \int \chi_e(|\mathbf{r} - \mathbf{r}'|)v_\tau(\mathbf{r}')\,\mathrm{d}\mathbf{r}' \tag{6.71}$$

and the pseudoatom is defined as the potential v_τ together with its screening cloud ρ_τ. The function $\rho_\tau(\mathbf{r})$ is spherically symmetric and does not depend on the positions of the atoms, only on the average density which enters the screening function.

The properties of pseudoatoms are a consequence of second-order perturbation theory with jellium as the starting point, which is expressed by (6.71) in the form of linear response theory. However the atoms are arranged at constant volume, the total charge density in the metal is given by the sum of the above pseudoatom charge densities. When the atoms move, the nuclei carry their screening clouds rigidly with them. In this respect they are very different to real atoms. The electron clouds around real atoms are constantly being distorted, that is polarised, by the influence of their environment, but pseudoatoms cannot be polarised. We might ask then, what happens if we apply an electric field to a pseudoatom? If a uniform electric field is applied to an isolated, macroscopic piece of metal, after an initial transient flow of electrons, all the redistribution of charge that takes place is localised near the surface so as to screen the interior. So we cannot imagine applying a uniform external field. We can, however, apply a *local* field to the Ith pseudoatom say, simply by moving one or more of its neighbours away from their equilibrium positions. If we do that, an electric field is created at the position of the nucleus of I which will therefore move in response. Contrast this to the behaviour of a real atom in an electric field. A neutral atom in an applied uniform electric field will not tend to move, because the force on the nucleus from the applied field is exactly compensated by the force from its own cloud of electrons which have polarised in response to the external field. In both cases, the force on the nucleus is given by the Hellmann–Feynman theorem (see Section 3.1). The difference is that pseudoatoms do not polarize.

We can think of the pairwise potential derived in this chapter as the interaction between a pair of pseudoatoms, depicted in Fig. 6.4. To be precise, the interaction of pseudoatom I with pseudoatom J is just the electrostatic attraction between the screening cloud of J and the ion core of I together with the direct Coulomb interaction:

$$V_{IJ} = \int v_{\tau(I)}(\mathbf{r})\rho_{\tau(J)}(\mathbf{r})\,\mathrm{d}\mathbf{r} + \frac{Z_I Z_J}{R_{IJ}}. \tag{6.72}$$

At first sight this expression looks asymmetric in I and J, which would be a very unsatisfactory state of affairs, violating Newton's Third Law, but we only have to

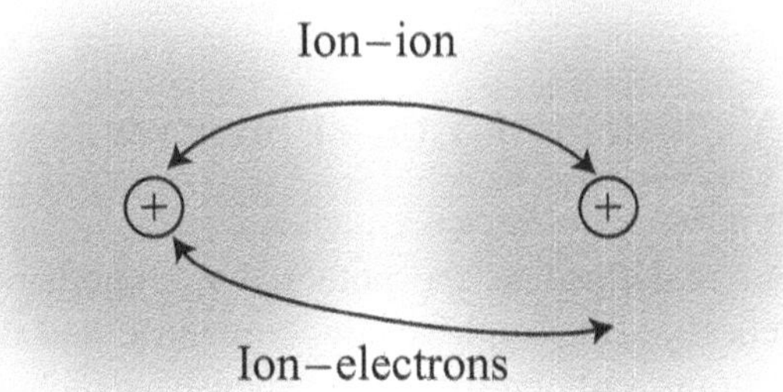

FIG. 6.4 The interaction between two pseudoatoms.

substitute (6.71) into (6.72) to obtain the symmetric form

$$V_{IJ} = \int v_{\tau(I)}(\mathbf{r})\chi_{\mathrm{e}}(|\mathbf{r}-\mathbf{r}'|)v_{\tau(J)}(\mathbf{r}')\,\mathrm{d}\mathbf{r}\,\mathrm{d}\mathbf{r}' + \frac{Z_I Z_J}{R_{IJ}}. \tag{6.73}$$

This picture is consistent with the Hellmann–Feynman theorem in the following way. As that theorem tells us, the force on atom I which we can associate with pseudoatom J is precisely the electrostatic force on ion I due to pseudoatom J, namely:

$$\frac{\partial V_{IJ}}{\partial R_{IJ}} = \int \frac{\partial v_{\tau(I)}(\mathbf{r})}{\partial R_{IJ}}\rho_{\tau(J)}(\mathbf{r})\,\mathrm{d}\mathbf{r} - \frac{Z_I Z_J}{R_{IJ}^2}. \tag{6.74}$$

It acts radially along the direction $\mathbf{R}_{IJ}$. Now consider the total energy change ΔE^{tot} when the distance between I and J is varied by moving atom I along some arbitrary path from $\mathbf{R}_I(A)$ to $\mathbf{R}_I(B)$. We can obtain ΔE^{tot} by integrating the force due to J along the path moved by I:

$$\Delta E^{\mathrm{tot}} = \int_{\mathbf{R}_I(A)}^{\mathbf{R}_I(B)} \frac{\partial V_{IJ}}{\partial R_{IJ}}\mathrm{d}R_{IJ} = \Delta V_{IJ}. \tag{6.75}$$

This will have to be done for the contributions to the force from all the other pseudoatoms too, but the contributions to the force and hence to the energy change are independently additive for each pseudoatom. Thus the energy of any rearrangement is just given by the pairwise interaction defined as above.

We can verify that the **r**-space form of the pairwise potential (6.73) can be derived directly from the **k**-space form (6.53), which for local pseudopotentials can be written in the form:

$$V_{IJ} = \frac{\Omega^2}{(2\pi)^3}\int \chi_{\mathrm{e}}(q)V_{\tau(I)}(q)V_{\tau'(J)}(q)\exp(-\mathrm{i}\mathbf{q}\cdot\mathbf{R}_{IJ})\mathrm{d}\mathbf{q} + \frac{Z_I Z_J}{R_{IJ}}. \tag{6.76}$$

This is a straightforward exercise for the reader. It is simply necessary to replace the functions of q by their Fourier transformations.

We have now seen how, at least when the pseudopotentials are local, the pairwise potential describes an electrostatic interaction between pseudoatoms. Notice however that it is not the *entire* elecrostatic interaction of the pseudotoms with each other. For one thing it excludes the electrostatic interaction between the screening clouds $\rho_{\tau(I)}$ and $\rho_{\tau(J)}$. And it only counts half of the electron–ion interaction. These excluded terms are just what we would expect from the earlier discussions of second-order perturbation theory, in particular they have a direct analogy with the classical image interaction discussed in Section 3.1.3.

The physical interpretation of the large on-site terms V_I is very similar. They represent half the electrostatic interaction of the screening clouds with their own ions.

The most mysterious aspect of the total energy in **r**-space (6.51) is perhaps the appearence of half the bulk modulus of the free electron gas, with a negative sign. What is a term like this doing in the total energy? At most one might expect to see it in an expression for the bulk modulus. The answer lies in the compressibility sum rule, specifically in the use of the compressibility sum rule in the form (4.102). As the nature of the V_{IJ} and V_I terms makes clear, all the electron–ion energy, including the $q = 0$ part of it, is treated by linear response theory, which produces energy changes of the form $\frac{1}{2}\chi_e \Delta V_{ext}^2$. The net outcome looks like a sum of simple electrostatic interactions between electrons and ions, each with a factor of 1/2. Because of the long wavelength limiting form of χ_e, in which we have to include the q^4 term as well as the q^2 term, these simple electrostatic terms hide a term which is half the jellium bulk modulus. However, this term has been smuggled into the $q = 0$ limit of the electrostatic energies for the sake of obtaining a convenient expression for the energy in real space, hence it must be subtracted out again to give the correct total energy.

This mysterious aspect of the pseudoatom picture is deeper than the above description suggests, for it sheds light on the paradox alluded to earlier (cf. Section 6.1), namely the discrepancy between the bulk modulus as calculated by the method of long waves and by homogeneous deformation. Let us carry out a thought experiment in which we compress a simple metal. Suppose that we pretend the screening function χ_e is independent of the jellium density ρ. This has two immediate consequences. It means that the bulk modulus term B^{jel}/ρ^2 in (4.102) is supposed to be constant, and it means that the pseudoatoms remain constant in shape and size as they are increasingly overlapped during the compression. Now consider the jellium terms in the total energy:

$$\tilde{E}^{jel} = E^{jel} - \frac{1}{V}\left[\frac{V^2 B^{jel}}{2}\right]. \tag{6.77}$$

The factor in square brackets is the part which came directly from the long wavelength limit of χ_e. Now if we evaluate the contribution of $\tilde{E}^{\mathrm{jel}}$ to the total bulk modulus we find, treating the contents of the square brackets as constant

$$\frac{d^2\tilde{E}^{\mathrm{jel}}}{dV^2} = \frac{d^2E^{\mathrm{jel}}}{dV^2} - \frac{2}{V^3}\left[\frac{V^2B^{\mathrm{jel}}}{2}\right] = \frac{B^{\mathrm{jel}}}{V} - \frac{2}{V^3}\left[\frac{V^2B^{\mathrm{jel}}}{2}\right] = 0. \tag{6.78}$$

During the compression, these pseudoatoms which we are forcing to be rigid do not change their self-energies V_I, so the second derivative of the total energy is entirely described by the pair potential. This explains why the pair potential alone *may* give quite a reasonable value for the bulk modulus. However in reality the density dependence of χ_e is significant. The pseudoatoms relax their form under compression or expansion, their self energy changes, the pairwise interaction function changes and $\tilde{E}^{\mathrm{jel}}$ has its full volume dependence. As a result the use of pair potentials treated as rigid functions may give a very poor estimate of the bulk modulus. Finnis (1974) estimated this to be an underestimate of about 50% in the case of aluminium, but the exact error is very dependent on the choice of pseudopotential.

6.7.1 The Energy of a Pseudoatom and the Local Density

The pseudoatom picture is so physically appealing that it is worth exploring it further. If it were not for the problems of the compressibility paradox and the difficulties of non-local pseudopotentials, it would probably have found greater favour as a basis for atomistic simulations. In this section I also point out some ways forward for simulation purposes.

It is convenient to write the total energy as a sum over atomic contributions,

$$E^{\mathrm{tot}} = \sum_I E_I. \tag{6.79}$$

To do this we need to introduce the jellium energy per particle, as in eq. (1.224), into (6.51). Although it is not unique, the obvous expression for E_I which is compatible with (6.51) is then

$$E_I = Z_I\left(\epsilon^{\mathrm{jel}} - \frac{1}{2\rho}B^{\mathrm{jel}}\right) + \frac{Z_I}{4\pi^3\rho}\int f_{\mathrm{F}}(\epsilon_{\mathbf{k}})\left\{V^{\mathrm{core}}_{\tau(I)}(\mathbf{k},\mathbf{k}) - V^{\mathrm{core}}_{\tau(I)}(\mathbf{k}_{\mathrm{F}},\mathbf{k}_{\mathrm{F}})\right\}\,d\mathbf{k}$$
$$+ V_I + \frac{1}{2}\sum_J V_{IJ}. \tag{6.80}$$

The most compact and general equations for V_I and V_{IJ} are (6.56) and (6.55). E_I has the form of a density dependent energy plus the sum of a pairwise potential which is also a function of density:

$$E_I = F_I(\rho) + \frac{1}{2}\sum_J V_{IJ}(R_{IJ}, \rho). \tag{6.81}$$

The difficulty lies in the fact that by virtue of its derivation from second-order perturbation theory, ρ is the average density of electrons within the large volume V. Physically, if a region of V is occupied by vacuum, as it would be if the system of interest is a large cluster or a thick slab, or material containing a large cavity, the interatomic forces far from the surfaces *should* be just as they are in infinite bulk material. This is because the effect of surfaces or other disturbances in a metal is screened out by the conduction electrons over microscopic distances. The appearance of the average density $\rho = N/V$ in the total energy, where V includes the volume occupied by vacuum, is clearly inappropriate. Second-order perturbation theory starting from a uniform jellium is completely inadequate to describe a system of which some parts are so very different to others, in particular any system with a free surface.

In such a case it would be more reasonable to define the density in terms of the volume actually occupied by material. However this is not a particularly useful strategy in view of the difficulty of defining at the atomic scale the average density of a system with a surface, and it certainly does not correct the error made by second-order perturbation theory in the neighbourhood of a surface. Even a single vacancy is too great a perturbation for the above expressions for the energy to work in general, as discussed by Evans and Finnis (1976).

A possible way forward is to define a *local* density, ρ_I, which depends on the local environment of atom I, and to use it in eq. (6.80) in place of the global density ρ. This has been proposed in different forms by several authors, including Finnis and Sachdev (1976), Evans and Kumaravadivel (1976), Rosenfeld and Stott (1987) and Finnis *et al.* (1998*c*). To be of any use, ρ_I must satisfy three criteria. First, it should be relatively local on the atomic scale, reflecting the short range of metallic screening. Second, it should reproduce as closely as possible the true average density of electrons in a perfect crystal, so that the above formulae for total energy still apply. Third, it should provide an accurate extrapolation of the total energy to inhomogeneous distributions of atoms, including point defects and surfaces. Finnis *et al.* (1998*c*) made a detailed study of some candidate functions with regard to the first and second of these criteria. If they are satisfied the model will at least predict bulk properties exactly as the original model with a global density, but it will avoid the compressibility paradox. These authors concluded that the most promising candidate was a sum of Gaussian functions over neighbouring atoms, just as proposed by Rosenfeld and Stott (1987). In the present notation, including the possibility of more than one element, the local density would have the form:

$$\rho_I = \bar{Z} \sum_J \frac{\alpha^3}{\pi^{3/2}} \exp(-\alpha^2 R_{IJ}^2). \tag{6.82}$$

The picture here is that each pseudoatom is associated with a Gaussian density, of fairly narrow width determined by a screening length $1/\alpha$ which is of order a few Angstroms. This satisfies the first criterion above. Now consider the case of a perfect crystal lattice in which all atoms occupy equivalent sites, and the volume per atom is Ω. The broader the Gaussian, the more accurately the summation can be replaced by an integral; the limit is given by:

$$\lim_{\alpha \to 0} \sum_J \frac{\alpha^3}{\pi^{3/2}} \exp(-\alpha^2 R_{IJ}^2) \cdot \Omega = \int \frac{\alpha^3}{\pi^{3/2}} \exp(-\alpha^2 r^2)\, d\mathbf{r} = 1. \tag{6.83}$$

Hence if we choose $\bar{Z}$ to be the mean number of electrons per atom, that is the weighted average valence, we have:

$$\lim_{\alpha \to 0} \rho_I = \frac{\bar{Z}}{\Omega} = \rho. \tag{6.84}$$

Now a compromise has to be made between the first two criteria. Unfortunately there is no theory to tell us exactly what the 'best' value of α should be, and this will have to be determined pragmatically. We need to choose α to be small enough for the summation to be a good approximation, but if the Gaussians are thereby too long ranged, they will (a) be unwieldy for the rapid atomistic simulations for which the model is intended and (b) fail to capture the local environment. Fortunately the compromise is surprisingly unproblematic. Take the simple example of a face-centred cubic crystal, with lattice parameter a. Try $\alpha = 2/a$. If the sum is extended to long range, the density is overestimated by a mere 0.5%. If it is truncated after third nearest neighbours (which are at a distance $1.2247a$) the density is overestimated by 0.33%. If we choose a smaller value of α we will obtain an even better approximation to ρ when the summation is extended to long range, while reducing the contribution of the nearest neighbours. For example with $\alpha = 1.6/a$, truncating the sum after third nearest neighbours now *underestimates* ρ by 2.2%. These results are very encouraging in terms of satisfying the first two criteria; it remains to be seen how well the third criterion can also be satisfied, given that there is some freedom in the choice of α.

7

TIGHT BINDING

7.1 Introduction

There is no single tight-binding theory or model, but several, depending on the kind of approximations that are made to the full DFT total energy. Their common feature is the use of a minimal local basis of local orbitals. This usually means at most just one s-orbital $|s\rangle$, three p-orbitals $\{|p_x\rangle, |p_y\rangle, |p_z\rangle\}$ and five d-orbitals $\{|d_{xy}\rangle, |d_{yz}\rangle, |d_{zx}\rangle, |d_{x^2-y^2}\rangle, |d_{3z^2-r^2}\rangle\}$ per atom. There is no distinction between principal quantum numbers in such a basis. Normally the orbitals are chosen to be real in r-space, as the labelling indicates. The symmetry of the orbitals in this case is described in Section 1.5.1 and depicted in Fig. 7.1. The valence wavefunctions built of these orbitals dominate the bonding in the vast majority of common materials. The transition metals (groups 1B–8B of the periodic table) are characterized as sd-bonded, since their cohesion arises mainly from the overlap of d-orbitals with each other and with the nearly free electron states. Most of the elements which are not transition metals are referred to as *sp-bonded.* This means that the wavefunctions which dominate their cohesion are, to a good approximation, linear combinations of overlapping, local, atomic-like s and p orbitals. The noble metals lie on the border. Their Fermi surfaces are very nearly free electron like, suggesting that they are sp-bonded. This is consistent with the band structures and densities of states which show a full d-band. However, detailed calculations show that the mixing or *hybridization* of s with d orbitals makes an important contribution to their wavefunctions and cohesive energy. A discussion of these characteristics with reference to particular elements is given by Pettifor (1995), chapter 7 .

The range of tight-binding approaches forms a loose heirarchy in terms of the level of approximations made. The most exact are *ab initio* methods, in which the only approximations are the treatment of exchange and correlation within density functional theory and the incompleteness of the basis. Then in the interests of rapid computation and simple physical insight, a sequence of further approximations can be made, notably

- ignore three-centre integrals
- parameterize two-centre integrals
- ignore non-spherical terms on an atom
- ignore inter-site Coulomb interactions
- ignore charge transfers
- assume orbitals are orthogonal
- make a real space representation as multi-ion potentials
- make a real space second-moment approximation

The further down this list we go, the simpler the model we can derive. The meaning of these approximations will become clear in the following sections.

7.1.1 Predicting the Past

Simple, non-self-consistent tight-binding models, as described in Section 7.2 were first used in the nineteen sixties to explain the trends in crystal structure and bonding of the transition metals going from left to right across the periodic table (Friedel, 1964, 1969; Cyrot-Lackmann, 1968; Heine, 1980). Applications to *sp*-bonded and *pd*-bonded compounds soon followed (Chadi and Cohen, 1975; Majewski and Vogl, 1986; Pettifor and Podloucky, 1986). The differences in energy between different crystal strucures are small, one or two orders of magnitude less than cohesive energies, so it is quite an impressive achievement for simple, non-self-consistent tight binding to be able to get even their signs right.

The quality of a model can be very loosely thought of as the ratio of its predictive power to its complexity:

$$Q = \text{predictive power/complexity.}$$

Predictive power is a common misnomer, since we are also referring here to quantities that have already been measured, or calculated by a more accurate method (hence the title of this section). It really means *transferability*, and describes how well the model performs when it used to calculate some physical quantity which was not involved in its parameterisation, as discussed in Chapter 5. In the remainder of this book we shall be making many more approximations, which can only be tested empirically by testing the transferability of the model. The approximations are probably interesting if they improve Q, and they may also be of practical use.

7.1.2 *Ab Initio* Tight Binding

If the Kohn–Sham wavefunctions are expanded in a minimal basis of orbitals, and the matrix-eigenvalue equation is solved, the whole thing without any adjustable parameters, the method can be referred to as first-principles or *ab initio* tight binding. The method need not be self-consistent for this description to be appropriate. In its most exact form, however, *ab initio* tight binding is fully self-consistent and the limited basis can be made more complete by adding more orbitals to it. In this sense then, practically any first-principles method using local orbitals could be described as *ab initio* tight-binding. I shall not describe any of the *ab initio* methods here; there are many technical details involved, they are constantly being developed further, and whole books have been written about some of them already. There is no 'best' method in general; like the choice of a model, it depends on the system size, the elements involved and most importantly on the questions you are asking. We will quickly pass over the *ab initio* methods to discuss the simplifications that can be made, but since it is a growing field of activity, a few brief remarks about them are appropriate.

Perhaps the most sophisticated approach is the tight-binding representation of the linear muffin-tin orbital method pioneered by Andersen and coworkers, which is described in detail in the book edited by Dreyssé (2000). Although this method is complicated, it includes within the algorithm a prescription for the exact form of the local orbitals which does not leave any arbitrary, that is to say empirical, choices to be made. Other approaches such as Sankey *et al.* (1991), Elstner *et al.* (1998) and Soler *et al.* (2002) require the orbitals to be constructed in advance. This may not be difficult, but there is no unique way to do it, so you can never be sure you have made the optimum choice for the problem in hand.

A general difficulty of local orbital methods when you have chosen a limited basis, is that there is usually no optimal and systematic way of including additional orbitals in order to make the basis more complete. This weakness of local orbital methods is conversely one of the great attractions of plane-waves as a basis: you know exactly how to increase the size of the basis set, simply by increasing the maximum wave-vector within which all plane waves are to be included. There are other advantages of the plane-wave basis which are discussed elsewhere; see for example the article by Lindan (2002).

Nevertheless, local orbital methods have proved to be powerful tools for first-principles atomistic simulation, because in certain cases they can deal with thousands if not yet millions of atoms. For the purpose of large-scale calculations, their great strength as opposed to those *ab initio* methods which use plane waves or other extended basis sets is that algorithms exist with which the computation of the total energy and forces can be done within a processor time which scales linearly as the number of atoms. For non-metallic systems such an algorithm is implemented for example in the codes SIESTA (Soler *et al.*, 2002) and CONQUEST

Bowler *et al.* (2002). Methods which exploit a linear scaling algorithm are usually called O(N) methods, pronounced *order N methods*. For metallic systems it is harder to come up with a successful O(N) method, fundamentally because of the longer range of the off-diagonal elements of the density matrix $\rho(\mathbf{r}, \mathbf{r}')$, nevertheless a useful scheme was devised in the tight-binding setting by Goedecker (1993). Besides the potential advantages of O(N), the great attraction of local orbital methods which is shared by the tight-binding models to be described below is their chemical picture of bonding between atoms. Good accounts of bonding from the point of view of local orbitals can be found in the books by Harrison (1980), Sutton (1993) or Pettifor (1995).

7.1.3 One-, Two- and Three-centre integrals

The task which is common to all tight-binding models is to solve the Kohn–Sham equation, that is the single-particle Schrödinger equation, in the form of the matrix eigenvalue problem, eq. (1.235). This involves constructing the matrix elements of the Hamiltonian

$$H_{I\mu J\nu} = \langle I\mu|\hat{H}|J\nu\rangle \equiv \langle I\mu|\hat{T}|J\nu\rangle + \langle I\mu|\hat{V}_{\text{eff}}|J\nu\rangle. \tag{7.1}$$

A large part of tight-binding modelling is concerned with how to approximate these quantities. Let us restrict ourselves to a local potential for the present purposes. With a choice of real orbitals, the kinetic energy part of (7.1) looks like

$$\begin{aligned}\langle I\mu|\hat{T}|J\nu\rangle &\equiv \int \langle I\mu|\mathbf{r}\rangle \left\{-\tfrac{1}{2}\nabla^2\right\} \langle \mathbf{r}|J\nu\rangle \, d\mathbf{r} \\ &\equiv \int \phi_{I\mu}(\mathbf{r}-\mathbf{R}_I) \left\{-\tfrac{1}{2}\nabla^2\right\} \phi_{J\nu}(\mathbf{r}-\mathbf{R}_J).\end{aligned} \tag{7.2}$$

These are integrals of a kind that depends only on the vector $\mathbf{R}_{IJ}$ and the particular orbitals, not on the environment of the atoms in the material. They are therefore called *two-centre integrals*. If $I = J$ they are simply *one-centre integrals* or *on-site* terms.

More difficult are the potential integrals, because the potential $V_{\text{eff}}(\mathbf{r})$ depends on the local environment. Even without considering the effect of self-consistency, if we just consider the $V_{\text{eff}}^{\text{in}}$ resulting from the superposition of free atom charge densities, we shall have to make some approximations at this point in order to derive simpler models. A key simplification is to write V_{eff} as a sum of atom-centred contributions

$$\hat{V}_{\text{eff}} = \sum_I \hat{V}_{\text{eff}}^{(I)}, \tag{7.3}$$

or

$$V_{\text{eff}}(\mathbf{r}) = \sum_I V_{\text{eff}}^{(I)}(\mathbf{r}-\mathbf{R}_I). \tag{7.4}$$

Then the integrals to be done are much less daunting than in the general case. There are one- and two-centre integrals with the same structure as for the kinetic energy. In addition there are *three-centre integrals*. Two- and three-centre integrals are explicit in the following representation of the potential, for $I \neq J$:

$$\langle I\mu|\hat{V}_{\text{eff}}|J\nu\rangle = \langle I\mu|\hat{V}_{\text{eff}}^{(I)}|J\nu\rangle + \langle I\mu|\hat{V}_{\text{eff}}^{(J)}|J\nu\rangle + \sum_{K\neq I,J} \langle I\mu|\hat{V}_{\text{eff}}^{(K)}|J\nu\rangle. \quad (7.5)$$

When the orbitals are on the same site we have the one-centre integrals,

$$\langle I\mu|\hat{V}_{\text{eff}}^{(I)}|I\nu\rangle \quad (7.6)$$

and also two-centre integrals of a different kind, in which the potential from site J overlaps with the orbitals on site I:

$$\langle I\mu|\hat{V}_{\text{eff}}^{(J)}|I\nu\rangle, \quad (7.7)$$

which together give the following form for the on-site matrix elements of the potential:

$$\langle I\mu|\hat{V}_{\text{eff}}|I\nu\rangle = \langle I\mu|\hat{V}_{\text{eff}}^{(I)}|I\nu\rangle + \sum_{J\neq I} \langle I\mu|\hat{V}_{\text{eff}}^{(J)}|I\nu\rangle. \quad (7.8)$$

The tight-binding models we shall be describing make the approximation that three-centre integrals can be ignored. At the same time, one- and two-centre integrals are not calculated explicitly, with chosen orbitals, rather they are parameterized. This means that the functional form of a particular orbital is not assumed, but rather the matrix elements are modelled directly as specific functions of the atomic positions. It is hard to quantify the errors in these approximations of decomposing V_{eff} into a sum of atom-centred potentials and neglecting three-centre integrals, because the basis orbitals and the decomposition of the potential are not uniquely prescribed. There is no doubt some optimal choice which would minimize the errors. The neglect of three-centre integrals can to some extent be compensated during the fitting of parameters to describe the one and two-centre integrals, but it nevertheless remains probably the most serious approximation made by empirical tight-binding models Foulkes (1993), and the most difficult to circumvent.

It will be easiest to start by considering the simplest non-self-consistent model, with a view to adding complications later, rather than the more logical approach of starting with the most sophisticated self-consistent model and stripping it down by successive approximations. The following section will also establish some of the terminology to be used later.

7.2 Non-self-consistent Tight Binding

The starting point for a simple, non-self-consistent tight-binding model is the first-order functional defined by (3.48). Let us use the density matrix introduced in Section 1.7.2 together with eq. (1.175) to write this functional in the form:

$$E^{(1)}[\rho] = \mathrm{Tr}\,\hat{\rho}\hat{H}_{\mathrm{in}} + E_{\mathrm{xc}}^{\mathrm{in}} - \rho^{\mathrm{in}} V_{\mathrm{xc}}^{\mathrm{in}} - E_{\mathrm{H}}^{\mathrm{in}} + E_{ZZ}. \tag{7.9}$$

As in Section 4.5, I am using the shorthand notation in which, when the context is clear, the product of a charge density and a potential is to be integrated over. Thus in (7.9):

$$\rho^{\mathrm{in}} V_{\mathrm{xc}}^{\mathrm{in}} \equiv \int \rho^{\mathrm{in}}(\mathbf{r}) V_{\mathrm{xc}}^{\mathrm{in}}(\mathbf{r})\, d\mathbf{r} \equiv \mathrm{Tr}\,\hat{\rho}^{\mathrm{in}} \hat{V}_{\mathrm{xc}}^{\mathrm{in}}. \tag{7.10}$$

In theory, the main task in non-self-consistent tight binding is to minimize $E^{(1)}$ with respect to ρ within the space of feasible test densities. As we have seen, this problem is equivalent to solving the one-particle Schrödinger equation

$$\hat{H}^{\mathrm{in}} |\psi_n\rangle = \epsilon_n |\psi_n\rangle \tag{7.11}$$

for the eigenvalues ϵ_n. As described in Section 1.10, in the chosen local basis, the Schrödinger equation is cast as a matrix eigenvalue problem, eq. (1.235):

$$\sum_{J\nu} H_{I\mu J\nu}^{\mathrm{in}} C_{J\nu}^{n} = \epsilon_n \sum_{J\nu} S_{I\mu J\nu} C_{J\nu}^{n}. \tag{7.12}$$

Neglecting three-centre integrals means that the Hamiltonian matrix is easy to construct if we know the functions $H_{I\mu J\nu}^{\mathrm{in}}(\mathbf{R}_{IJ})$. We shall refer to these functions as *hopping integrals*, because from the elementary ideas of time-dependent perturbation theory (e.g. Landau and Lifshitz, 1965, p. 136), they are related to the probability of an electron hopping from one orbital to another. Like the overlap matrix elements $S_{I\mu J\nu}(\mathbf{R}_{IJ})$ they depend only on the vector joining the atoms I and J. Here and in the remainder of the chapter the formulae will be written as if they refer to a finite system rather than a periodic system. As we saw previously, the treatment of a periodic system leads to the same equations in which the matrix elements of H and S are replaced by their Bloch sums, with the introduction of an extra quantum number, the $\mathbf{k}$-vector, see eqs. (1.243)–(1.245). For generality and simplicity of notation I will omit $\mathbf{k}$ labels. This means that that all the formulae can be taken over as they stand for periodic systems, with the understanding that a $\mathbf{k}$-summation will have to accompany an n-summation and the H and S matrices are complex Hermitian if $\mathbf{k} \neq 0$.

In addition, we have to calculate the remaining part of (7.9) that does not depend on ρ but on the choice of input density ρ^{in}. If we can do all this, then (7.9) gives us the Harris–Foulkes approximation to the energy, which is accurate to first-order in the wavefunction, the charge density and effective potential.

The first term of (7.9) we have already referred to as the *band energy*. Using the expression (1.183) for the trace it can be written as

$$E^{\text{band}} = \text{Tr}\hat{\rho}\hat{H}^{\text{in}} = \sum_{I\mu J\nu} \rho^{I\mu J\nu} H^{\text{in}}_{J\nu I\mu}. \tag{7.13}$$

It can be calculated by summing the lowest N eigenvalues, corresponding to the occupied eigenstates. We will also define the *bond energy* as the intersite part of the band energy:

$$E^{\text{bond}} = \sum_{I\mu J\nu, I\neq J} \rho^{I\mu J\nu} H^{\text{in}}_{J\nu I\mu}. \tag{7.14}$$

This is a quantity of great physical significance, since it is representing an important part of the total energy, in fact the main contribution to the cohesion of the atoms, as a sum over bonds. The individual bond energies for each pair of atoms and orbitals can be defined as

$$E^{\text{bond}}_{I\mu J\nu} = \rho^{I\mu J\nu} H^{\text{in}}_{J\nu I\mu}. \tag{7.15}$$

The factor $\rho^{I\mu J\nu}$ is the *bond order* introduced in Section 1.7.3, eq. (1.188). The strength of the bonding is therefore directly related to the bond orders and the hopping integrals in a way which could hardly be simpler. If only the bond orders could be represented as functions of R_{IJ} like the hopping integrals we would have an interatomic potential! This *can* be done with some further simplifying assumptions, as we shall see in Section 7.12.1. The standard approach however is to calculate the bond orders or the sum of one electron energies by solving the Schrödinger equation.

We now need to discuss several points in more detail, before considering how to actually go about solving (7.12). First, there is the question of how to calculate the hopping integrals. Second, there is the question of how to treat the overlap or **S** matrix, raised already in Section 1.5.2. A common strategy is to make the further simplification of *orthogonal tight binding,* in which the **S** matrix becomes the unit matrix. This needs some justification, since a calculation of **S** with atomic orbitals gives interatomic matrix elements which are of order unity, the size of the diagonal elements. For completeness I will develop the formalism for non-orthogonal orbitals and indicate where the assumption of orthogonal orbitals simplifies matters. Third, there is the question of how to treat the remaining terms in the total energy which do not depend on ρ. They are normally lumped together and approximated by the sum of a repulsive pairwise potential V_{IJ}, and we need to examine why this is at all reasonable.

Historically $V_{IJ}(R)$ has usually been represented as a Born–Mayer potential of the form $A\exp(-pR)$ or an inverse power law AR^{-m}, where A, p and m are treated as parameters to be fitted. The resulting potentials are steeply repulsive, typical values of m lying in the range 4 for s–p bonded semiconductors to 10 for d-bonded transition metals. However, as we shall see, there are very different approaches, according to whether the on-site terms in the density matrix are lumped in with the fitted pair potential or treated explicitly as in the band energy. Finally, notice that in parameterizing the hopping integrals we discard any information about the r-space forms of the orbitals $\langle \mathbf{r}|I\mu\rangle$, so we can no longer calculate $\rho(\mathbf{r})$. Hence we need to investigate how it is possible, without explicitly knowing $\rho(\mathbf{r})$, to apply the Hellmann–Feynman theorem in order to obtain the interatomic forces. I discuss these matters in turn in Sections 7.3–7.6.

7.3 Slater–Koster Parameters

7.3.1 The Symmetries of Two-centre Integrals

At first sight the number of Hamiltonian parameters to be fitted is daunting. In a purely d-band model we would have 5 orbitals on each site and $5 \times 5 = 25$ hopping integrals $H^{\text{in}}_{I\mu,J\nu}(\mathbf{R}_{IJ})$. In a full s, p, d model we would have 9 orbitals on each site, hence $9 \times 9 = 81$ matrix elements of the Hamiltonian between each pair of atoms. In the two-centre approximation, the number of these hopping integrals which are actually independent can be greatly reduced by symmetry. Slater and Koster (1954) showed how to do this in a classic paper on the theory of tight binding. Their method is based on the formulae for rotating the spherical harmonics from one set of Cartesian axes to another. A detailed example should make the idea clear. For reference, Fig. 7.1 shows the non-vanishing hopping integrals in the standard reference frame described below, in which the number of independent elements has been reduced to 14 by choosing the z-axis to lie along $\mathbf{R}_{IJ}$. Consider first just the p-orbitals. Suppose we want to know the hopping integral $H^{\text{in}}_{Ip_x Jp_y}(\mathbf{R}_{IJ})$, where the vector $\mathbf{R}_{IJ}$ has direction cosines l, m, n. Slater and Koster designated this hopping integral $E_{x,y}$ so in their notation:

$$E_{x,y} = \int \phi_{Ip_x}(\mathbf{r}-\mathbf{R}_I)\left\{-\tfrac{1}{2}\nabla^2_{\mathbf{r}} + V^{(I)}_{\text{eff}}(\mathbf{r}-\mathbf{R}_I) + V^{(J)}_{\text{eff}}(\mathbf{r}-\mathbf{R}_J)\right\}\phi_{Jp_y}(\mathbf{r}-\mathbf{R}_J)\,d\mathbf{r}. \tag{7.16}$$

There are similar expressions for $E_{x,z}$ and the remaining seven hopping integrals involving pairs of p-orbitals. Now let us switch to rotated Cartesian coordinates in which the z-axis is along the bond $\mathbf{R}_{IJ}$; call it the z' direction. It does not matter in which orthogonal directions the x' and y' axes lie. In these coordinates a little thinking reveals that there are only two independent and non-vanishing hopping integrals between p-orbitals, namely $E_{z'z'}$ and $E_{x'x'} = E_{y'y'}$. They are conventionally called $pp\sigma$ and $pp\pi$, respectively (the fourth and third entries down the diagonal of the 9×9 matrix depicted in Fig. 7.1). The fact that the

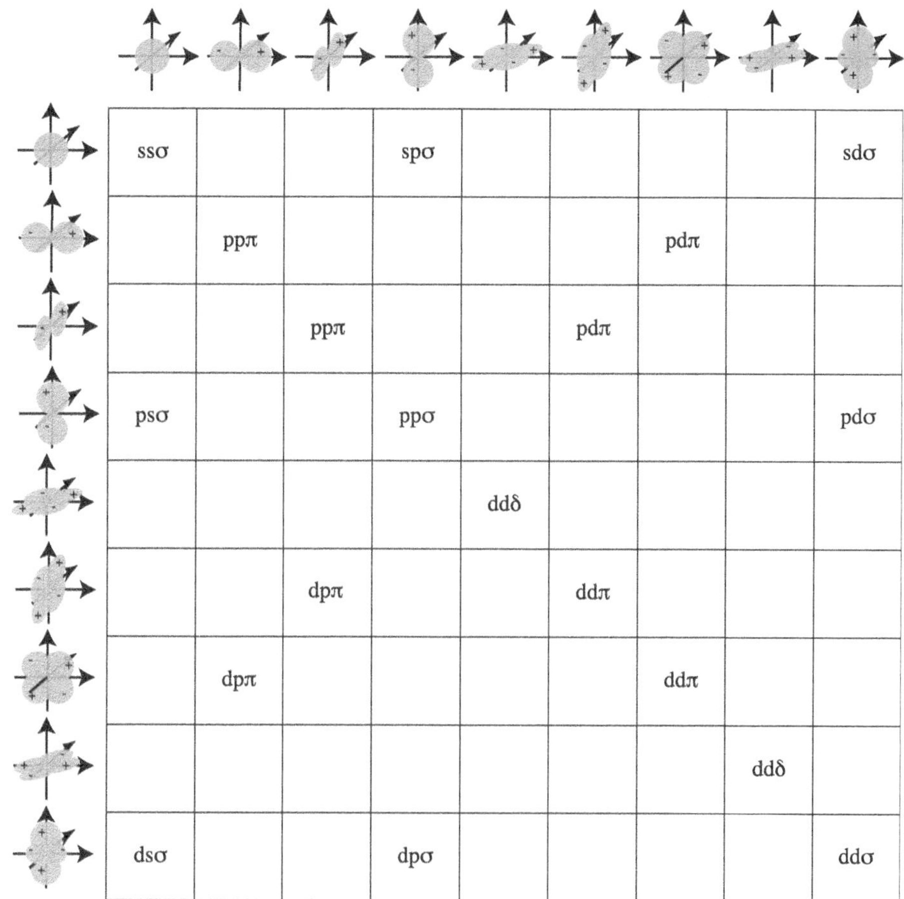

FIG. 7.1 The independent two-centre hopping integrals between atoms on the z-axis.

other entries in the 3×3 block (between rows 2-4 and columns 2-4) vanish can be seen by imagining the shape of the orbitals in which there are equal positive and negative lobes along the positive and negative axes, so in all but the $pp\sigma$ and $pp\pi$ orientations we find there are cancelling positive and negative contributions to the integral. Now returning to the original axes we realize that ϕ_{Ip_x} can be expressed as a linear combination of $\phi_{Ip_{x'}}, \phi_{Ip_{y'}}, \phi_{Ip_{z'}}$

$$\phi_{Ip_x} = a_{xx}\phi_{Ip_{x'}} + a_{xy}\phi_{Ip_{y'}} + a_{xz}\phi_{Ip_{z'}} \tag{7.17}$$

and similarly

$$\phi_{Ip_y} = a_{yx}\phi_{Ip_{x'}} + a_{yy}\phi_{Ip_{y'}} + a_{yz}\phi_{Ip_{z'}}. \tag{7.18}$$

Hence the hopping integral $E_{x,y}$ can be expressed as the following linear combination of $pp\sigma$ and $pp\pi$:

$$E_{x,y} = a_{xz}a_{yz}\, pp\sigma + (a_{xx}a_{yx} + a_{xy}a_{yy})\, pp\pi. \tag{7.19}$$

The coefficients can be expressed in terms of the direction cosines (l, m, n) of $\mathbf{R}_{IJ}$ with the result

$$E_{x,y} = lm(pp\sigma) - lm(pp\pi). \tag{7.20}$$

Slater and Koster (1954) worked out expressions of this kind for all the hopping integrals between s, p and d orbitals and presented them in table 1 of their paper, which has been invaluable for practical calculations. We shall also refer to these special hopping integrals as *bond integrals*. Between s orbitals there is of course just one bond integral $ss\sigma$. Between an s and a p orbital there is also only one, $sp\sigma$, however notice that

$$sp\sigma \neq ps\sigma. \tag{7.21}$$

If atoms I and J are of the same type then $sp\sigma = -ps\sigma$, but if they are different, $sp\sigma$ and $ps\sigma$ are completely independent. There is only one non-vanishing interaction of an s with the d orbitals, namely the one between an s and the orbital with symmetry $3z^2 - r^2$. This is designated $sd\sigma$. If the atoms are identical $sd\sigma = ds\sigma$. The symmetries of all the hopping integrals in the specially chosen orientation with R_{IJ} along the z-axis are illustrated in Fig. 7.1. This figure shows which of the 81 matrix elements of H are non-vanishing in this special orientation. If atoms I and J are of the same type you can see by inspecting the pattern of positive and negative lobes on the orbitals that the following symmetries apply:

$$\begin{aligned} sp\sigma &= -ps\sigma \\ sd\sigma &= ds\sigma \\ pd\sigma &= -dp\sigma \\ pd\pi &= -dp\pi. \end{aligned} \tag{7.22}$$

These sign changes are a fruitful source of error for anyone writing their first tight-binding code (which is probably the best way to proceed if you really want to understand this stuff), simply because the matrix $H_{I\mu J\nu}$ is only symmetric under the *simultaneous* interchange of I with J and μ with ν. Now consider the symmetry of the two-centre Hamiltonian operator in the definition of the hopping integrals. That is the factor between the orbitals in (7.16). It is independent of the orbitals and symmetric under rotations about the axis of the bond; that is it has the same symmetry as a constant scalar. This means that that same symmetries apply to $S_{I\mu J\nu}$ as to $H_{I\mu J\nu}$. Hence Slater and Koster's table can be used for generating all 81 elements of $\mathbf{H}$ (or $\mathbf{S}$) given the independent bond integrals (or overlap integrals) and the direction cosines of $\mathbf{R}_{IJ}$. There are 11 independent bond (or overlap) integrals when the atoms at $\mathbf{R}_I$ and $\mathbf{R}_J$ are of the same type and 15 when they are different.

Not surprisingly σ-integrals tend to be larger in magnitude than π- or δ-integrals. In fact the theory of canonical band structures (Andersen, 1973) predicts the ratio

$$dd\sigma : dd\pi : dd\delta = -6 : 4 : -1. \tag{7.23}$$

7.3.2 Distance Dependence of the Bond Integrals

Each of the bond integrals is a function not only of (l, m, n) but also of the distance R_{IJ}. This did not concern Slater and Koster, who were interested in band-structures in crystals with fixed atomic positions. However, in order to calculate forces we need to know how the bond integrals change continuously with atomic positions. With empirical tight binding, all the bond integrals, as well as the S-matrix elements in non-orthogonal tight binding, are given an R-dependence, where R stands for a general interatomic distance.

There are various strategies for describing the R-dependence of the bond integrals. In earlier work (Ducastelle, 1970; Allan and Lannoo, 1976) a bond integral $\beta(R)$ was assumed to vary as

$$\beta(R) = \beta_0 \exp(-qR), \tag{7.24}$$

where β_0 and q are constant. This was convenient to use but did not address the physical fact that q must depend on the type of orbitals involved. Later work suggested that a power law

$$\beta(R) = \beta_0 R^{-n} \tag{7.25}$$

should be more appropriate, see, e.g., Harrison (1980). The simplest argument for this form refers to the width of an s–p band. In a tight-binding model this must be proportional to the bond-integrals. On the other hand, in a free electron metal it is proportional to the Fermi energy and therefore inversely proportional to R^{-2} (see Section 1.9). Hence, we expect the bond integrals to vary as R^{-2} in s–p bonded materials. There are more sophisticated arguments suggesting that the d–d bond integrals should vary as R^{-5} (Heine, 1980).

Now for practical purposes you might think of parameterizing the bond integrals according to an inverse power law (7.25) and fitting values of β_0 and n to curves of bandwidth versus volume calculated by a fully self-consistent band structure method. The main problem about that is how do you truncate the functions? Even R^{-5} is a long ranged function compared to the exponential, and if we truncate it say at $R = R_c$ after first or second nearest neighbours, a discontinuity is introduced which could be disastrous. The disaster would occur if the functions were used in a simulation in which the distance between some pair of atoms moved through R_c. In a molecular dynamics calculation this would lead at best to non-conservation of energy and at worst to an instability as the numerical integration tried to deal with a delta function in the force. The partial answer is to connect $\beta(R)$ smoothly to zero over a range from say R_1 to R_2. A cubic polynomial or spline is a convenient way to do this since it contains 4 parameters and there are 4 conditions. The spline has to match the function $\beta(R_1)$and its derivative at $R = R_1$ and its value and its first derivative have to vanish at $R = R_2$.

A more flexible approach is not bound by the inverse power law, which is anyway only an approximation. A useful functional form was devised by Goodwin *et al.* (1989), and is referred to as the GSP function. They were motivated to smooth the cutoff by multiplying the inverse power law by an exponential. At the same time they wanted a function that could be fitted easily to experimental or calculated data. The GSP function $f_{\text{GSP}}(R)$ describes the bond integrals by the form

$$\beta(R) = \beta(R_0) f_{\text{GSP}}(R), \tag{7.26}$$

where

$$f_{\text{GSP}}(R) = (R_0/R)^n \exp\{-n(R/R_c)^{n_c}\} \exp\{n(R_0/R_c)^{n_c}\}. \tag{7.27}$$

R_0 is the nearest neighbour distance, at which $f_{\text{GSP}} = 1$, and R_c is an additional parameter to control the decay rate of the function. If R_c is chosen to be large compared to R_0, the simple inverse power law is recovered. The same function is useful for parameterizing the repulsive potential which will be described below. The GSP function is commonly accompanied by a cubic spline, as described previously, to bring the bond integral smoothly to zero at a prescribed cutoff distance R_2.

In summary, rather drastic approximations have been introduced to describe the R-dependence of the bond integrals, because they have to be cut off to zero at short range. This is partly for practical reasons, in order to speed up the computation of total energy and forces. However, we should not ignore the fact that the two-centre integrals beyond first or second neighbours would really be less important than the three-centre integrals which we are neglecting, so it is not worth the effort to compute them. Most of the bonding energy and its dependence on structure has been captured with the short-ranged, two-centre integrals, as the abundant applications in the literature testify.

7.4 The Repulsive Energy

When we talk about *the repulsive energy* in the context of tight binding, there are historically two possible interpretations, which differ significantly. In some of the literature, especially early papers on tight binding, but also for example Esfarjani and Kawazoe (1998, eq. (15)), the repulsive energy is defined as the remainder of the energy in (7.9) besides the band energy. We shall call this $E_{\text{band}}^{\text{rep}}$:

$$E_{\text{band}}^{\text{rep}} = E_{\text{xc}}^{\text{in}} - \rho^{\text{in}} V_{\text{xc}}^{\text{in}} - E_{\text{H}}^{\text{in}} + E_{ZZ}. \tag{7.28}$$

The *tight-binding band model* expresses $E_{\text{band}}^{\text{rep}}$ as a sum of pairwise repulsive potentials V_{IJ} which are empirical in form and fitted to experimental values of the lattice constant, cohesive energy and bulk modulus, thus (7.9) becomes:

$$E_{\text{band}}^{(1)}[\rho] = \text{Tr}\, \hat{\rho}\hat{H}^{\text{in}} + \frac{1}{2}\sum_{I \neq J} V_{IJ}. \tag{7.29}$$

This approach has been superseded by the tight-binding *bond* model described below, nevertheless I introduce it here because the band model has been a widely used form of empirical tight binding. Some of the most important applications of it are cited in the present bibliography or references in the books by Pettifor and Sutton.

There is no rigorous justification for the pairwise form of $E_{\text{band}}^{\text{rep}}$ and it is usually taken as a working assumption. However, a partial justification is possible along the following lines. In all the models discussed in this book that are based on the first- or second-order HKS functionals, with the exception of the pairwise potentials in simple metals, Section 6, our starting density ρ^{in} will be a superposition of spherical, atom- or ion-like charge densities. This helps justify the pairwise representation, since two of the terms in (7.28) are *exactly* pairwise in form, namely E_{H}^{in} and E_{ZZ}. E_{H}^{in} contains the constant self-energy of atomic charges plus the repulsive electrostatic interaction between spherical atomic electron densities. This must be at least a little less than the repulsion between ions, E_{ZZ}, because the electron densities overlap to some extent. Hence the difference $E_{ZZ} - E_{\text{H}}^{\text{in}}$ is a net repulsive potential, which can be represented exactly by a V_{IJ}. Notice that the terms E_{H}^{in} and $\rho^{\text{in}} V_{\text{xc}}^{\text{in}}$ enter here with the *opposite* sign to their physical contribution to the total energy. As discussed in Section 2.4.3, this is because they have already been counted twice in the band energy, which sums the energy of each electron disregarding the fact that the electon-electron interactions are thereby counted twice. Unfortunately I have found no very good arguments why the first two terms in $E_{\text{band}}^{\text{rep}}$ might be well described by pairwise potentials, except in special cases. One such case is when the electron density is rather uniform. Then we could decompose the density as

$$\rho^{\text{in}}(\mathbf{r}) = \bar{\rho}^{\text{in}} + \sum_I \Delta\bar{\rho}_I^{\text{in}}(\mathbf{r}), \tag{7.30}$$

where $\bar{\rho}^{\text{in}}$ is the average density obtained by superimposing the free atom charge densities and the second term sums the contributions from each atom. We can now make a Taylor expansion of any functional of ρ, stopping at second order. The first-order terms vanish by the definition of the average charge, and the second-order terms are pairwise, e.g.,

$$E_{\mathrm{xc}}[\rho^{\mathrm{in}}] = E_{\mathrm{xc}}[\bar{\rho}^{\,\mathrm{in}}] + \frac{1}{2}\int \frac{\partial^2 E_{\mathrm{xc}}}{\partial\rho\partial\rho'} \Delta\bar{\rho}_I^{\mathrm{in}}(\mathbf{r})\Delta\bar{\rho}_J^{\mathrm{in}}(\mathbf{r}')\,\mathbf{dr}\,\mathbf{dr}'. \tag{7.31}$$

Similar expressions would apply to the exchange and correlation potential. These terms are exactly pairwise, but the expression goes wrong when the density is very different from $\bar{\rho}^{\,\mathrm{in}}$, for instance outside a surface. Any approximation based on $\rho^{\mathrm{in}} \approx \bar{\rho}^{\,\mathrm{in}}$ is unsatisfactory at free surfaces, or perhaps even at vacancies. An analogous expansion can be made if we can write the total charge density as a uniform part and a part which is rather localised on atoms (such as the core charges), as described by Moriarty (1988).

Another approach is discussed by Sutton *et al.* (1988), which uses a many-body approximation to the correlation energy (Lindholm and Lundqvist, 1985), and a different approximation to the exchange energy. This has the advantage of approximating the difference between $E_{\mathrm{xc}}^{\mathrm{in}}$ and the exchange and correlation energy of free atoms, which means that it will not systematically break down when the atoms are at surfaces or otherwise have a lower coordination, but its disadvantage is that it represents an uncontrolled approximation to E_{xc}.

7.5 The Tight-Binding Bond Model

7.5.1 Basic Ideas

The alternative definition of the repulsive energy is used in the *tight-binding bond model* (TBBM) which I now describe. The repulsive energy is here also represented as a sum of pairwise potentials, but they are different in principle from those in the tight-binding band model. The TBBM approximates directly the *cohesive* energy, defined in Section 5.1, eq. (5.1), rather than the total energy. It was implicit in the work of Friedel (1964, 1969), elaborated in detail from a point of view closer to the present one by Sutton *et al.* (1988), and further developed and applied by Pettifor (1987, 1995). Following the notation of Sutton *et al.* we define the total binding energy of the material E_{B} as the difference between its total energy and the total energy of the same atoms when they are single and separated in free space. It is a negative quantity, given in terms of the number of atoms and the cohesive energy (5.1) by

$$E_{\mathrm{B}} = -N_{\mathrm{a}} E^{\mathrm{coh}} \equiv E^{\mathrm{tot}}(\text{condensed matter}) - E^{\mathrm{tot}}(\text{free atoms}). \tag{7.32}$$

The energy of the free atoms is given in an obvious notation by

$$E^{\mathrm{tot}}(\text{free atoms}) = \sum_I \left\{ \mathrm{Tr}\,\hat{\rho}^I \hat{H}^I - \rho^I V_{\mathrm{eff}}^I + \rho^I V_{\mathrm{ext}}^I + E_{\mathrm{xc}}^I + \tfrac{1}{2}\rho^I V_{\mathrm{H}}^I \right\}. \tag{7.33}$$

We are going to write down and simplify an expression for E_{B}. Before doing so, two comments on the free atom quantities are relevant. First, notice how the densities ρ^I, the Hartree potentials and the ionic potentials of free atoms are additive, in the sense that:

$$\rho^{\mathrm{in}}(\mathbf{r}) = \sum_I \rho^I(\mathbf{r} - \mathbf{R}_I),$$
$$V_{\mathrm{H}}^{\mathrm{in}}(\mathbf{r}) = \sum_I V_{\mathrm{H}}^I(\mathbf{r} - \mathbf{R}_I), \tag{7.34}$$
$$V_{\mathrm{ext}}(\mathbf{r}) = \sum_I V_{\mathrm{ext}}^I(\mathbf{r} - \mathbf{R}_I).$$

However, unless we make a further approximation, the effective potentials of free atoms, V_{eff}^I, cannot be superimposed in this way. In other words, the dissection of $V_{\mathrm{eff}}^{\mathrm{in}}$ into atomic contributions $V_{\mathrm{eff}}^{(I)}$ as in (7.3) certainly does *not* imply that $V_{\mathrm{eff}}^{(I)} = V_{\mathrm{eff}}^I$. This is because the exchange-correlation potential, unlike the Hartree potential, is not a linear function of the density.

The second comment is that we have made a significant error by writing the energy of free atoms in the form (7.33) because we are neglecting the *spin polarization*. Although this is a perfectly good approximation for non-magnetic atoms, it is a poor approximation for most free atoms, in which the energy is lowered by the electrons aligning their spins and occupying separate orbitals. This is the phenomenon which leads to *Hund's rules* for the electronic structure of atoms. In the non-spin-polarized density functional theory on which the derivations in this book are based, the cohesive energy of materials is always overestimated because of it. All we can do without an explicit description of the electron spins and their interactions is to note that a constant term should always be subtracted from the energy of free atoms, or added to the binding energy, to correct for it. In the models we are dealing with this term does not appear explicitly, but it would have to be taken into account if you wanted to reproduce the experimental binding energy by fitting some parameters of the model. How to model it explicitly as a function of atomic coordination and interatomic spacing is a matter for further research.

Now the binding energy to first-order can be obtained by subtracting (7.33) from (7.9). We make use of the definition of the bond energy to write the result in the form:

$$E_{\mathrm{B}}^{(1)} = E^{\mathrm{bond}} + \sum_{I\mu\nu} \left\{ \rho^{I\mu I\nu} H_{I\nu I\mu}^{\mathrm{in}} - \rho_I^{I\mu I\mu} H_{I\mu I\mu}^I \right\} - \sum_I \rho^I (V_{\mathrm{eff}}^{\mathrm{in}} - V_{\mathrm{eff}}^I)$$
$$+ E_{\mathrm{xc}}^{\mathrm{in}} - \sum_I E_{\mathrm{xc}}^I + \sum_{I \neq J} \rho^I V_{\mathrm{ext}}^J + \frac{1}{2} \sum_{I \neq J} \rho^I V_{\mathrm{H}}^J + E_{ZZ}. \tag{7.35}$$

We have used the additivity expressions (7.34) to write the electrostatic interactions in a pairwise way, e.g.:

$$E_{\rm H}^{\rm in} = \frac{1}{2}\rho^{\rm in} V_{\rm H}^{\rm in} = \frac{1}{2}\sum_{I} \rho^{I} V_{\rm H}^{I} + \frac{1}{2}\sum_{I \neq J} \rho^{I} V_{\rm H}^{J}. \tag{7.36}$$

This makes the on-site electrostatic terms cancel between the condensed system and the free atoms. The Hamiltonian of the free atom is spherically symmetric, which makes all the off-diagonal matrix elements of H^I vanish between orbitals having different angular momentum quantum numbers L. In a tight-binding basis or in a basis of atomic orbitals this always the case.

It is time now to discuss the purpose of all this rearrangement of the terms, which so far has introduced nothing new into the expression for the energy. The idea is that we want to express all the terms besides the *bond* energy as either a constant or a pairwise repulsive energy. This is a reasonable approximation for one main reason. In reality, the on-site elements of the density matrix, which describe the total charge 'on an atom' in the system (leaving aside for a moment the O-matrix terms) are relatively insensitive to the local arrangements of the atoms. This is most obviously true in a metallic environment, in which any fluctuations in charge are strongly screened by conduction electrons, so that the true electronic distribution on atoms in a metal is rather constant. Charge transfers between atoms are inhibited by an increase in the Coulomb energy. *The charge transfers are not inhibited in this way unless we do a self-consistent calculation.* In the non-self-consistent case we are considering, the calculated ρ, which is called $\rho^{\rm out}$ in the discussion of the Harris–Foulkes functional (Section 3.4.1), has variable on-site components $\rho^{I\mu I\nu}$ which will tend to exaggerate the charge transfers. These on-site terms are a large contribution to the total energy, and a major source of errors in the band model when comparing the energies of different crystal structures for example. In the tight-binding *bond* model they are absorbed into the constant or the repulsive potential, thereby eliminating their spurious influence. This is formulated mathematically in the discussion of self-consistent tight binding which comes later.

7.5.2 Development of the Model

It makes sense to pick out another part of the total binding energy for special consideration, namely the *promotion energy*, $E^{\rm prom}$. The idea is that the $\rho^{I\mu I\nu}$ are not so close to the corresponding terms in free atoms, but closer to those of atoms that have been prepared for their role in solids by promoting certain electrons into higher energy orbitals. The archetypical example is silicon, in which the free atoms have the s^2p^2 ground state. In solid silicon, the bonding is best described in terms of prepared atoms in which say Δn_p s-electrons have been promoted to the empty p-orbital. In this case the promotion energy is just $\Delta n_p(\epsilon_p - \epsilon_s)$. The more recent development of these ideas by Fähnle and coworkers (Börnsen *et al.*, 1999; Bester and Fähnle, 2001) suggests the following general definition:

$$E^{\text{prom}} = \sum_{I\mu} \left\{ q_{I\mu} H^{I}_{I\mu I\mu} - \rho_I^{I\mu I\mu} H^{I}_{I\mu I\mu} \right\}. \tag{7.37}$$

This has the the property we would expect of the promotion energy from the discussion above. By including the S-matrix we ensure that all the 'promoted' charge is counted, including bond charge, since the Mulliken charge associated with orbital μ on atom I is given by (see eq. (1.187)):

$$q_{I\mu} = \sum_{J\nu} \rho^{I\mu J\nu} S_{J\nu I\mu}. \tag{7.38}$$

In most applications of the TBBM orthogonal orbitals have been assumed ($\mathbf{S} = \mathbf{1}$ or $\mathbf{O} = \mathbf{S} - \mathbf{1} = \mathbf{0}$). This requirement is strictly speaking incompatible with the short-ranged hopping integrals that are also imposed, and amounts to the neglect of terms in the overlap matrix $\mathbf{O}$. The approximation is also made that the on-site elements of H^{in} are identical to those of the free atoms. With these assumptions the promotion energy is just equal to the second term in eq. (7.35):

$$E^{\text{prom}} \approx \tilde{E}^{\text{prom}} = \sum_{I\mu\nu} \left\{ \rho^{I\mu I\nu} H^{\text{in}}_{I\nu I\mu} - \rho_I^{I\mu I\mu} H^{I}_{I\mu I\mu} \right\}. \tag{7.39}$$

We now write eq. (7.35) in the form:

$$E^{(1)}_{\text{B}(S=1)} = E^{\text{bond}} + \tilde{E}^{\text{prom}} + \tilde{E}^{\text{rep}}, \tag{7.40}$$

where the tilde on $\tilde{E}^{\text{prom}}$ is a reminder that this term is not the most satisfactory definition of promotion energy, (7.37), although it is a substitute for it. Further approximations are made now. First, it is assumed that $\tilde{E}^{\text{prom}}$ is constant. Recall our main reason for carving out $\tilde{E}^{\text{prom}}$ for special treatment was to avoid the poor approximation that would result if it were calculated with ρ^{out}, and even keeping it fixed is an improvement on that approximation, which was made in the tight-binding *band* model. Second, $\tilde{E}^{\text{rep}}$ is modelled as a pairwise sum, which can be partially justified as discussed previously:

$$\begin{aligned} \tilde{E}^{\text{rep}} &= -\sum_I \rho^I (V^{\text{in}}_{\text{eff}} - V^I_{\text{eff}}) \\ &\quad + E^{\text{in}}_{\text{xc}} - \sum_I E^I_{\text{xc}} + \sum_{I \neq J} \rho^I V^J_{\text{ext}} + \frac{1}{2} \sum_{I \neq J} \rho^I V^J_{\text{H}} + E_{ZZ} \\ &\approx \frac{1}{2} \sum_{IJ} V_{IJ}. \end{aligned} \tag{7.41}$$

Naturally, when the pair potential is parameterized, some of the errors in the other approximations are absorbed in the fitting.

With these approximations, the on-site Hamiltonian matrix elements $H^{\text{in}}_{I\mu I\nu}$ no longer explicitly enter the formula for the binding energy, but they are required in order to calculate the bond orders in the bond energy (7.14). In the bond energy model, the coupling of different angular momenta on a site would create on-site, off-diagonal elements of $H^{\text{in}}_{I\mu I\nu}$ which do not vanish for $\mu \neq \nu$. This effect is neglected, and only diagonal elements $H^{\text{in}}_{I\mu I\mu}$ are retained. Off-diagonal elements will be considered when we discuss self-consistent tight binding. The diagonal elements are normally fitted to calculated band structures or fitted, together with the other adjustable parameters, to properties such as the lattice parameter and elastic constants of the equilibrium crystal structure. However, the insight that the atoms should be neutral provides an additional constraint on the parameters, which can only be met if we allow $H^{\text{in}}_{I\mu I\mu}$ to be adjusted atom by atom so as to keep $\sum_\mu \rho^{I\mu I\mu}$ constant. This adjustment has to be done iteratively and repeated for each configuration of the atoms. In effect, it introduces an elementary form of self-consistency into the calculation, while neglecting the second-order terms in the energy. We shall be able to justify this in more detail when we discuss self-consistent tight binding in Section 7.7.

7.5.3 A Closer Look at the Energy in a Non-Orthogonal Basis

The above scheme has been implemented and applied to many systems within the framework of an orthogonal basis. However, an interesting modification suggests itself in the more realistic case of non-orthogonal orbitals ($\mathbf{O} \neq \mathbf{0}$) which deserves to be more widely known. In what we have defined as the approximate promotion energy, (7.39), the term $\rho^{I\mu I\mu} H^{\text{in}}_{I\mu I\mu}$ represents the charge on atom I interacting with the input potential of atom I, but it is missing the bond charge. This results in a systematic error with respect to our aim that this model should embody atomic charge neutrality. More precisely we can say that it spoils the assumption that this term is a constant irrespective of the configuration of the atoms, since clearly the bond charge will decrease and the on-site elements $\rho^{I\mu I\mu}$ will increase as we pull the atoms apart for example. The constancy of this term is also desirable if we choose not atomic charge densities to make the input charge but *ionic* charge densities, which would be the natural assumption for modelling an ionic material. We can improve matters by adding in the bond charge to the charge counted in (7.35), which we do if we replace $\rho^{I\mu I\mu}$ by $\sum_{J\nu} \rho^{I\mu J\nu} S_{J\nu I\mu}$. Now we have to subtract again the overlap term we have just added to preserve the same total energy, which we do by replacing E^{bond} by a quantity which we will call the *covalent energy* E^{cov}, defined by

$$E^{\text{cov}} = \sum_{I\mu J\nu, I\neq J} \rho^{I\mu J\nu} (H^{\text{in}}_{J\nu I\mu} - O_{J\nu I\mu} H^{\text{in}}_{I\mu I\mu}). \tag{7.42}$$

The term in parenthesis is equivalent to the hopping integral (h) introducted by Pettifor (1995, p. 53) for constructing the bond energy when the orbitals are non-orthogonal. This covalent energy was also defined by Bester and Fähnle (2001). In detail, the way to proceed is to replace $\tilde{E}^{\rm prom}$ in eq. (7.40) by $E^{\rm prom}$, but this does not quite do the job because of the difference between $H^{\rm in}_{I\mu I\nu}$ and the free atom Hamiltonians $H^{I}_{I\mu I\nu}$. Making due corrections for these terms we find after a little manipulation which can be left as an exercise for the reader:

$$E_{\rm B}^{(1)} = E^{\rm cov} + E^{\rm prom} + E^{\rm cfs} + E^{\rm M\mathit{q}} + \tilde{E}^{\rm rep}, \tag{7.43}$$

where

$$E^{\rm cfs} = \sum_{I\mu\nu} \rho^{I\mu I\nu} \left(H^{\rm in}_{I\nu I\mu} - \delta_{\mu\nu} H^{\rm in}_{I\mu I\mu} \right) \tag{7.44}$$

and

$$E^{\rm M\mathit{q}} = \sum_{I\mu} q_{I\mu} \left(H^{\rm in}_{I\mu I\mu} - H^{I}_{I\mu I\mu} \right). \tag{7.45}$$

Besides a new definition of the quantity which describes the chemical bonding, which is now $E^{\rm cov}$, and a more satisfactory definition of the promotion energy, these manipulations have made explicit two new kinds of terms. First, following the promotion energy we have a term $E^{\rm cfs}$ which depends only on the interaction of the *off-diagonal* on-site elements of the Hamiltonian with the corresponding elements of the charge. This term in the energy depends on the *crystal field* splitting of the atomic energy levels. In the free atoms there are only diagonal elements on a site, $H^{I}_{I\mu I\nu} = \delta_{\mu\nu} H^{I}_{I\mu I\mu}$ by virtue of the spherical symmetry, a restriction that does not apply in the solid. Likewise, the charge on an atom can polarize, which introduces the terms $\rho^{I\mu I\nu}$ for $\mu \neq \nu$ on a site. This is the case for example when the same wavefunction contains coefficients of both s and p orbitals on an atom, in which case its atomic charge has a dipolar character, being skewed along the axis of the p orbital. This term in the binding energy can be thought of as a polarization energy (Bester and Fähnle, 2001), since it represents the interaction between the non-spherical charge and the non-spherical potential. It is normally neglected, or swept into $E^{\rm rep}$. However, in principle a more accurate model might be constructed by including this term with a parameterized form for the non-spherical terms $H^{\rm in}_{I\mu I\nu}$.

Like $E^{\rm cfs}$ the second new term $E^{\rm M\mathit{q}}$ is accommodated by sweeping it into the 'catch-all' repulsive energy, which we now distinguish as $E^{\rm rep}$ without the tilde. This term represents the change in energy of atom-centred charges when they move from the environment of a free atom to the environment of an atom embedded in a solid. It is not unlike a Madelung energy of the electronic charges, although it is double counting the charge–charge interactions and it does not originate from the electrostatics of point charges. There must be a lot of cancellation between this term and the first term in the original $\tilde{E}^{\text{rep}}$, that is we expect:

$$\sum_{I\mu} q_{I\mu} \left(H^{\text{in}}_{I\mu I\mu} - H^{I}_{I\mu I\mu} \right) - \sum_{I} \rho^{I} (V^{\text{in}}_{\text{eff}} - V^{I}_{\text{eff}}) \approx 0, \tag{7.46}$$

so it is probably not too bad an approximation just to absorb this term into a redefined, pairwise E^{rep}, at least no worse than the approximations already made! Our final preferred expression for the total binding energy in the TBBM is therefore:

$$E^{(1)}_{\text{B}} = E^{\text{cov}} + E^{\text{prom}} + E^{\text{rep}}. \tag{7.47}$$

This reduces to (7.40) in the case of orthogonal orbitals.

Finally, it is worth mentioning that the rearrangement from an expression with E^{bond} describing the chemical bonding to the expression with E^{cov} has brought another subtle but quite satisfying benefit, of greatest value for interpreting the binding energy in self-consistent calculations. This is that it makes the binding energy invariant to a uniform shift ΔV_{ext} in the electrostatic potential. With the full DFT expression for the binding energy, the invariance of the total binding energy to a shift ΔV_{ext} is guaranteed because the system is neutral: a rise in energy of the electrons is compensated by a lowering in energy of the nuclei. However in a tight binding model this cancellation is not built in because there is now no term that can represent the interaction of the nuclei with an external potential. This possibility was sacrificed when we introduced a pairwise description of E^{rep}. If we add the shift ΔV_{eff} to our input potential within H^{in}, it changes E^{bond} (eq. 7.14) by:

$$\Delta E^{\text{bond}} = \sum_{I\mu J\nu, I \neq J} \rho^{I\mu J\nu} \langle J\nu | \Delta V_{\text{eff}} | I\mu \rangle = \Delta V_{\text{eff}} \sum_{I\mu J\nu} \rho^{I\mu J\nu} O_{J\nu I\mu}. \tag{7.48}$$

It would be disturbingly unphysical to describe the strength of chemical bonding by a term which appears to depend on an arbitrary constant shift in the background electrostatic potential! There is no such shift in E^{cov}, because the shifts in its two parts exactly cancel. In electronic structure calculations on infinite systems the zero of electrostatic energy is arbitrary, and it is desirable that this arbitrariness does not enter the description of the strength of a bond. The use of E^{cov} for describing the strength of the bonding avoids this problem. This was in fact the original motivation of Börnsen *et al.* (1999) for introducing E^{cov}.

The shift need not be in the external potential for this difficulty to arise, it can be in the effective potential. At the surface of a material there is always a dipole layer which shifts the mean electrostatic potential within the material relative to the distant vacuum level by an amount which depends on the details of the surface structure. It is a major contribution to the work function of the surface. From the point of view of the atoms deep within the solid, this shift is equivalent to adding an external potential. On the other hand, in fairness to the users of E^{bond}, there is no such arbitraryness in the zero of energy as long as one is only working with H^{in}, and not trying to construct a bond energy from the self-consistent H. The external and Hartree potential of the input Hamiltonian must be constructed by superimposing the electrostatic potentials of isolated neutral free atoms, disregarding any real shifts caused by the rearrangement of electrons at surfaces.

Exercise
Starting from the density functional formula (7.35) which expresses the first-order binding energy in terms of the bond energy, derive the formula (7.43) for the first-order binding energy in terms of the covalent energy.

7.6 Hellmann–Feynman Forces

7.6.1 The Pitfall of Incomplete Bases

We have made various approximations for the energy and we expect to be making similar approximations to the interatomic forces. This is true with one important caveat. Whatever we do we must ensure that the expression we use for the force on an ion reproduces the derivative of the expression we use for the total energy with respect to that ion's position. Otherwise our model would be the basis of a perpetual motion machine. That is to say, a simulation would not necessarily conserve energy, since the path integral of the forces would not equal the change in energy along the path. Consistency between the total energy and the forces in this sense is a basic requirement of any model.

At first you might think this condition should be easy enough to satisfy with any reasonable model. But a simple example illustrates the kind of difficulty that can arise. Think of a single hydrogen atom in free space, with one electron in an s-orbital. For the purpose of the argument let us neglect the motion of the proton and think of it as a heavy classical particle. Now apply a uniform electric field $\mathbf{E}$. What is the force on the proton? The answer of course is zero. Otherwise the neutral hydrogen atom would start to accelerate! From the point of view of the Hellmann–Feynman theorem the direct force due to the external field is exactly compensated by the force due to the electron, which arises because the electron distribution has been shifted off-centre by the field. In other words the polarization of the electron cloud is essential in making the resultant force on the nucleus vanish. Now if we imagine calculating the energy and charge distribution of the hydrogen atom, the standard procedure would be to expand the wave function $|\psi\rangle$ in an orbital basis set, with variable coefficients, and to minimize the energy functional with respect to the expansion coefficients. In this case the total energy has a very simple form. There is no exchange and correlation since we are dealing with a single electron, so the only contributions are from the kinetic energy $\langle\psi| - \frac{1}{2}\nabla^2|\psi\rangle$ and from the electrostatic interactions between the electron and the nucleus $\langle\psi| - Z/r|\psi\rangle$ and between the electron and the external field $\langle\psi|\mathbf{E}\cdot\mathbf{r}|\psi\rangle$. Looking back to the derivation of the Hellmann–Feynman theorem, we see it is only satisfied if the charge density has complete variational freedom. That is, the charge $\delta\rho(\mathbf{r})$ induced when the perturbation is applied (in this case by moving the proton) must be within the space of charge variations within which the density functional was minimized. This would clearly not be the case for example if we solved for the ground state of the hydrogen atom by varying a spherically symmetric wave function. That procedure gives the exact answer in the absence of an electric field but it does not 'see' an electric field at all. We need a basis which includes higher angular momentum components p- d-, etc. in order to do the job, because they have to represent the polarized charge density that is not spherically symmetric about the proton.

In general, we see that the Hellmann–Feynman theorem can only be applied if the electron density has been variationally minimised with respect to variations with a basis having a certain degree of completeness. While the error in the energy is of second-order in the error in the wave functions, the error in the Hellmann–Feynman force is of first-order. This is a well-known problem in the compution of forces using first-principles methods. It can be solved by augmenting the standard electrostatic force due to $\rho(\mathbf{r})$ by a term involving the first derivatives of the basis functions; a term commonly referred to as the *Pulay force*, after the author who discussed their significance (Pulay, 1969).

In empirical tight binding we have abandoned the explicit electron density altogether, since we have abandoned any reference to the explicit $\mathbf{r}$-dependence of the basis orbitals. We do not have to introduce the complication of Pulay forces into empirical tight-binding models, and something like the Hellmann–Feynman theorem still applies. However, as we shall see an extra term does appear in the case of non-orthogonal orbitals, which is a bit subtle, involving the derivatives of the overlap matrix $\mathbf{S}$. The starting point for deriving a formula for the force will be a model energy functional of the type introduced above, which is a function of the atomic positions and the wave function coefficients, $E^{\text{tot}}(\{\mathbf{R}_I\}, \{C^n_{I\mu}\})$. To obtain an expression which satisfies our requirement for consistency we have to start with our chosen energy functional and differentiate it without further approximations. The following derivation is also valid for self-consistent tight-binding models, which are based on the second-order functional.

7.6.2 The Force on an Ion

If we make a small change in the position of an ion I the energy change is given by

$$\delta E^{\text{tot}} = \left.\frac{\partial E^{\text{tot}}}{\partial \mathbf{R}_I}\right|_{\{C^n_{I\mu}\}} \cdot \delta\mathbf{R}_I + \sum_{nJ\nu} \left.\frac{\partial E^{\text{tot}}}{\partial C^n_{J\nu}}\right|_{\{\mathbf{R}_I\}} \delta C^n_{J\nu}. \tag{7.49}$$

The notation $\partial E^{\text{tot}}/\partial \mathbf{R}_I$ stands for the vector gradient of E^{tot} with respect to the position of atom I:

$$\partial E^{\text{tot}}/\partial \mathbf{R}_I \equiv (\partial E^{\text{tot}}/\partial R_{Ix}, \partial E^{\text{tot}}/\partial R_{Iy}, \partial E^{\text{tot}}/\partial R_{Iz}).$$

The second term in eq. (7.49) stands for the change in the total energy when we change the wavefunctions at fixed atomic positions. At first sight we might think that by the variational principle this term should vanish. That would be wrong, because $\{\delta C^n_{j\nu}\}$ which are induced by the movement of atom I are not arbitrary changes in the space of valid wave function coefficients at fixed $\{\mathbf{R}_I\}$; they are such as to keep the energy functional minimized under the constraint that the eigenfunctions remain orthogonal as the positions change. That means the coefficients $\{C^n_{I\mu}\}$ must satisfy simultaneously a matrix eigenvalue equation, (1.235), and the orthogonality condition which involves the S-matrix (1.123).

A clearer understanding of the situation is reached by going back to first-principles. Consider variations of the functional E^{tot} in the space of $\{C^n_{I\mu}\}$ at constant $\{\mathbf{R}_I\}$. The orthogonality condition (1.123) defines a surface in this space, which would be the unit sphere if the orbitals were orthogonal ($O = 0$). The normal to this surface at every point on it is a vector in the space; let us call it $\boldsymbol{v}^n(\{C^n_{I\mu}\})$. Its components $v^n_{I\mu}$ are the derivatives of the quantity we can denote by $\mathbf{C}^{n*}\mathbf{S}\mathbf{C}^n$, where

$$\mathbf{C}^{n*}\mathbf{S}\mathbf{C}^n \equiv \sum_{I\mu J\nu} C^{n*}_{I\mu} S_{I\mu J\nu} C^n_{J\nu} \tag{7.50}$$

and

$$v^n_{I\mu} = \frac{\partial}{\partial C^n_{I\mu}} \mathbf{C}^{n*}\mathbf{S}\mathbf{C}^n. \tag{7.51}$$

The only variations $\delta C^n_{I\mu}$ that are allowed must preserve the orthonormality condition, that is $\delta \mathbf{C}^n$ must be perpendicular to $\boldsymbol{v}^n$. Hence the variational principle states that for every $\delta \mathbf{C}^n$ satisfying:

$$\boldsymbol{v}^n \cdot \delta \mathbf{C}^n = 0 \tag{7.52}$$

it must also be true that

$$\frac{\partial E^{\text{tot}}}{\partial \mathbf{C}^n} \cdot \delta \mathbf{C}^n = 0. \tag{7.53}$$

If it were not so, we could lower the energy by making the appropriate small change $\delta \mathbf{C}^n$. In order to be true for all $\delta \mathbf{C}^n$, eqs. (7.52) and (7.53) require that the vector $\partial E^{\text{tot}}/\partial \mathbf{C}$ with components $\partial E^{\text{tot}}/\partial C^n_{J\nu}$ is *parallel* to $\boldsymbol{v}^n$, which can be expressed as

$$\frac{\partial E^{\text{tot}}}{\partial C^n_{I\mu}} = f_n \epsilon_n \frac{\partial}{\partial C^n_{I\mu}} \mathbf{C}^{n*}\mathbf{S}\mathbf{C}^n, \tag{7.54}$$

where I have anticipated that the constant of proportionality turns out to be the product of the occupancy and the eigenvalue. I have made use here of a theorem of linear algebra which says that if a vector is orthogonal to all the vectors in a particular subspace then, if it is not zero, it must lie in the complementary subspace. If you think about this in three dimensions it is intuitively obvious. Taking the derivative in (7.54) gives us directly the matrix eigenvalue equation (1.235), where the Hamiltonian is specified by

$$\frac{\partial E^{\mathrm{tot}}}{\partial C^{n*}_{I\mu}} = f_n \sum_{J\nu} H_{I\mu J\nu} C^n_{J\nu}. \tag{7.55}$$

Equation (7.55) introduces an important general formula, which *defines* the Hamiltonian matrix given any description of the energy as a function of the wavefunction coefficients. Thus within the second-order energy functional the Hamiltonian is itself a function of the wavefunction coefficients.

A couple of subtleties may puzzle the astute reader. One is the way we seem to differentiate with respect to $C^n_{I\mu}$ or its complex conjugate $C^{n*}_{I\mu}$ as if they were independent variables, which they clearly are not. It is easy to prove that this is legitimate if we change variables to the real and imaginary parts of $C^n_{I\mu}$. The other subtlety is that we seem to have ignored differentials of f_n. The short answer to this is that we are working at $T = 0\,\mathrm{K}$, where f_n is a constant, either 2 or 0. A more satisfactory answer is that the derivative of the *free* energy with respect to the f_n vanishes at finite temperature, but this only works if we include an entropy term in the functional. Indeed the vanishing of this derivative is just the condition which generates the Fermi distribution. The proof is left as an exercise, and although it is rather straightforward we are not going to carry through the extra complication of temperature dependence.

Now we are going to use (7.54) to evaluate the $\mathbf{C}^n$-derivative in (7.49). We have

$$\sum_{nJ\nu} \left.\frac{\partial E^{\mathrm{tot}}}{\partial C^n_{J\nu}}\right|_{\{\mathbf{R}_I\}} \delta C^n_{J\nu} = \sum_n f_n \epsilon_n \left\{ \delta \mathbf{C}^{n*} \mathbf{S} \mathbf{C}^n + \mathbf{C}^{n*} \mathbf{S} \delta \mathbf{C}^n \right\}. \tag{7.56}$$

Making use of the orthogonality condition we can eliminate the $\delta\mathbf{C}^n$ by writing:

$$\delta \mathbf{C}^{n*} \mathbf{S} \mathbf{C}^n + \mathbf{C}^{n*} \mathbf{S} \delta \mathbf{C}^n = \delta\{\mathbf{C}^{n*} \mathbf{S} \mathbf{C}^n\} - \mathbf{C}^{n*} \delta \mathbf{S} \mathbf{C}^n = -\mathbf{C}^{n*} \delta \mathbf{S} \mathbf{C}^n. \tag{7.57}$$

With this substitution, the change in total energy (7.49) becomes

$$\delta E^{\mathrm{tot}} = \left.\frac{\partial E^{\mathrm{tot}}}{\partial \mathbf{R}_I}\right|_{\{C^n_{I\mu}\}} \cdot \delta \mathbf{R}_I - \sum_{nJ\nu J'\nu'} f_n \epsilon_n C^{n*}_{J\nu} \frac{\partial S_{J\nu J'\nu'}}{\partial \mathbf{R}_I} C^n_{J'\nu'} \cdot \delta \mathbf{R}_I. \tag{7.58}$$

Notice that only the terms with $J = I$ or $J' = I$ are non-zero. This can also be expressed in terms of the expansion coefficients of the Green function (1.209), as in Sutton *et al.* (1988, eq. (2.38)):

$$\delta E^{\text{tot}} = \left.\frac{\partial E^{\text{tot}}}{\partial \mathbf{R}_I}\right|_{\{C^n_{I\mu}\}} \cdot \delta \mathbf{R}_I + \sum_{J\nu\nu'} \frac{4}{\pi} \int_{-\infty}^{\infty} \mathrm{d}\epsilon f_{\mathrm{F}}(\epsilon)\epsilon \,\mathrm{Im}\, G^{J\nu' I\nu}(\epsilon) \frac{\partial S_{I\nu J\nu'}}{\partial \mathbf{R}_I} \cdot \delta \mathbf{R}_I. \tag{7.59}$$

Hence the force on atom I is:

$$\mathbf{F}_I = -\left.\frac{\partial E^{\text{tot}}}{\partial \mathbf{R}_I}\right|_{\{C^n_{I\mu}\}} - \sum_{J\nu\nu'} \frac{4}{\pi} \int_{-\infty}^{\infty} \mathrm{d}\epsilon f_{\mathrm{F}}(\epsilon)\epsilon \,\mathrm{Im}\, G^{J\nu' I\nu}(\epsilon) \frac{\partial S_{I\nu J\nu'}}{\partial \mathbf{R}_I}. \tag{7.60}$$

This is the central result of this section.

With this expression for the forces, energy conservation is guaranteed when we integrate them along any trajectory. As a consequence, in order to calculate the force on atom I within the first-order functionals, having obtained the occupied wavefunctions, we need to know the derivatives:

$$\frac{\partial H^{\text{in}}_{I\mu J\nu}(\mathbf{R}_{IJ})}{\partial \mathbf{R}_I}, \quad \frac{\partial E^{\text{rep}}}{\partial \mathbf{R}_I} \quad \text{and} \quad \frac{\partial S_{I\nu J\nu'}}{\partial \mathbf{R}_I}.$$

If $\mathbf{O} = \mathbf{0}$ ($\mathbf{S} = \mathbf{1}$) we are left with an expression for the force which is very similar in form to that of the Hellmann–Feynman theorem derived previously, namely we have to differentiate the total energy with respect to an ion position at constant wave function coefficients. For example, the force in the orthogonal TBBM is given by:

$$\mathbf{F}_I = -2 \sum_{\mu J\nu, J\neq I} \rho^{I\mu J\nu} \frac{\partial H^{\text{in}}_{J\nu I\mu}}{\partial \mathbf{R}_I} - \sum_{J\neq I} \frac{\partial V_{IJ}}{\partial \mathbf{R}_I}. \tag{7.61}$$

Non-orthogonality adds the term in the gradients of $\mathbf{S}$ exactly as in eq. (7.60).

In practice, there is some lengthy but straighforward algebra to do in order to obtain these derivatives from the expressions for $H^{\text{in}}_{J\nu I\mu}$ in terms of the bond length and direction cosines. Like many standard formulae and functions, this is now embodied in well-tested computer codes, see for example Horsfield *et al.* (1996*a*,*b*) and no one needs to repeat the whole exercise.

7.7 Self-consistent Tight-Binding

7.7.1 The Self-consistent Charge Transfer Model

The idea of all self-consistent tight-binding models is to include the second-order term in the HKS functional, as in eq. (3.32). In the $\mathbf{r}$-representation we recall this was given by

$$E_2 = \frac{1}{2}\int C_{\rm in}(\mathbf{r},\mathbf{r}')\delta\rho(\mathbf{r})\delta\rho(\mathbf{r}')\,d\mathbf{r}\,d\mathbf{r}'. \tag{7.62}$$

Our first task will be to describe E_2 in the basis of local orbitals. By differentiating it with respect to the wavefunction coefficients we will then obtain a matrix Schrödinger equation analogous to (3.33). In whatever basis it is expressed this matrix equation includes an effective potential which depends on the wavefunctions and it must therefore be solved by an iterative process. The price to be paid for the extra accuracy of second-order is therefore considerably longer computation time. Another disadvantage of including second-order terms is that there are more parameters to be determined. It is easy to say that in situations when charge transfer is important the second-order functional will be necessary, but there are no hard and fast rules, since a degree of charge transfer may be built into the input charge density, as in the ionic model. However, when a species needs to be described in very different environments within the same system, we expect to find the first-order functional inadequate. Approaches similar to the one I present here were worked out independently by several groups at about the same time (Finnis *et al.*, 1998*a*,*b*; Elstner *et al.*, 1998; Esfarjani and Kawazoe, 1998; Schelling *et al.*, 1998), where examples are given.

Notice the strategy we shall follow is to first construct a model for the energy functional including the second-order term and then differentiate it in order to generate a model for the effective potential $V_{\rm eff}$, eq. (3.34), and to get an expression for the forces. If we attempted to construct a model of $V_{\rm eff}$ or the forces directly we would risk the inconsistency of incomplete bases discussed above.

To understand the principles, we start with the simplest description of the second-order term. We assume that the induced charges $\delta\rho$ are given by the changes in the Mulliken charges $q_I - q_I^{\rm in}$ given by eq. (1.187), concentrated at points at the atomic sites. The second-order term now takes the form:

$$E_2 = \frac{1}{2}\sum_I U_I \delta q_I^2 + \frac{1}{2}\sum_{IJ, I\neq J} U_{IJ}\delta q_I \delta q_J \,. \tag{7.63}$$

where

$$\delta q_I = (q_I - q_I^{\rm in}). \tag{7.64}$$

We refer to this as the *self-consistent charge transfer model* (SCTM). The on-site energy changes are described by a parameter U_I, called the *Hubbard U*, which we suppose is constant for a given species. The intersite terms are assumed to be just Coulomb interactions:

$$U_{IJ} = 1/R_{IJ}, \tag{7.65}$$

which at short range may be damped by the effect of overlapping the atomic charge distributions. This is consistent with a local density approximation for the exchange and correlation energy, in which $C_{\rm in}(\mathbf{r},\mathbf{r}') = 1/|\mathbf{r}-\mathbf{r}'|$ when $\mathbf{r} \neq \mathbf{r}'$.

Now the new terms in the effective potential are obtained by differentiating E_2, following eq. (7.55). Thus the action of the total effective potential on a wavefunction, analogous to eq. (3.34), is specified by

$$f_n \sum_{J\nu} V^{(2)}_{\text{eff}\, I\mu J\nu} C^n_{J\nu} = f_n \sum_{J\nu} V^{\text{in}}_{\text{eff}\, I\mu J\nu} C^n_{J\nu} + \frac{\partial E_2}{\partial C^{n*}_{I\mu}}. \tag{7.66}$$

This equation is completely general, since it does not depend on the particular form of the second-order term. We now have to evaluate

$$\frac{\partial E_2}{\partial C^{n*}_{I\mu}} = \sum_{I'} \frac{\partial E_2}{\partial q_{I'}} \frac{\partial q_{I'}}{\partial C^{n*}_{I\mu}}. \tag{7.67}$$

This is a single term in orthogonal tight binding, because the charge on a site depends only on the wavefunction coefficients on the same site, but in general the q_I are given by eq. (1.187), from which there are additional overlap terms. Differentiating (1.187):

$$\frac{\partial q_I}{\partial C^{n*}_{I\mu}} = f_n C^n_{I\mu} + \frac{1}{2} f_n \sum_{J\nu} C^n_{J\nu} O_{I\mu J\nu} \tag{7.68}$$

and

$$\frac{\partial q_J}{\partial C^{n*}_{I\mu}} = \tfrac{1}{2} f_n C^n_{J\nu} O_{I\mu J\nu} \quad \{J \neq I\}. \tag{7.69}$$

Combining the preceding equations we find the Hamiltonian:

$$\begin{aligned} H_{I\mu J\nu} = H^{\text{in}}_{I\mu J\nu} &+ \Big(U_I \delta q_I + \sum_{I' \neq I} U_{II'} \delta q_{I'}\Big) \delta_{IJ} \delta_{\mu\nu} \\ &+ \frac{1}{2}\Big(U_I \delta q_I + U_J \delta q_J + \sum_{I' \neq I} U_{II'} \delta q_{I'} + \sum_{I' \neq J} U_{JI'} \delta q_{I'}\Big) O_{I\mu J\nu}. \end{aligned} \tag{7.70}$$

In the literature you will find the orthogonal form of this expression, in which only the diagonal elements of the input Hamiltonian are shifted by the charge transfers. In any case the only new parameter introduced in this second-order model is the Hubbard U.

It is straightforward to get expressions for the forces, especially in the orthogonal form. We apply the general expression (7.60), noting the derivative at constant wave function coefficients implies differentiating at constant q_I in the orthogonal case. Thus we can write down the result directly:

$$\mathbf{F}_I = -\sum_{\mu\nu} \rho^{I\mu I\nu} \frac{\partial H^{\text{in}}_{I\nu I\mu}}{\partial \mathbf{R}_I} - 2 \sum_{\mu J\nu, J\neq I} \rho^{I\mu J\nu} \frac{\partial H^{\text{in}}_{J\nu I\mu}}{\partial \mathbf{R}_I} - \sum_{J\neq I} \frac{\partial V_{IJ}}{\partial \mathbf{R}_I} - \sum_J \delta q_I \delta q_J \frac{\mathbf{R}_{IJ}}{R^3_{IJ}}. \tag{7.71}$$

Compared to the TBBM there are two extra terms. First, we are now treating all the charge explicitly, so the diagonal elements are included. In practice up to now crystal field splitting of H^{in} on a site has been ignored, and $\mu \neq \nu$ terms are therefore set to zero. It is also common to assume the on-site terms of H^{in} are constant and omit the first term altogether. The final term is simply the electrostatic repulsion between the transferred charges $\{\delta q_I\}$. This comes from the first term of (7.60), the derivative at constant wavefunction coefficients. With a non-orthogonal basis additional terms in $\partial \mathbf{O}/\partial \mathbf{R}_I$ would have to be included.

Exercise
Derive the extra terms (involving $\partial \mathbf{O}/\partial \mathbf{R}_I$) that have to be included in (7.71) if the basis is not orthogonal.

7.7.2 Local Charge Neutrality

In the light of the above self-consistent model we can revisit the tight-binding bond model (TBBM) introduced in Section 7.5 and give it a bit more theoretical support. Recall that to implement the TBBM you have to do an iterative calculation in order to determine self-consistent on-site matrix elements $H_{I\mu I\mu}$ such as to make all the δq_I vanish. The values of these on-site matrix elements do not appear in the TBBM expression for the total energy.

Imagine 'switching on' the Hubbard Us starting with all $U_I = 0$. For this purpose the U_{IJ} are irrelevant; they can have arbitrary values, including zero. As the U_I get larger, the δq_I get smaller. Either by linear response theory, or by simply differentiating the functional with respect to δq_I, including the constraint of charge conservation, and setting the result to zero, you should be able to convince yourself that for sufficiently large U_I, δq_I must be inversely proportional to U_I. The product $U_I \delta q_I$ therefore tends to a constant, which is precisely the correction to the diagonal elements of the Hamiltonian that you find in the TBBM by iterative calculation. In this limit the second-order term tends to zero in proportion to U_I^{-1}. We can therefore think of the TBBM as the limit of the above SCTM when the U_I tend to infinity.

7.7.3 Including Atomic Polarization

The self-consistent charge transfer model assumes by the approximation (7.63) for E_2 that the atoms are point monopoles of charge. To extend the tight-binding model to encompass ionic materials, it is desirable to include a description of dipolar and quadrupolar distortions of the atoms induced by their mutual electrostatic interactions. A practical motivation for doing so is the indication that such distortions are important in establishing the sequence of energies of various phases of alumina (Wilson *et al.*, 1996*a*) and zirconia (Wilson *et al.*, 1996*b*) by the shell model and its extensions, which I desribe more fully in Chapter 9.

To proceed, we now write down an exact expression for the second-order energy (7.62) in the local basis, which we will later approximate. First, inserting the expressions (1.179) for the charge densities we have

$$E_2 = \frac{1}{2} \sum_{I\mu I'\mu'} \sum_{J\nu J'\nu'} \int C(\mathbf{r}, \mathbf{r}') \Big(\langle \mathbf{r}|I\mu\rangle \rho^{I\mu I'\mu'} \langle I'\mu'|\mathbf{r}\rangle - \langle \mathbf{r}|I\mu\rangle \rho_I^{I\mu I\mu} \langle I\mu|\mathbf{r}\rangle \Big)$$

$$\cdot \Big(\langle \mathbf{r}'|J\nu\rangle \rho^{J\nu J'\nu'} \langle J'\nu'|\mathbf{r}'\rangle - \langle \mathbf{r}'|J\nu\rangle \rho_J^{J\nu J\nu} \langle J\nu|\mathbf{r}'\rangle \Big) \, \mathrm{d}\mathbf{r}\, \mathrm{d}\mathbf{r}'. \tag{7.72}$$

The integrals define all the matrix elements of the kernel C, in terms of which we can now write E_2 as

$$E_2 = \frac{1}{2} \sum_{I\mu I'\mu'} \sum_{J\nu J'\nu'} C_{I\mu I'\mu' J\nu J'\nu'} \left(\rho^{I\mu I'\mu'} - \rho_I^{I\mu I\mu} \right) \left(\rho^{J\nu J'\nu'} - \rho_J^{J\nu J\nu} \right). \tag{7.73}$$

The C-matrix is real and symmetric with respect to interchange of $I\mu$ with $I'\mu'$ or $J\nu$ with $J'\nu'$. It also has the symmetries that derive from the symmetry $C(\mathbf{r}, \mathbf{r}') = C(\mathbf{r}', \mathbf{r})$, namely $C_{I\mu I'\mu' J\nu J'\nu'} = C_{J\nu J'\nu' I\mu I'\mu'}$. Using these symmetries, we can derive the contribution of the induced charge density $\delta\rho$ to the effective potential as we did for the SCTM, by differentiating with respect to $C_{J\mu}^{n*}$, with the result:

$$H_{I\mu J\nu} = H_{I\mu J\nu}^{\mathrm{in}} + \sum_{I'\mu' J'\nu'} C_{I\mu J\nu I'\mu' J'\nu'} \left(\rho^{I'\mu' J'\nu'} - \rho_{I'}^{I'\mu' I'\mu'} \right). \tag{7.74}$$

This general approach has taken us beyond the SCTM in two significant ways, as you can see by comparing the full version of the Hamiltonian (7.74) with its simplified, SCTM version (7.70). For our present purpose the most significant new feature is the addition of on-site terms with $\mu \neq \nu$, which couples orbitals of different symmetry on the same atom; the coupling is induced by the fields from charges on other atoms. Coupling between different orbitals on an atom is the physical mechanism for distorting its spherical shape. Thus a matrix element between s and p orbitals on a site allows a dipole moment to develop. Or, for example, a transfer of electrons from p_z to p_x and p_y orbitals would create a quadrupole. If an external field did not create such matrix elements, a free atom or ion would not react to it. It would also be possible in principle to construct on-site elements of $H^{\text{in}}_{\mu\neq\nu}$ as a short range effect of the breaking of spherical symmetry on the atoms when the charge densities are overlapped, but this has not been done in any empirical tight-binding models of which I am aware.

It is worth pointing out that the purely covalent effects of the input Hamiltonian H^{in} alone, without any splitting of the diagonal elements, will create 'notional' dipoles and higher multipoles by mixing the different angular momentum components on a site, given that the symmetry of that site is sufficiently low. Such multipoles are rather like ghosts, because they do not generate any electrostatic interaction energy unless we include terms from the second-order functional, as in (7.74).

The second new feature of the general Hamiltonian (7.74) is that inter-site matrix elements ($I \neq J$) are induced by the induced charges. The effect of these has not been investigated. In the applications to date, only the modification of on-site matrix elements by Coulomb interactions has been considered, as I will now describe. There is plenty of scope for future investigations to include other terms. For example the form of (7.74) is sufficiently general to include the exchange integrals of HF theory and thereby exclude self-interactions. There is also no reason in principle why the local density approximation has to be made, since non-local correlation effects could also be built into the parameterization. However, all I want to do at this stage is to pick out the C-matrix elements that induce the formation of dipoles and quadrupoles and that quantify their electrostatic interactions as point multipoles. This will involve a minimum number of new parameters to be fitted.

As in the SCTM we will assume all exchange and correlation effects are local terms, to be included in on-site Hubbard Us. The simplest strategy is to retain the form of the SCTM for the monopoles on the atoms, but generalise it to include the new multipole terms which (7.74) introduces. For this purpose we make use of the results in Section 1.5.5. Thus from (1.139) we write

$$E_2 = \frac{1}{2}\sum_I U_I q_I^2 + \frac{1}{2}\sum_{ILJL', I\neq J} Q_L(\mathbf{R}_J) B_{LL'}(\mathbf{R}_{IJ}) Q_{L'}(\mathbf{R}_I). \tag{7.75}$$

We have already used the monopoles $Q_0(\mathbf{R}_I) \equiv \delta q_I$, and $B_{00}(\mathbf{R}_{IJ}) = 1/R_{IJ}$. Now with a little more effort we can include the dipoles and quadrupoles, $Q_{1\kappa}(\mathbf{R}_I)$ and $Q_{2\kappa}(\mathbf{R}_I)$. Remember the B coefficients just describe the electrostatic interactions between the multipoles. To restrict the number of parameters we omit the overlap contributions to the multipoles for $L > 0$ and write (1.128) in the form:

$$Q_L(\mathbf{R}_I) = \sum_{\mu\nu} \rho^{I\mu I\nu} \int \langle I\nu|\mathbf{r}\rangle R_L(\mathbf{r}) \langle \mathbf{r}|I\mu\rangle \, d\mathbf{r} \quad \{L > 0\}. \tag{7.76}$$

Since the input atoms are spherically symmetric, there are no input contributions to be subtracted for calculating the induced multipoles except for $L = 0$.

The integral in (1.128) or (7.76) is a constant depending only on the type of atom and the three indices μ, ν and L, let us give it a name by defining

$$D^I_{\nu L\mu} = \int \langle I\nu|\mathbf{r}\rangle R_L(\mathbf{r})\langle\mathbf{r}|I\mu\rangle\,\mathrm{d}\mathbf{r}. \tag{7.77}$$

Because the orbitals are real the Ds satisfy

$$D^I_{\mu L\nu} = D^I_{\nu L\mu}. \tag{7.78}$$

In addition there are useful symmetries, usually called *selection rules*, which arise from the angle integrals in (7.77), because they are integrals of the products of three spherical harmonics. These are the Gaunt integrals we met in Section 1.5.5, eq. (1.134). Explicitly, using the form (1.106) for the local orbitals, we see that the D-parameters are proportional to the Gaunt coefficients:

$$D_{\mu L\nu} = \Delta_{l'll''} C_{L'LL''}, \tag{7.79}$$

where

$$\Delta_{l'll''} = \int r^2 f_\mu(r) r^l f_\nu(r)\,\mathrm{d}r \tag{7.80}$$

and L', L and L'' are the angular momentum quantum numbers standing for the pairs $l'\kappa'$, $l\kappa$ and $l''\kappa''$. L' and L'' correspond to μ and ν respectively. I have dropped the atom index I for simplicity, on the understanding that we define the origin to be at atom I, and that the particular form of the orbitals described by μ and ν will depend on the species of atom I.

One selection rule is that they are only non-vanishing if the three angular momenta l', l, and l'' satisfy the triangle inequality. Furthermore the sum of the ls must be even. Such rules are best proved by the theory of group representations, with reference to the properties of symmetrized product groups, although when only s and p orbitals are involved they can be seen easily 'by inspection'. Assuming just one principal quantum number per orbital angular momentum in the basis, the symmetries of the Gaunt coefficients drastically limits the number of extra parameters represented by the Ds. Without these symmetries there would be hundreds of them! For a given species of atom with s, p and d orbitals in the basis only the following combinations of $l'll''$ give independent, non-vanishing Gaunt coefficients: sss, psp, dsd, spp, ppd, sdd, pdp, ddd, pfd, and dgd. The sss, psp and dsd D-parameters are both unity, because (see 1.114) $R_{00} = 1$. So there will be at most an additional two parameters per element needed to describe multipoles with an sp basis: one (spp) to describe the magnitude of dipoles, and another (pdp) to describe the magnitude of quadrupoles. If d-orbitals are included in the basis we have the possibility of generating octapoles (f-symmetry) and hexadecapoles

(g-symmetry) by including pfd and dgd parameters respectively, but it seems unlikely that these higher multipoles will be used in practical models. It is probably more important to consider the contributions which d-orbitals can make to dipoles, via the ppd parameter, and perhaps to quadrupoles, via the three parameters sdd, pdp and ddd.

The Hamiltonian matrix elements can be obtained as before, and the more general form of (7.70) including multipoles is straightforward to derive:

$$H_{I\mu J\nu} = H^{\text{in}}_{I\mu J\nu} + U_I \delta q_I \delta_{IJ} \delta_{\mu\nu} + \sum_{I' \neq I} \sum_{LL'} D^l_{\mu L\nu} B_{LL'}(\mathbf{R}_{II'}) Q_{L'}(\mathbf{R}_{I'}) \delta_{IJ}$$

$$+ \frac{1}{2}\Big(U_I \delta q_I + U_J \delta q_J + \sum_{I' \neq I} U_{II'} \delta q_{I'} + \sum_{I' \neq J} U_{JI'} \delta q_{I'} \Big) O_{I\mu J\nu}. \quad (7.81)$$

The form of the Hamiltonian shows explicitly how fields and field gradients induce polarisation of the ions via the D-parameters, each of which is the product of a Gaunt coefficient and a coupling strength $\Delta_{l'll''}$. Remember the Gaunt coefficients are purely mathematical functions, completely independent of the physical system; all the system-specific physics is in the atomic coupling strengths $\Delta_{l'll''}$.

The spherical tensor components of p and d character have a physical interpretation as electric fields and field gradients (quadrupoles of potential). An electric field is a p component of potential, which couples s and p orbitals with a strength proportional to Δ_{spp}. A quadrupolar field is a d component of potential which couples p orbitals with a strength proportional to Δ_{pdp} and d orbitals with a strength proportional to Δ_{ddd}, besides coupling s orbitals to d orbitals with a strength proportional to Δ_{sdd}.

7.8 Moments of the Density of States

It has been known since Friedel's work that the main factors governing the relative energies of different crystal structures are encoded within the sum of the energies of the occupied states, the band energy, or the bond energy. This focusses attention on the form of the density of states and the results of Section 1.8, since:

$$E^{\text{band}} = \sum_n f_n \epsilon_n = 2 \int_{-\infty}^{\infty} f_{\text{F}}(\epsilon) \epsilon D(\epsilon)\, \mathrm{d}\epsilon. \quad (7.82)$$

Many models based on tight-binding attempt in some way to approximate the density of states, leading to an approximation to the bond energy in the tight-binding bond model. We shall restrict ourselves for simplicity to orthogonal orbitals, since most models have not taken on the complexities of non-orthogonality, but note that this is not an absolute restriction of the simple formalism to be presented, since any set of local orbitals can be transformed into an orthogonal set at the price of making the matrix elements longer in range.

To derive useful results the global density of states $D(\epsilon)$ is best written as a sum of local densities of states, so again from Section 1.8 we have

$$E^{\text{bond}} = 2\sum_{J\nu}\int_{-\infty}^{\infty} f_{\text{F}}(\epsilon)(\epsilon - H^{\text{in}}_{J\nu J\nu})D_{J\nu}(\epsilon)\,\text{d}\epsilon. \tag{7.83}$$

A very attractive idea was developed by Françoise Cyrot–Lackmann and her colleagues from Friedel's original concept of modelling the density of states. This was to characterise the density of states in terms of its *moments*, defined by:

$$\mu_p^{J\nu} = \int_{-\infty}^{\infty} \epsilon^p D_{J\nu}(\epsilon)\,\text{d}\epsilon. \tag{7.84}$$

By expressing $D_{J\nu}(\epsilon)$ in terms of the eigenstates (see Section 1.8) we can also write (7.84) in the form of a weighted sum over the eigenvalues to the power p:

$$\mu_p^{J\nu} = \sum_n |\langle n|J\nu\rangle|^2 \epsilon_n^p. \tag{7.85}$$

In a tight-binding model the density of states is only non-vanishing within a finite range of energy, usually called the bandwidth, so the moments are all finite quantities. You might notice that would certainly not be the case if we had a complete basis, with eigenstates of arbitrarily high energy! The great thing about the moments is that they have a very simple expression in terms of the Hamiltonian. The total p^{th} moment is given by:

$$\mu_p = \sum_n \epsilon_n^p = \sum_n \langle n|\hat{H}^p|n\rangle \equiv \text{Tr}\,\hat{H}^p. \tag{7.86}$$

Similarly we can write the moments of the local density of states from (7.85) as

$$\mu_p^{J\nu} = \sum_n \langle n|J\nu\rangle\langle J\nu|\hat{H}^p|n\rangle. \tag{7.87}$$

Recognising the identity operator $\sum_n |n\rangle\langle n|$ we therefore obtain the expression:

$$\mu_p^{J\nu} = \langle J\nu|\hat{H}^p|J\nu\rangle. \tag{7.88}$$

It is satisfying to find that this agrees with the expression for the moments of the total density of states in the sense that from (7.86) and (7.88) we have

$$\mu_p = \sum_{J\nu} \mu_p^{J\nu}. \tag{7.89}$$

The value of this simple expression (7.88) for $\mu_p^{J\nu}$ lies in the fact that it is a real space expression which can be evaluated *without* solving any Schrödinger equations! Furthermore we can obtain a more explicit expression by inserting the identity operator $\sum_{I\mu} |I\mu\rangle\langle I\mu|$ between each of the factors $\hat{H}$; we thereby write the moment as a sum over paths, each path consisting of a series of p links between two local orbitals, starting and finishing with orbital $J\nu$:

$$\mu_p^{J\nu} = \sum_{J^{(1)}\nu^{(1)}\ldots J^{(p-1)}\nu^{(p-1)}} \langle J\nu|\hat{H}|J^{(1)}\nu^{(1)}\rangle\langle J^{(1)}\nu^{(1)}|\hat{H}|J^{(2)}\nu^{(2)}\rangle \\ \cdots\langle J^{(p-2)}\nu^{(p-2)}|\hat{H}|J^{(p-1)}\nu^{(p-1)}\rangle\langle J^{(p-1)}\nu^{(p-1)}|\hat{H}|J\nu\rangle. \quad (7.90)$$

In simple cases this can be evaluated 'by hand', for example for the second moment, about which we shall have more to say in Section 7.10, but normally such path counting is done in a computer program.

The idea of the moments approach is to calculate just the lowest few moments and with them to reconstruct somehow the density of states. In the earlier work this was simply done by making an ansatz for the density of states, such as a Gaussian function, and fitting it to the calculated moments. Well, you cannot get very far with a Gaussian except to fit the first and second moments, and to introduce more complicated functions to be fitted becomes a very arbitrary process. The general problem of recovering a function from a knowledge of its moments exercised mathematicians for some time, and is full of subtleties which it would be beyond the scope of this book, and my competence, to discuss, see for example Shohat and Tamarkin (1950). One attractive approach is the *maximum entropy* method, which various authors have investigated in this context (Mead and Papanicolaou, 1984; Brown and Carlsson, 1985; Glanville *et al.*, 1988). It is based on Baysian probability theory and its virtue is that no information is introduced apart from the moments themselves, which eliminates any spurious structure in the reconstructed density of states. However, a particularly efficient way of generating and using moments turned out to be the *recursion method* which deserves a section on its own.

7.9 The Recursion Method

The recursion method was developed by Haydock *et al.* (1972, 1975) in the first instance as a systematic way of approximating the local density of states in a general configuration of atoms. Their clear papers give a good historical perspective of its early development. Comprehensive references on the theory of the recursion method and its applications are the voume devoted to tight binding in the Solid State Physics series (Heine, 1980) and the volume of Proceedings of a special conference on the subject, edited by Pettifor and Weaire (1985). I will explain it in outline here. Mathematically, the method is almost identical to the Lanzcos algorithm, but its purpose is rather different. It is closely related to the moments method as we shall see. The general idea is to obtain approximate densities of states, or more precisely Green function matrix elements, without calculating eigenvalues or requiring periodic boundary conditions and the machinery of k-space. This has made it an attractive starting point for deriving multi-ion interatomic potentials, such as the bond-order potentials (BOPs) (see Section 7.12).

The method proceeds in two distinct stages. The first stage begins with a particular orbital, let us say $|I\mu\rangle$, and generates from it a new set of mutually orthogonal states called the recursion orbitals. For generality we shall call the starting orbital, which is also the zeroth recursion orbital, $|0\rangle$ (not to be confused with the ground state wavefunction for which the same symbol was used) and we shall call the orthonormal recursion orbitals to be generated $|1\rangle, \ldots, |i\rangle$, etc. The first recursion orbital is proportional to the state $|\tilde{1}\rangle$ obtained by the following operations:

$$|\tilde{1}\rangle = \hat{H}|0\rangle - a_0|0\rangle. \tag{7.91}$$

I use the tilde to denote that the recursion orbital so defined has not yet been normalised. The coefficient a_0 is determined by the condition that $|\tilde{1}\rangle$ must be orthogonal to $|0\rangle$. Thus multiplying (7.91) to the left by $\langle 0|$ we have

$$a_0 = \langle 0|\hat{H}|0\rangle. \tag{7.92}$$

The process on the right-hand side of (7.91), namely operating $\hat{H}$ on $|0\rangle$, then orthogonalizing the resulting state to $|0\rangle$, has generated a state which we now normalize in the usual way by introducing the factor b_1 such that

$$|1\rangle = \frac{1}{b_1}|\tilde{1}\rangle, \tag{7.93}$$

and b_1 is calculated from

$$b_1^2 = \langle\tilde{1}|\tilde{1}\rangle. \tag{7.94}$$

It is also significant that b_1 satisfies the relation

$$b_1 = \langle 1|\tilde{1}\rangle = \langle 1|\hat{H}|0\rangle = \langle 0|\hat{H}|1\rangle. \tag{7.95}$$

The subsequent recursion orbitals are constructed by repeated operation of $\hat{H}$ and orthogonalization, using the *recursion* formula

$$|\widetilde{i+1}\rangle = \hat{H}|i\rangle - a_i|i\rangle - b_i|i-1\rangle. \tag{7.96}$$

The b coefficients are defined after each recursion step by the normalization of the resulting orbital, exactly as b_1 was determined:

$$b_i^2 = \langle\tilde{i}|\tilde{i}\rangle. \tag{7.97}$$

The a coefficients are then given by

$$a_i = \langle i|\hat{H}|i\rangle. \tag{7.98}$$

The recursion formula (7.96), together with the defining equations for the a and b coefficients have some wonderful properties. At first sight it seems as if, by the above definition of a_i, we have only orthogonalised each orbital $|i+1\rangle$ to the preceding orbital $|i\rangle$. But the definition of b_i ensures that we have actually orthogonalised it to *all* the preceeding orbitals! The proof is an exercise for the reader, or it can be found in the literature mentioned above. Now we have generated a new orthonormal basis, we can express the Hamiltonian as a matrix within it. The second delightful property of the recursion formula is that the matrix elements of $\hat{H}$ in the recursion basis only couple the consecutive orbitals i and $i+1$. This is obvious from the fact that acting with $\hat{H}$ on $|i\rangle$ only produces components on $|i-1\rangle$, $|i\rangle$ and $|i+1\rangle$. In other words the matrix is *tridiagonal*. It has the form:

$$\mathbf{H} = \begin{pmatrix} a_0 & b_1 & 0 & 0 & 0 & \cdots \\ b_1 & a_1 & b_2 & 0 & 0 & \cdots \\ 0 & b_2 & a_3 & b_3 & 0 & \cdots \\ \vdots & \vdots & \ddots & \ddots & \ddots & \cdots \\ \vdots & \vdots & \vdots & \ddots & \ddots & \ddots \end{pmatrix}. \tag{7.99}$$

In effect, we have mapped the original problem onto the problem of a one dimensional chain of orbitals, each of which only interacts with its nearest neighbours, as indicated in Fig. 7.2.

The Hamiltonian matrix elements in this chain are

$$\begin{aligned} H_{00} &= a_0 \\ H_{01} &= H_{10} = b_1 \\ H_{i,i+1} &= H_{i+1,i} = b_{i+1}, \quad \{i > 0\}. \end{aligned} \tag{7.100}$$

The a_{i-1} and b_i are determined up to a chosen maximum i we shall denote by $\mathcal{L}$. This completes the first stage of the recursion method, requiring $\mathcal{L}$ operations of the Hamiltonian. The second stage is to extract the required approximation to the density of states from the coefficients that have been generated. As we have seen, to get the local density of states $D_{I\mu}(\epsilon)$ all we need is the Green function matrix element $G_{I\mu I\mu} = G_{00}$, from which

$$D_{I\mu}(\epsilon) = -\frac{1}{\pi} \lim_{\eta \to 0} \operatorname{Im} G_{I\mu I\mu}(\epsilon + i\eta). \tag{7.101}$$

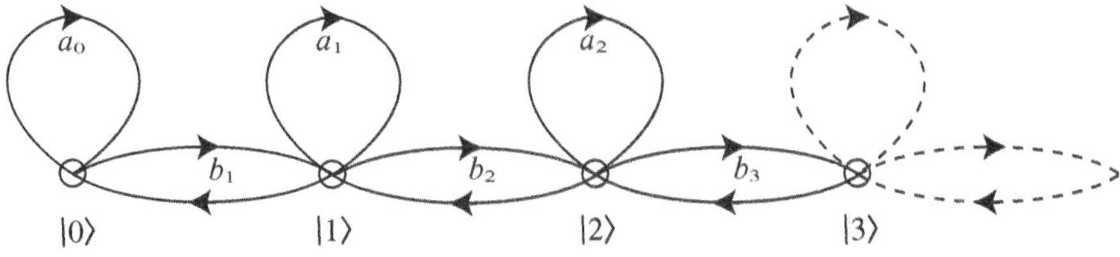

FIG. 7.2 The one-dimensional chain created by recursion.

In the recursion basis this is now also equal to the diagonal element of the Green function matrix at the start of our one dimensional chain. As shown by Haydock *et al.* (1972), this matrix element has the form of an infinite continued fraction:

$$G_{00}(\epsilon) = \cfrac{1}{\epsilon - a_0 - \cfrac{b_1^2}{\epsilon - a_1 - \cfrac{b_2^2}{\epsilon - a_2 - \cfrac{b_3^2}{\epsilon - \cdots}}}} \tag{7.102}$$

I would like to give the heuristic 'proof' that this is so, because it illustrates the physics of the relationship between the moments of the density of states and its continued fraction representation. The idea is to try to expand $\hat{G}$ from eq. (1.211) in powers of $\hat{H}/\epsilon$:

$$\hat{G}(\epsilon) = \frac{1}{\epsilon}\left(1 + \frac{\hat{H}}{\epsilon} + \frac{\hat{H}^2}{\epsilon^2} + \frac{\hat{H}^3}{\epsilon^3} + \cdots\right). \tag{7.103}$$

Taking the diagonal elements of $\hat{G}$ gives us what looks like a moments expansion, so for G_{00} :

$$G_{00}(\epsilon) = \frac{1}{\epsilon}\left(1 + \frac{\mu_1^0}{\epsilon} + \frac{\mu_2^0}{\epsilon^2} + \frac{\mu_3^0}{\epsilon^3} + \cdots\right), \tag{7.104}$$

where, in the recursion basis:

$$\mu_i^0 = \sum_{j^{(1)}\ldots j^{(i-1)}} \langle 0|\hat{H}|j^{(1)}\rangle\langle j^{(1)}|\hat{H}|j^{(2)}\rangle \cdots \langle j^{(i-1)}|\hat{H}|0\rangle. \tag{7.105}$$

Compared to (7.90), the paths in (7.105) are on the one-dimensional chain, with only nearest neighbour interactions, which makes them a lot easier to deal with than the three-dimensional problem. The next step is a linked cluster summation, familiar in diagramatic perturbation theory if you like that kind of thing. If you do not, skip this bit. We first define 'irreducible paths' $\bar{\mu}_i^0$, which exclude from (7.105) all the paths which at any intermediate step return to state $|0\rangle$. Equation (7.104) is then equivalent to an expansion of:

$$G_{00}(\epsilon) = \frac{1}{\epsilon\left(1 - \Sigma_{(i>0)}\, \bar{\mu}_i^0/\epsilon^i\right)}. \tag{7.106}$$

We can now repeat exactly the same trick on the summation in the denominator. First pick out the first term, $\bar{\mu}_1^0$, which we recognize as a_0:

$$G_{00}(\epsilon) = \frac{1}{\epsilon - a_0 - \epsilon\,\Sigma_{(i>1)}\, \bar{\mu}_i^0/\epsilon^i}. \tag{7.107}$$

Now the summation can be factored into the two steps from $|0\rangle$ to $|1\rangle$ and back, which from (7.95) is b_1^2, times the sum of all paths leaving from and returning to $|1\rangle$, but not visiting $|0\rangle$. This takes us to:

$$G_{00}(\epsilon) = \frac{1}{\epsilon - a_0 - (b_1^2/\epsilon)\left(1 + \Sigma_{(i>0)}\, \mu_i^{>1}/\epsilon^i\right)}. \tag{7.108}$$

The paths of i steps represented by $\mu_i^{>1}$ start and finish at $|1\rangle$ and can revisit it on the way. The meaning of the '>' is that they cannot go back to $|0\rangle$. We have now generated the first level of the continued fraction. Let us define the last factor in the denominator as

$$G_{11}^{>}(\epsilon) = \frac{1}{\epsilon}\left(1 + \sum_{i>0} \mu_i^{>1}/\epsilon^i\right)$$

Effectively, by comparison with (7.104), this is the diagonal element of the Green function of the chain as if the first orbital were missing and we were simply starting with $|1\rangle$. We can now treat it exactly as we did (7.104), which will generate the second level in the continued fraction:

$$G_{00}(\epsilon) = \frac{1}{\epsilon - a_0 - (b_1^2/(\epsilon - a_1 - b_2^2 G_{22}^{>}(\epsilon)))}. \tag{7.109}$$

And so on. There is a heirarchy defined by:

$$G_{ii}^{>}(\epsilon) = \frac{1}{\epsilon - a_i - b_{i+1}^2 G_{i+1\,i+1}^{>}(\epsilon)} \qquad \{i \geq 0, G_{00}^{>} \equiv G_{00}\}. \tag{7.110}$$

We now see that successive operations of $\hat{H}$ on the starting state correspond to the generation of successive levels in the continued fraction representing G_{00}, the ith level being characterised by a_{i-1} and b_i. Furthermore, at each level i of the continued fraction we see we are adding a further two moments of the density of states to the expansion of G_{00}, one with a_{i-1} and another with b_i. I stress that the above 'proof' is very far from rigourous, for example the series (7.104) does not even converge in the energy range of interest. The later work of Haydock and Nex put the formulae on a much sounder footing. Nevertheless, the formal manipulations we have just made seem to be telling us some real physics, in the sense that the crudest description of G_{00} is established by its second moment $(a_0^2 + b_1^2)$ and this description is successively refined as we go down the continued fraction, incorporating information about the topology and potential from atomic sites further and further away from $|0\rangle$.

Explicit expressions for the moments in terms of the continued fraction coefficients or vice versa can be obtained; the first five are given by:

$$\begin{aligned}
\mu_1 &= a_0 \\
\mu_2 &= a_0^2 + b_1^2 \\
\mu_3 &= a_0^3 + 2a_0 b_1^2 + a_1 b_1^2 \\
\mu_4 &= a_0^4 + 3a_0^2 b_1^2 + 2a_0 a_1 b_1^2 + a_1^2 b_1^2 + b_1^4 + b_1^2 b_2^2 \\
\mu_5 &= a_0^5 + 4a_0^3 b_1^2 + 3a_0 b_1^4 + 3a_0^2 a_1 b_1^2 + 2a_0 a_1^2 b_1^2 \\
&\quad + 2a_1 b_1^4 + a_1^3 b_1^2 + 2a_0 b_1^2 b_2^2 + 2a_1 b_1^2 b_2^2 + a_2 b_1^2 b_2^2. \qquad (7.111)
\end{aligned}$$

The general procedure is to calculate the top left matrix element of H^p, which is precisely μ_p. This is easy to do since H is a tridiagonal matrix, which makes it straightforward to construct the sequence of matrices $H^2, H^3, \ldots$. We can always get twice as many moments as the highest order b_i coefficient we have. Once the desired moments have been obtained in terms of the a_i and b_i it is easy to generate the inverse formulae for a_i and b_i in terms of the moments, which will be needed for constructing BOPs. This is done by replacing a_i and b_i step by step from $i = 0$ by their expressions in terms of the μ_p up the heirarchy of equations of which (7.111) shows the first five. In this way just from the first four of (7.111) we get:

$$\begin{aligned}
b_1 &= \sqrt{\mu_2 - \mu_1^2} \\
a_1 &= \frac{\mu_3 - \mu_1^3 - 2\mu_1 b_1^2}{b_1^2} \qquad (7.112) \\
b_2 &= \frac{\sqrt{\mu_4 - \mu_1^4 - 3\mu_1^2 b_1^2 - 2\mu_1 a_1 b_1^2 - a_1^2 b_1^2 - b_1^4}}{b_1}.
\end{aligned}$$

The recursion method is more than just an efficient way of calculating moments of the density of states. The continued fraction expression for the density of states itself is useful if care is taken to terminate it properly. This is rather a delicate mathematical question which it would take us too far afield to delve into, and the interested reader is referred to the literature (Heine, 1980; Haydock and Nex, 1984; Pettifor and Weaire, 1985). Two options have been commonly used. One is simply to terminate with $b_{\mathcal{L}+1} = 0$. This gives a rational approximation to G_{00}. Let us see how this works in a four level approximation, in which the continued fraction is truncated by taking $b_4 = 0$. We can simplify it to the quotient of a cubic and a quartic polynomial in ϵ. This is accomplished by first multiplying top and bottom of the tail

$$\frac{b_2^2}{\epsilon - a_2 - (b_3^2/\epsilon - a_3)}$$

by $\epsilon - a_3$ to clear the bottom fraction. Top and bottom of the new tail

$$\frac{b_1^2}{\epsilon - a_1 - (b_2^2(\epsilon - a_3)/(\epsilon - a_2)(\epsilon - a_3) - b_3^2)}$$

are rationalised in a similar way, and after one more such step we find:

$$G_{00}(\epsilon) = \frac{(\epsilon - a_1)(\epsilon - a_2)(\epsilon - a_3) - b_3^2(\epsilon - a_1) - b_2^2(\epsilon - a_3)}{(\epsilon - a_0)[(\epsilon - a_1)(\epsilon - a_2)(\epsilon - a_3) - b_3^2(\epsilon - a_1) - b_2^2(\epsilon - a_3)] - b_1^2[(\epsilon - a_2)(\epsilon - a_3) - b_3^2]}, \tag{7.113}$$

which can be written in the general form:

$$G_{00}(\epsilon) = \frac{\epsilon^3 + A_{00}\epsilon^2 + B_{00}\epsilon + C_{00}}{(\epsilon - \epsilon_1)(\epsilon - \epsilon_2)(\epsilon - \epsilon_3)(\epsilon - \epsilon_4)}. \tag{7.114}$$

The constants A_{00}, B_{00} and C_{00} in the numerator are easy to write down in terms of $a_0 \cdots b_3$. The energies $\epsilon_1 \cdots \epsilon_4$ are the roots of the quartic equation obtained by setting the denominator of (7.113) equal to zero, and are a little more difficult to obtain in closed form, although this can be done as shown by Pettifor and Oleinik (1999).

We can generalise the above procedure to any number of levels, so that the G_{00} can always be rationalized to the form:

$$G_{00}(\epsilon) = \frac{\mathcal{D}_1(\epsilon)}{\mathcal{D}_0(\epsilon)}, \tag{7.115}$$

where $\mathcal{D}_n(\epsilon)$ is a certain polynomial of order $\mathcal{L} + 1 - n$. On the way to this expression you climb through the heirarchy, starting on the right:

$$G_{11}^{>}(\epsilon) = \frac{\mathcal{D}_2(\epsilon)}{\mathcal{D}_1(\epsilon)}, \quad G_{22}^{>}(\epsilon) = \frac{\mathcal{D}_3(\epsilon)}{\mathcal{D}_2(\epsilon)}, \quad \ldots, \quad G_{\mathcal{L}\mathcal{L}}^{>}(\epsilon) = \frac{1}{\epsilon - a_{\mathcal{L}}}. \tag{7.116}$$

From the definitions above, these polynomials satisfy the recurrence relations:

$$\begin{aligned} \mathcal{D}_{\mathcal{L}+1} &= 1 \\ \mathcal{D}_{\mathcal{L}} &= \epsilon - a_{\mathcal{L}} \\ \mathcal{D}_i &= (\epsilon - a_i)\mathcal{D}_{i+1} - b_{i+1}^2 \mathcal{D}_{i+2}. \end{aligned} \tag{7.117}$$

The same result can be obtained more directly from the theory of determinants. The polynomial $\mathcal{D}(\epsilon)$ is simply the determinant of the matrix $\epsilon\mathbf{1} - \mathbf{H}$, where $\mathbf{H}$ is the tridiagonal matrix representation of $\hat{H}$ in the recursion basis, eq. (7.99). The polynomial $\mathcal{D}_1(\epsilon)$ is the determinant of the submatrix obtained by deleting the first row and column of $\epsilon\mathbf{1} - \mathbf{H}$. The polynomial $\mathcal{D}_2(\epsilon)$ is the determinant of the submatrix obtained by deleting the first two rows and columns of $\epsilon\mathbf{1}-\mathbf{H}$, and so on.

Finally we can express the result in the form of partial fractions:

$$G_{00} = \sum_{i=1}^{\mathcal{L}} \frac{w_i}{\epsilon - \epsilon_i}. \tag{7.118}$$

Now from the definition of the delta function in eq. (1.197) we can write the local density of states to this level of approximation as a weighted sum of four delta functions:

$$D_0(\epsilon) = -\frac{1}{\pi} \lim_{\eta\to 0} \operatorname{Im} G_{00}(\epsilon + i\eta) = \sum_{i=1}^{\mathcal{L}} w_i \delta(\epsilon - \epsilon_i). \tag{7.119}$$

Notice that the energies ϵ_i are not eigenvalues of the Hamiltonian, but they are sampling the spectrum of its eigenvalues. An exception is the special case of a small molecule with only four basis orbitals altogether, in which the truncation of the continued fraction is not an approximation and the subsequent formulae are exact. We will run through an example in the next section.

The discrete approximation to the continued fraction, and hence to densities of states, can be very useful, but it is sometimes even more useful to have these as continuous functions of ϵ. This can be accomplished by a technique that was much used in the early days of the recursion method (Haydock *et al.*, 1972, 1975). It exploits the fact that in an infinite system, with a single band of states, the coefficients tend to constant values a_∞ and b_∞. The a_i and b_i coefficients are set to these values at some level called the terminating level, beyond which there are assumed to be still an infinite number of identical levels. The continued fraction from the terminating level down is then replaced by a terminating function $t(\epsilon)$, which must satisfy

$$t(\epsilon) = \epsilon - a_\infty - \frac{b_\infty^2}{t(\epsilon)}. \tag{7.120}$$

This is a quadratic equation for t with the solution

$$t(\epsilon) = \frac{\epsilon - a_\infty}{2} \left\{ 1 - \left(1 - \frac{4b_\infty^2}{(\epsilon - a_\infty)^2} \right)^{1/2} \right\}. \tag{7.121}$$

This terminator immediately defines a continuous range (the *bandwidth*) over which the density of states is non-vanishing, it is just the range over which G_{00} has a non-vanishing imaginary part and corresponds to where the square root in (7.121) has a negative argument, namely $a_\infty - 2b_\infty < \epsilon < a_\infty + 2b_\infty$. Some ingenuity has gone into making the best estimate of these parameters. Other terminators for the continued fraction such as the splicing method of Luchini and Nex (1987) have been proposed. Physically, the choice of a terminating function corresponds to replacing the actual chain of orbitals beyond the maximum length by some effective medium. If this is a vacuum, the chain is truncated and its energy levels are delta functions. Otherwise, with a terminator such as the square root one, these delta functions are smeared into a continuum, which corresponds to a physical model in which the finite chain is coupled to an infinite system within which its eigenstates are resonances in a continuum of states.

Besides these two methods of obtaining a density of states from the recursion coefficients I should mention two other interesting ways of processing them, but refer you to the literature for details. First, there is an elegant link between the mathematics of continued fractions and orthogonal polynomials which Nex and Haydock exploited to devise an efficient and accurate quadrature scheme for generating integrals of arbitrary functions over the density of states (Nex, 1984; Haydock and Nex, 1984, 1985). Second, there is the method of maximum entropy I mentioned previously, which directly reconstructs the smoothest possible density of states from its moments.

The recursion method can also be used to develop approximations to the *off*-diagonal matrix elements of $\hat{G}$ in a local basis. These are the ingredients of bond orders, which can be used in the TBBM. The key is to recognise that the starting state $|0\rangle$ is arbitrary. It does not have to be a single orbital but can be a linear combination of orbitals. In particular if we started with the combination

$$|0^+\rangle = \frac{1}{\sqrt{2}}(|\mu\rangle + |\nu\rangle) \tag{7.122}$$

we would generate $\langle 0^+|\hat{G}|0^+\rangle$, which we denote for short G_{00}^+, where

$$G_{00}^+ = \tfrac{1}{2}G_{\mu\mu} + \tfrac{1}{2}G_{\nu\nu} + G_{\mu\nu}, \tag{7.123}$$

similarly from

$$|0^-\rangle = \frac{1}{\sqrt{2}}(|\mu\rangle - |\nu\rangle) \tag{7.124}$$

we would generate

$$G_{00}^- = \tfrac{1}{2}G_{\mu\mu} + \tfrac{1}{2}G_{\nu\nu} - G_{\mu\nu} \tag{7.125}$$

and from

$$|0^c\rangle = \frac{1}{\sqrt{2}}(|\mu\rangle + \mathrm{i}|\nu\rangle) \tag{7.126}$$

we would generate

$$G_{00}^c = \tfrac{1}{2}G_{\mu\mu} + \tfrac{1}{2}G_{\nu\nu}. \tag{7.127}$$

The superfix c here stands for complex.

Hence the off-diagonal element can be calculated in principle from the combinations

$$G_{\mu\nu} = G_{00}^{+} - G_{00}^{c}, \tag{7.128}$$

or

$$G_{\mu\nu} = \tfrac{1}{2}\left(G_{00}^{+} - G_{00}^{-}\right). \tag{7.129}$$

Another useful starting orbital, introduced by Aoki and Pettifor (1993), is the combination

$$|0^{\lambda}\rangle = \tfrac{1}{\sqrt{2}}(|\mu\rangle + e^{i\psi}|\nu\rangle), \tag{7.130}$$

Defining $\lambda = \cos\psi$, this gives us the diagonal element

$$G_{00}^{\lambda} = \tfrac{1}{2}G_{\mu\mu} + \tfrac{1}{2}G_{\nu\nu} + \lambda G_{\mu\nu} \tag{7.131}$$

from which we get a key result for the off-diagonal elements of the Green function matrix:

$$\langle I\mu|\hat{G}|J\nu\rangle = \left.\frac{\partial G_{00}^{\lambda}}{\partial\lambda}\right|_{\lambda=0}. \tag{7.132}$$

The evaluation point $\lambda = 0$ is purely a matter of convenience; since G_{00}^{λ} is a linear function of λ, its derivative is independent of λ.

The latter expression is particularly useful in the development of analytic approximations for bond orders as we shall see. There are still many difficulties, some of which have only recently been overcome, concerning the most efficient way to use recursion coefficients to estimate the integral of the energy over a density of states or to estimate a bond order. Not least is the familiar problem of formulating an expression for the forces which is compatible with that for the total energy. We shall be touching on some of the approaches in the following sections.

7.9.1 Block Recursion

I mention here a refinement of the recursion method called *block recursion* or *matrix recursion*, or the block Lanzcos algorithm. Because it is rather technical, I refer the reader to the literature which is detailed enough for anyone who really needs to understand it. The papers by Jones and Lewis (1984) and Inoue and Ohta (1987) describe the idea well, and subsequent papers by Nex (1989) and Godin and Haydock (1991) describe how it has been implemented. Important refinements and applications to BOPs have since been introduced by Ozaki *et al.* (2000).

Typically for applied mathematics, it was devised in response to a difficulty with the conventional recursion method as I have described it. The problem is that unless results are perfectly converged with the number of levels you can find that they do not exactly respect the symmetry of the Hamiltonian, which becomes obvious when dealing with perfect crystals. Things like the total density of states on a site for example should not depend on the choice of the z-axis, which defines the orientation of the atomic orbitals. When a density of states is summed over a subset of the orbitals on a site which span a particular representation of the full rotation group, such as the p-orbitals or the d-orbitals, then it must be invariant under rotations. These properties are guaranteed by the block recursion technique, in which all the densities of states associated with the subset of orbitals on a site are calculated in one sweep of recursion operations. The idea is that instead of defining a starting orbital, a starting *vector* of orbitals is defined, such as the subset of p-orbitals on a site. Then the operations of the Hamiltonian on orbitals become operations on vectors of orbitals, so the role previously played by Hamiltonian matrix elements is now played by blocks of Hamiltonian matrix elements. The a_i and b_i coefficients naturally become matrices. All the other relationships, such as the expression for the bond energy in terms of the Green function can generalised in a natural way. Remember that there are two ways in principle to calculate a bond energy, either from the trace of a bond order times the Hamiltonian using (7.15), or by integrating the energy times the density of states obtained from the diagonal elements of G. A pleasing result of block recursion is that these two ways give identical numerical results at whatever level the recursion is terminated, something that is not guaranteed by scalar recursion due to loss of symmetry.

7.10 Second-Moment Models

7.10.1 General Ideas

Second moment models are the simplest level of approximation within a tight-binding framework. They have been very widely used for simulation, and they will also serve to illustrate some of the ideas of tight-binding theory without too much heavy algebra.

The local density of states $D_I(\epsilon)$ is a total for all the n_I^{orb} orbitals on site I. For a pure d-orbital model of transition metals for example we have $n_I^{\text{orb}} = 5$. The zeroth moment is the normalisation of D_I:

$$n_I^{\text{orb}} = \int_{-\infty}^{+\infty} D_I(\epsilon)\,\mathrm{d}\epsilon. \tag{7.133}$$

The first moment a_{0I} is the centre of gravity of the density of states and is specified by

$$0 = \int_{-\infty}^{+\infty} (\epsilon - a_{0I}) D_I(\epsilon)\,\mathrm{d}\epsilon. \tag{7.134}$$

The second moment μ_2^I is a direct measure of the width of the density of states, which we shall call σ_I, but only if a_{0I} is taken as the zero of energy, as we can see from the defining relations

$$\mu_2^I = \int_{-\infty}^{+\infty} \epsilon^2 D_I(\epsilon)\, d\epsilon \tag{7.135}$$

and

$$\sigma_I = \int_{-\infty}^{+\infty} (\epsilon - a_{0I})^2 D_I(\epsilon)\, d\epsilon \equiv \int_{-\infty}^{+\infty} \epsilon^2 D_I(a_{0I} + \epsilon)\, d\epsilon. \tag{7.136}$$

We can anticipate that the choice of energy zero so as to make $a_{0I} = 0$ will be very convenient. It will not necessarily be the same energy zero for all atoms. Following the ideas of moments, in particular (7.90), we shall find that μ_2^I can be written as the sum of the squares of bond integrals to nearest neighbours of I, and eventually takes the form:

$$\mu_2^I = \sum_J \beta_{IJ}(R_{IJ}), \tag{7.137}$$

where the functions $\beta_{IJ}(R)$ depend on the type of atoms at $\mathbf{R}_I$ and $\mathbf{R}_J$ but not on the orientation of $\mathbf{R}_{IJ}$. I now describe the way this is derived and how it is useful.

7.10.2 The TBBM with a Gaussian Density of States

The earliest approach (Cyrot-Lackmann, 1968; Ducastelle, 1970; Allan and Lannoo, 1976) was to assume a Gaussian local density of states on atom I:

$$D_I(\epsilon) = \frac{n_I^{\text{orb}}}{\sqrt{2\pi\sigma_I}} \exp\left(-\frac{(\epsilon - a_{0I})^2}{2\sigma_I}\right). \tag{7.138}$$

Other one-parameter model densities of states are even simpler, for example two delta functions, which is the density of states obtained by truncating the continued fraction after b_1^2/ϵ, or the rectangular density of states introduced by Friedel, which is just a constant within the band and zero outside it. The first is clearly not so attractive if one is looking at trends with varying ϵ_F, but there is no obvious reason why a Gaussian should be any better than the rectangular model. In the light of maximum entropy theory one can argue that the Gaussian form introduces the least extraneous information over and above the second moment. However, the argument is not a strong one for a model which is such a gross simplification of the real problem. Let us simply accept it as a model and follow through the consequences.

The best way to apply the Gaussian density is within the framework of the *bond* model (see Section 7.5). Besides being more realistic for metals than the band model, the TBBM solves the problem of where to put the centre of the Gaussian

a_{0I}, which is the first moment of the density of states. In the early days it was *assumed* that $a_{0I} = 0$. However, if we assume local charge neutrality, the condition that the Fermi energy ϵ_F is constant everywhere fixes the position of the centre of gravity a_{0I} of each D_I, as follows.

Another standard (but dubious!) assumption is that splitting of the on-site energy levels can be neglected, so all the diagonal elements of $\hat{H}$ or $\hat{H}^{\text{in}}$ on a site are assumed to be the same, although they may differ from site to site. In other words the centre of gravity of each orbital density of states $D_{I\mu}$ is implicitly assumed to be the same for each site orbital μ, and given by

$$\int_{-\infty}^{+\infty} \epsilon D_{I\mu}(\epsilon)\, d\epsilon = H_{I\mu I\mu} = a_{0I}. \tag{7.139}$$

The total number of electrons on the site is given by

$$Z_I = q_I = \sum_\mu q_{I\mu} = 2\int_{-\infty}^{\epsilon_F} D_I(\epsilon)\, d\epsilon = 2\int_{-\infty}^{\epsilon_F - a_{0I}} D_I(\epsilon + a_{0I})\, d\epsilon. \tag{7.140}$$

Given the constant value of ϵ_F, we understand that although it turns out we shall never need to calculate a_{0I} its value is determined by the upper limit of the integral which fixes the total charge Z_I in eq. (7.140). The idea is that if q_I is less than Z_I, there will be a flow of electrons into the orbitals on I, which will tend to raise the on-site matrix elements $H_{I\mu I\mu}$. The mechanism of this raising in self-consistent models is the Hubbard U, but in the TBBM it is implicit that U is infinite, so $q_I = Z_I$.

We have for E^{bond}, which is the same as E^{cov} for orthogonal orbitals (see (7.14) and (7.42)):

$$E^{\text{bond}} = \sum_I \left\{ 2\int_{-\infty}^{\epsilon_F} \epsilon D_I(\epsilon)\, d\epsilon - \sum_\mu q_{I\mu} H_{I\mu I\mu} \right\}. \tag{7.141}$$

Substituting (7.140) into (7.141) gives us after a little manipulation:

$$E^{\text{bond}} = \sum_I 2\int_{-\infty}^{\epsilon_F - a_{0I}} \epsilon D_I(\epsilon + a_{0I})\, d\epsilon. \tag{7.142}$$

The integrand is now independent of a_{0I}, the centre of gravity of the density of states on I. Although the Gaussian form of $D_{I\mu}$ is useful to fix ideas and get an explicit formula, in fact eqs. (7.139–7.141) do not depend on the form of the density of states. The Gaussian model is one of the simplest for which we can explicitly evaluate the bond energy by doing the integral in (7.142), which gives

$$E^{\text{bond}}_{\text{Gauss}} = -2\sum_I n_I^{\text{orb}} \sqrt{\frac{\mu_2^I}{2\pi}} \exp\left(-\frac{(\epsilon_F - a_{0I})^2}{2\mu_2^I} \right). \tag{7.143}$$

For the purpose of calculating the second moment we choose to measure the energy in each density of states from its centre of gravity, the local value of a_{0I}, because

μ_2^I is thereby the standard width of the Gaussian. Remember that ϵ_F is the same on all sites although $\epsilon_F - a_{0I}$ may vary from site to site as required to satisfy the condition of local charge neutrality. In fact $\epsilon_F - a_{0I}$ has to vary proportionally to $\sqrt{\mu_2^I}$. We can see this by writing the charge neutrality integral (7.140) in terms of the dimensionless energy $x = \epsilon/\sqrt{2\mu_2^I}$:

$$q_I = 2n_I^{\mathrm{orb}} \frac{1}{\sqrt{\pi}} \int_{-\infty}^{(\epsilon_F - a_{0I})/\sqrt{2\mu_2^I}} \exp(-x^2)\,\mathrm{d}x. \tag{7.144}$$

This equation tells us that the upper limit of the integral has to be the same for all atoms of the same type. Thus we reach the significant conclusion that the exponential in (7.143) is a constant and the bond energy is strictly proportional to $\sum_I \sqrt{\mu_2^I}$. The result does not depend on the specific assumption of a Gaussian density of states, but on the general assumption that the local density of states has the form

$$D_I(\epsilon) = \frac{n_I^{\mathrm{orb}}}{\sqrt{\mu_2^I}} F\left(\frac{(\epsilon - a_{0I})}{\sqrt{\mu_2^I}}\right) \tag{7.145}$$

where the function F is normalized such that:

$$\int_{-\infty}^{+\infty} F(x)\,\mathrm{d}x = 1. \tag{7.146}$$

In words, the argument also holds for any function $D_I(\epsilon)$ that is normalized to the same value n_I^{orb} and whose shape variation is determined by its second moment alone. That is, a function that is free only to stretch or compress uniformly along the energy axis while its height scales uniformly to preserve its integral (Ackland *et al.*, 1988). These are key results in the theory of second-moment models, referring to the TBBM. The proportionality to $\sum_I \sqrt{\mu_2^I}$ is not true in the *band* model except in the special case of an exactly half-filled band and when all atoms have the same value of a_{0I}, two conditions that together ensure $\epsilon_F = a_{0I}$ so that the exponent is constant. In the *bond* model these restrictions are not necessary.

Now we turn to the calculation of μ_2^I. It is the sum over the orbital contributions

$$\mu_2^I = \sum_\mu \mu_2^{I\mu} \tag{7.147}$$

and by applying (7.90) it can be expressed as a sum over the neighbours J of I:

$$\mu_2^I = \sum_\mu \langle I\mu|\hat{H}^2|I\mu\rangle = \sum_{\mu J\nu} H_{I\mu J\nu} H_{I\mu J\nu} = \sum_{\mu J\nu} H_{I\mu J\nu}^2. \tag{7.148}$$

By our choice of energy zero, together with the neglect of splitting of the on-site energy levels, which is an approximation, we have ensured that the $J = I$ terms vanish. A further simplification has to do with the fact that the result is independent of the orientation of the Cartesian axes used to describe the orbitals. This is an example of what is usually called *rotational invariance* and must be true for all scalar quantities in a consistent model. In this case each of the terms:

$$\mu_2^{IJ} = \sum_{\mu\nu} H_{I\mu J\nu}^2(\mathbf{R}_{IJ}) \tag{7.149}$$

is rotationally invariant. Notice that here it is the *orbitals* that we are rotating, not the bond $\mathbf{R}_{IJ}$; if we rotated all together, the invariance would be trivial! The proof of this statement is widely applicable, so let us take a moment to understand it. A rotation of the coordinate system transforms the orthormal set of orbitals $|\mu\rangle$ into the orthonormal set $|\mu'\rangle$, which is related to the original set by a unitary matrix $\mathbf{B}$ with elements $B_{\mu\mu'}$ such that

$$|\mu'\rangle = \sum_{\mu} B_{\mu'\mu}|\mu\rangle, \qquad \langle\mu'| = \sum_{\mu} B_{\mu'\mu}^*\langle\mu|. \tag{7.150}$$

Unitary means that the inverse of $\mathbf{B}$ is equal to its transposed complex conjugate $\mathbf{B}^{*T}$, which follows from the othonormality of the sets of orbitals:

$$\langle\mu'|\nu'\rangle = \delta_{\mu'\nu'} = \sum_{\mu\nu} B_{\mu'\mu}^* B_{\nu'\nu}\langle\mu|\nu\rangle = \sum_{\mu} B_{\mu\mu'}^{*T} B_{\nu'\mu} \equiv (\mathbf{BB}^{-1})_{\nu'\mu'}. \tag{7.151}$$

Hence

$$B_{\mu\mu'}^{*T} = B_{\mu'\mu}^* = (B^{-1})_{\mu\mu'}. \tag{7.152}$$

In our present case, incidentally, the matrix $\mathbf{B}$ would be real, since we work with a real basis, and a real matrix that satisfies the unitary property is called *orthogonal*. Now for any Hermitian operator $\hat{A}$ we have:

$$\begin{aligned}
\sum_{\mu'}\langle\mu'|\hat{A}|\mu'\rangle &= \sum_{\mu\nu\mu'} B_{\mu'\mu}^* B_{\mu'\nu}\langle\mu|\hat{A}|\nu\rangle \\
&= \sum_{\mu\nu\mu'} (B^{-1})_{\mu\mu'} B_{\mu'\nu}\langle\mu|\hat{A}|\nu\rangle \\
&= \sum_{\mu\nu} \delta_{\mu\nu}\langle\mu|\hat{A}|\nu\rangle \\
&= \sum_{\mu}\langle\mu|\hat{A}|\mu\rangle.
\end{aligned} \tag{7.153}$$

This looks very like the invariance of a trace which we have met before, except that the set of orbitals μ belonging to one atom only spans a subspace of the complete

Hilbert space. In similar fashion we can prove:

$$\sum_{\mu} |\mu\rangle\langle\mu| = \sum_{\mu'} |\mu'\rangle\langle\mu'|. \tag{7.154}$$

The point of this digression into the properties of unitary matrices is that we can apply these properties within the expression

$$\mu_2^{IJ} = \sum_{\mu} \langle I\mu|\hat{H} \sum_{\nu} |J\nu\rangle\langle J\nu|\hat{H}|I\mu\rangle,$$

to show that it is invariant when we rotate the sets $\{|I\mu\rangle\}$ and $\{|J\nu\rangle\}$. We can even rotate these two sets independently. We can therefore choose the orientation of the orbitals to be along the bond direction, according to Fig. 7.1, which has two major benefits. Firstly it reduces the number of bond integrals, and secondly it eliminates the angular dependence of the individual terms, since we have terms such as $dd\sigma(R_{IJ})^2$, depending on the bond length R_{IJ}, in place of general terms $H_{I\mu J\nu}(\mathbf{R}_{IJ})^2$ that depend on the direction of $\mathbf{R}_{IJ}$. For example in the pure d-orbital case we would have

$$\mu_2^{IJ} = dd\sigma(R_{IJ})^2 + 2dd\pi(R_{IJ})^2 + 2dd\delta(R_{IJ})^2, \tag{7.155}$$

which is far more convenient than the general expression. This completes the proof of (7.137), in which in the d-orbital case the function of the bond length is given by

$$\beta_{IJ}(R) = dd\sigma(R)^2 + 2dd\pi(R)^2 + 2dd\delta(R)^2. \tag{7.156}$$

The above analysis is background to the simplest possible form of second-moment model which looks like this:

$$E^{\text{tot}} = \sum_{I} E_I \tag{7.157}$$

where

$$E_I = -\sqrt{\phi_I} + \frac{1}{2}\sum_{J} V_{IJ}(R_{IJ}). \tag{7.158}$$

The repulsive term is pairwise as usual, and the bond energy looks like minus the square root of the function ϕ_I which is proportional to the second moment:

$$\phi_I = A_I \sum_{J} \beta_{IJ}(R_{IJ}). \tag{7.159}$$

You might be puzzled that the suffices IJ appear both on β and on R. Really on β they are a shorthand for the *type* of the atoms at sites I and J, because the bond integrals will depend on the elements involved even when the orbital indices are the same.

The Gaussian model for the bond energy (7.143) gives us an explicit expression for the prefactor A_I:

$$A_I = 2n_I^{\text{orb}}\sqrt{\frac{1}{2\pi}}\exp\left(-\frac{(\epsilon_{\text{F}} - a_{0I})^2}{2\mu_2^I}\right), \tag{7.160}$$

where the exponential, you remember, is constant for this species of atom by the assumption that its charge is constant. However, because of the crudeness of the Gaussian density of states this form is never used in practice. Instead both A_I and β_{IJ} are treated as quantitites to be fitted. For example $\beta_{IJ}(R)$ might be fitted to a GSP function, or, as it was by (1984b),to a polynomial.

Exercise
Derive a formula for the bond energy analogous to (7.143), assuming a rectangular density of states. This is the *Friedel model.* Notice the parabolic dependence of the bond energy on the filling of the band, ie. the number of electrons or the Fermi energy. Even this simple model explains why the strongest cohesion in transition metals occurs in the middle of the series, when the density of states is around half full.

7.10.3 An Effective Pairwise Potential

The second-moment model is a many-body description of the energy, in the sense that it cannot be expressed as a sum of pairwise potentials alone. This is a quality it shares with the description of simple metals discussed in Chapter 6. Neither can these models be expressed in terms of pairwise interactions plus three and four body interactions, since the quantity ϕ_I depends on the distances of *all* the nearest neighbours of I, of which there are twelve in close packed metals. Having said that, for deformations or rearrangements of the atoms which do not change the local density of atoms much, we can expand ϕ_I about a reference value and recover an effective pair potential description. Denoting reference values by a superscipt 0 and expanding the square root to first-order we have:

$$E_I = -\sqrt{\phi_I^0} - \frac{\phi_I - \phi_I^0}{2\sqrt{\phi_I^0}} + \frac{1}{2}\sum_J V_{IJ}(R_{IJ}) = -\frac{1}{2}\sqrt{\phi_I^0} - \frac{\phi_I}{2\sqrt{\phi_I^0}} + \frac{1}{2}\sum_J V_{IJ}(R_{IJ}). \tag{7.161}$$

Now the effective pairwise potential is defined by

$$V_{IJ}^{\text{eff}}(R_{IJ}) = -\left(\frac{A_I}{2\sqrt{\phi_I^0}} + \frac{A_J}{2\sqrt{\phi_J^0}}\right)\beta_{IJ}(R_{IJ}) + V_{IJ}(R_{IJ}). \tag{7.162}$$

This has the desired property that the total energy is given by a constant plus the sum of the effective pairwise potential:

$$E^{\text{tot}} = -\sum_I \sqrt{\phi_I^0} + \frac{1}{2} \sum_{IJ, I \neq J} V_{IJ}^{\text{eff}}(R_{IJ}). \tag{7.163}$$

This formula may give some insight in comparing total energies but it is not a good basis for the calculation of interatomic forces. The reason is that if our reference structure is in equilibrium, the exact forces will be of first order in deviations of the positions $\mathbf{R}_I$ from the reference. But forces of this order would also be obtained by differentiating the second and higher order terms in an expansion of E^{tot} in powers of $\phi_I - \phi_I^0$. So the forces neglected in (7.163) are of the same order in the displacements of atoms as the forces included.

7.11 Fourth-Moment Models

Although second-moment models continue to be widely used by virtue of their simplicity and speed of computation, their limitations are rather easily exposed. See for example Brown and Carlsson (1985). A well-studied example is the trend in crystal structures observed as we look from left to right across the series of transition metals. The first, second and third transition rows of the periodic table correspond to the filling of the $3d$-, $4d$- and $5d$-orbitals. In the solids these orbitals are broadened into d-bands and mixed or *hybridised* with s-orbitals. In a d-band there is room for ten electrons per atom, and in an s-band only for two, so the d-band character dominates the density of states of transition metals.

The FCC transition metals such as Ni, Pt, Pd, including the noble metals Cu, Ag, Au have rather similar densities of states. Their different Hamiltonian matrix elements do introduce differences, and fix the energy scale, but the topology of the FCC lattice, which controls the moments, is clearly reflected in the similar multiple peak structure of their densities of states. The typical FCC density of states is also rather similar to that of the hexagonal close packed (HCP) transition metals Ti and Zr because the nearest neighbours in HCP and FCC lattices are similarly disposed. Notice that the FCC and HCP lattices are characteristic of metals to the right and left of the transition series respectively. In the middle of the transition series we find the high-melting point metals with the body-centred cubic crystal structure, such as V, Nb, Cr, Mo, W. Their density of states is characterised by a dip in the middle, near the Fermi energy, with broad peaks to lower and higher (unnocupied) energy.

The coincidence of a dip in the density of states with the position of the Fermi energy leads to the strong bonding and stability of the BCC transition metals. This can be explained by comparing the bond energies E^{bond} for different shaped densities of states as a function of band filling or Fermi energy. For a metal with a half-filled band such as Mo, a BCC-style density of states shifts the centre of gravity of the lower, occupied part of the band to lower energy with respect to the centre of the band, as compared to an FCC-style density of states. Now there is no way that a second-moment model can distinguish differences of this kind between two different densities of states, so we can be sure that a second-moment model cannot explain the relative stability of the BCC versus FCC or HCP structures, although it can reproduce it! It is possible for a second-moment model to be constructed, as in Finnis *et al.* (1984), such that it correctly stabilizes the BCC versus the FCC structure, but this cannot be for the correct physical reason, which means that the model cannot be trusted to correctly describe other small energy differences between different structures. The stabilisation of BCC within a second-moment model takes advantage of the fact that the six second neighbours in BCC are only slightly further away than the eight first neighbours, so by making the first well in $V_{IJ}^{\text{eff}}(R)$ rather broad all fourteen neighbours can lie within it, as opposed to just twelve in the competing FCC structure.

You can see that to get a dip in the middle of a density of states it is necessary and sufficient to take account of μ_4. To get a feeling for the relationship, consider a symmetric density of states ($\mu_3 = 0$) consisting of three delta functions located at $\epsilon = 0$ and $\epsilon = \pm\epsilon_1$, with weights w_0 and w_1. We have $\mu_2 = 2w_1\epsilon_1^2$ and $\mu_4 = 2w_1\epsilon_1^4$. The *shape* of the density of states must be described by dimensionless parameters, so the appropriate parameter for representing the effect of μ_4 is $\mu_4/\mu_2^2 = (2w_1)^{-1}$. This relation describes how *decreasing* μ_4 is associated with *increasing* w_1, which to preserve the normalisation must take weight away from the central peak w_0. More generally, the trend towards a dip in the centre of the density of states, otherwise called a *bimodal* distribution, is associated with a reduction in μ_4, as shown schematically in Fig. 7.3. The topological explanation of the stability of the BCC structure lies in the configuraton of paths that result in a smaller μ_4.

The recursion method together with (7.111), or alternatively direct path counting, can be used to obtain μ_4^I from a given Hamiltonian matrix. The topology of all the paths that contribute is sketched in Fig. 7.4. I have omitted for brevity the atom labels in these sketches, and show only the paths that have a distinct topology. Thus path (b) of Fig. 7.4 really stands for two situations, either the *loops* are on the starting atom, for which the density of states is being evaluated, or they are on a neighbour to it. Likewise on path (e), the loop may be on the starting atom or one of its neighbours. In path (f), the starting atom may be in the middle or at one of the ends. If the zero of energy is chosen as usual to be a_0, then the terms represented by *loops* starting and ending at the starting atom vanish. Usually the approximation is made that the loops from all the other atoms can also be ignored; that is, all the on-site elements of the Hamiltonian on all the atoms are the same. In that case the only non-vanishing contributions are from paths of type (d), (f) and (g).

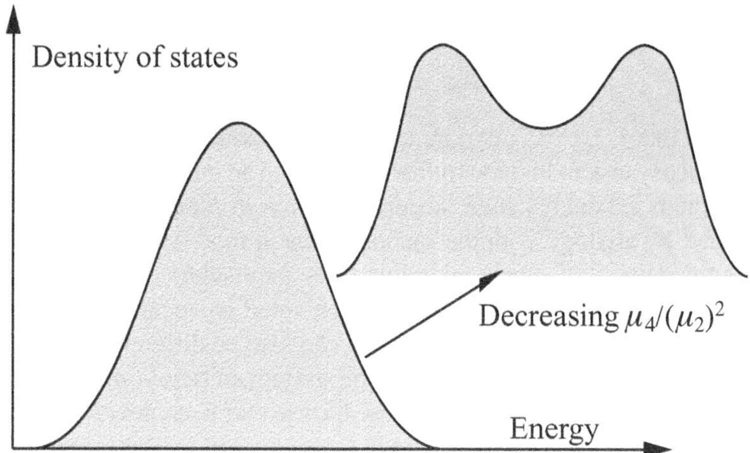

FIG. 7.3 The trend in the shape of a density of states as its fourth moment is varied.

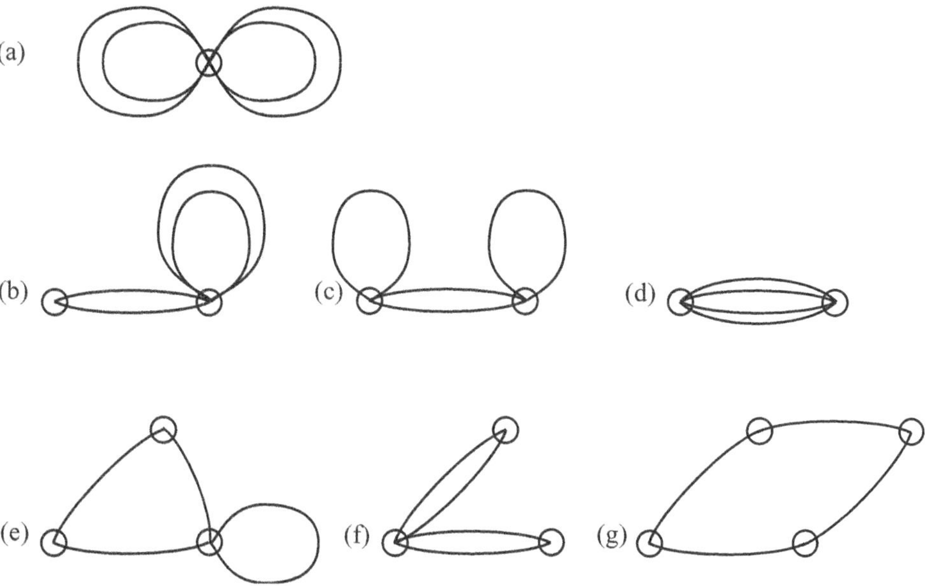

FIG. 7.4 Paths which contribute to the fourth moment of the local density of states. (a) 1-atom, (b)–(d) 2-atom, (e)-(f) 3-atom, (g) 4-atom.

A large contribution to μ_4^I comes from the reducible paths, representing terms that are the product of paths that enter μ_2^J, where J is a neighbour of I, including I itself. These are very naturally dealt with in the recursion approach. Notice from eq. (7.111) that when a_1 is chosen to be zero, $\mu_4 = \mu_2^2 + b_2^2\mu_2$. This tells us that the important part of the fourth moment for structural purposes is

$$\mu_4 - \mu_2^2 \equiv b_2^2 \mu_2. \tag{7.164}$$

The other part that is μ_2^2 has no more information about the structure than μ_2.

It is not an easy task to include fourth moments in an interatomic potential model to simulate the bond energy more accurately, even in pure transition metals. The obvious route, by analogy with the second-moment model, would be to construct a model of the density of states, including three parameters, which could be fitted to μ_2, μ_3 and μ_4. The moments could be obtained by recursion, using (7.111). Unfortunately there is no simple model that can also be differentiated analytically to give consistent forces. As we shall see, the strategy of *bond-order potentials* was a response to this difficulty, although these have so far been developed in analytic form only for certain sp-bonded systems.

A simple alternative has been proposed and applied by Carlsson (1990, 1991). The idea is to capture the essential μ_4-dependence of the energy by adding to a second-moment model a term linear in that part of μ_4 that is not already taken into account by the second moments. We might think of this as the first term in a Taylor expansion of the bond energy in deviations of the moments from some 'reference values'. The reference value of μ_4 should certainly not be zero, because any conceivable density of states which has a non-zero μ_2 will also have a non-zero μ_4. The simplest represention of the density of states that gets the second moment right is just the sum of two delta functions, obtained by truncating the continued fraction after the first level:

$$D_{I\mu}(\epsilon) = \tfrac{1}{2}\left\{\delta(\epsilon - b_1) + \delta(\epsilon + b_1)\right\}. \tag{7.165}$$

Its fourth moment is just the square of its second moment. This provides then a kind of reference for writing the new term in the bond energy as

$$E^{\text{bond}(4)} = B \sum_I \left\{\mu_4^I - (\mu_2^I)^2\right\} / (\mu_2^I)^{3/2}. \tag{7.166}$$

B is a constant which in practice is one of the fitting parameters. The factor of $(\mu_2^I)^{3/2}$ appears in the denominator for dimensional reasons, to make the quantity an energy. The picture is that $(\mu_2^I)^{1/2}$ sets the basic scale of energy for the density of states, and this is multiplied in (7.166) by the dimensionless fourth-moment parameter which we deduced from recursion, namely $\mu_4^I/(\mu_2^I)^2 - 1$. A more sophisticated treatment, taking account of matrix recursion, is given by Carlsson (1991), who applies it successfully to the tungsten (100) surface. A more empirical version of this fourth-moment model has been developed as an addition to the embedded atom model by Foiles (1993).

While there are a large number of contributions to the fourth-moment parameter, Carlsson (1991) suggested that those coming from paths of type (g) in Fig. 7.4

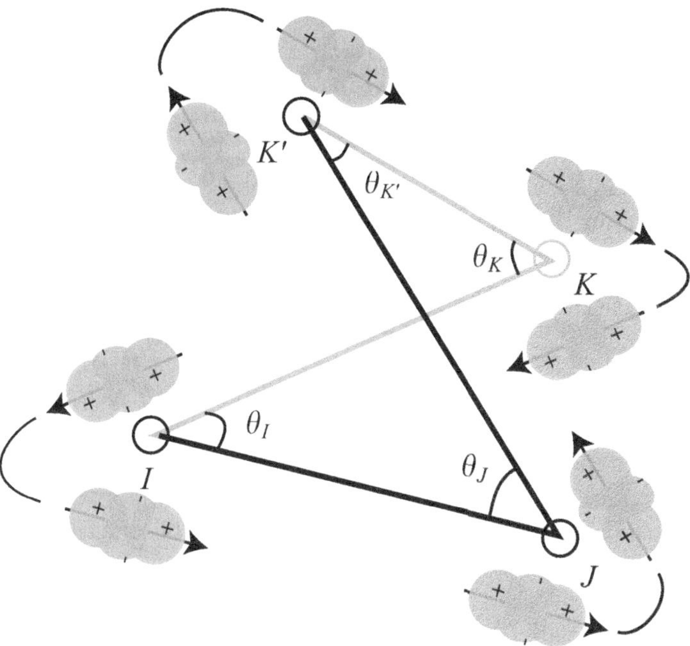

FIG. 7.5 Sketch to define the angles and the rotation of the sigma orbitals in a 4-body contribution to μ_4.

are the main ones responsible for the stability of the BCC structure in the middle of the transition series of the periodic table. If only the 'sigma' type interactions are included, in this case meaning $dd\sigma$ matrix elements, a four-body term has a particularly simple form. Referring now to Fig. 7.5, suppose the path starts with atom I to atom J, and this bond defines the z-axis for the orbitals. The first factor in the path is then $dd\sigma(R_{IJ})$. Now the σ orbital on J that participated in this bond is wrongly orientated with respect to the next step to atom K'. It has to be resolved into its five components with respect to the new orientation, which has the σ orbital along $\mathbf{R}_{JK'}$. Of these components we have decided to consider only the sigma orbital in the new direction. The other components will introduce π-bonds and δ-bonds into the path, but we are not following these now. The sigma component (or projection) along $\mathbf{R}_{JK'}$ is given by $P_2(\cos\theta_J)$, as can be shown from the rules for rotating spherical harmonics, where $P_2(x) = \frac{1}{2}(3x^2 - 1)$ is the Legendre polynomial of order 2. Incidentally, if these were p-orbitals rather than d-orbitals, then the component would just be $P_1(\cos(\pi - \theta_J)) = \cos(\pi - \theta_J)$, which is the usual projection of a 3-vector onto a given direction. Due to the projection onto the new direction, the step from J to K' introduces a factor $dd\sigma \cdot P_2(\cos\theta_J)$. The same process of rotation or projection has to be done at the next atoms in the path, until the final rotation brings the last orbital on I back to its original orientation. Altogether the contribution to μ_4 is

$$dd\sigma(R_{IJ})dd\sigma(R_{JK'})dd\sigma(R_{K'K})dd\sigma(R_{KI})$$
$$\times P_2(\cos\theta_J)P_2(\cos\theta_{K'})P_2(\cos\theta_K)P_2(\cos\theta_I).$$

Carlsson (1991) gives values of the angular factors involved in BCC and FCC lattices, which show that the quadrilaterals in the BCC lattice would give smaller contributions than those in the FCC lattice. This offers a simple real space explanation of the stability of the BCC lattice. However, as a quantitative scheme, the introduction of fourth-moment effects in the linearised way of (7.166) is only accurate for a limited range of structures.

Furthermore to explain more subtle energy differences, such as the HCP-FCC difference, higher moments are required. In semiconductors, with fourfold coordination based on the diamond structure, rings of six atoms are the shortest paths which do not retrace any steps. Because of the structural importance of these rings, we might expect that at least six moments of the density of states must be dealt with in order to describe the relative energies of different structures in a physically based model. Reconstruction of the densities of states with the maximum entropy method in this case leads to a rather disappointing conclusion; to achieve convergence of the bond energy, obtained by integration over the density of states, you actually seem to need many more moments than six (Kress and Voter, 1995). Bond-order potentials provide an alternative way of exploiting path counting information, which recovers the physics of structural energy differences in semiconductors using only about six moments.

7.12 Bond-order Potentials

7.12.1 A Multi-atom Expansion of the Bond Energy

It would be excellent to have a model for the bond order in the form of an analytic function of bond lengths and angles. With such a function the bond energy could be evaluated simply by multiplying it by the Hamiltonian matrix elements, the procedure would be much faster computationally than the calculation and processing of densities of states, and forces could be derived for use in large-scale $O(N)$ atomistic simulation. This has now been achieved for some useful practical cases, and although the basic theory looks deceptively straightforward, the derivation of an accurate scheme of approximation even in simple systems has proved surprisingly difficult. Consequently the literature of the last decade, over which the story of BOPs has unfolded, is rather difficult to get to grips with. Since Pettifor's seminal paper (Pettifor, 1989) there has been a chain of theoretical progress, leading to potentials for silicon, carbon and hydrocarbons that embody the physics of single, double and triple bonds as well as the formation of radicals. Models for intermetallic alloys were developed at the same time (Pettifor and Aoki, 1991), going beyond the second-moment level of approximation. As I write, further developments and applications to semiconductor compounds and transition metals are still underway. I will outline in this section the general theory of BOPs, and work through a simple example in the next section. This should help you to access the literature and enable you to follow future progress.

Historically there are two main ideas that set the ball rolling. The first is the identity (1.203) from which in an orthogonal basis we have

$$\epsilon\langle I\mu|\hat{G}(\epsilon)|I\mu\rangle = \sum_{J\nu}\langle I\mu|\hat{G}(\epsilon)|J\nu\rangle\langle J\nu|\hat{H}|I\mu\rangle + 1. \tag{7.167}$$

Taking imaginary parts and integrating up to the Fermi energy, the left-hand side gives the integral of the single electron energy over the local density of states, and the right-hand side becomes a sum of bond orders times Hamiltonian matrix elements. This identity invited calculations of the off-diagonal elements of G and stimulated development of the inter-site TBBM as an alternative to the on-site method. The other idea is that these off-diagonal elements are within our grasp via a real space method such as the recursion method, by taking the difference between two densities of states as in eq. (7.129). This difference is not just a mathematical trick, because the two densities in this equation refer to *bonding* and *antibonding* orbitals, and the equation is expressing the bond order as the difference in population of these orbitals. In the case of diatomic molecules it is a familiar idea that the bond order is related to the strength of a bond. Both are maximised with respect to the number of electrons when the bonding orbitals are occupied and the antibonding orbitals are empty, and the bonding weakens when electrons occupy antibonding orbitals. The idea of using bonding and antibonding orbitals as starting orbitals, and to generate moments from them, is basic to the bond order approach.

Subtracting the two densities of states in eq. (7.129) looks easy enough, but it was soon discovered that the resulting bond orders converge very slowly with the numbers of exact moments of the constituent bonding and antibonding densities of states. A calculation of the band energy from the on-site density of states converges much faster. Much of the effort of the ten years following the first papers on this topic was directed to finding ways to improve the cancellation of errors in the difference between the Green functions: $\langle 0^+|\hat{G}|0^+\rangle - \langle 0^-|\hat{G}|0^-\rangle = 2\langle I\mu|\hat{G}(\epsilon)|J\nu\rangle$. A key step (Aoki, 1993; Aoki and Pettifor, 1993), see also Pettifor and Oleinik (1999), was the discovery that there is an exact expansion for these inter-site Green functions that takes the form of a power series in the terms $\langle I\mu|\hat{H}^p|J\nu\rangle$, for which we adopt Pettifor's notation ζ_{p+1}. These terms can be evaluated by enumerating the paths of length p steps linking the two sites. They have been called *interference* terms. Without them, the bonding or antibonding diagonal elements $\langle 0^+|\hat{G}|0^+\rangle$ or $\langle 0^-|\hat{G}|0^-\rangle$ would be equal to the corresponding on-site Green function denoted G_{00}

where

$$G_{00}(\epsilon) = \tfrac{1}{2}\left(\langle I\mu|\hat{G}(\epsilon)|I\mu\rangle + \langle J\nu|\hat{G}(\epsilon)|J\nu\rangle\right), \tag{7.168}$$

as indeed they become in the limit that the two sites are far apart.

It is instructive to derive this expansion, but to do so we shall need a couple of new results and definitions. Firstly note that the moments of the density of states on the orbital $|0^\lambda\rangle$ defined by eq. (7.130) are given as follows:

$$\mu_p^\lambda = \langle 0^\lambda|\hat{H}^p|0^\lambda\rangle = \tfrac{1}{2}(\mu_p^{I\mu} + \mu_p^{J\nu}) + \lambda\zeta_{p+1}, \tag{7.169}$$

That is, these rather peculiar moments are just the average of standard on-site moments plus the interference term multiplied by λ. The interference term is thus given by the λ-derivative of μ_p^λ:

$$\frac{\partial\mu_p^\lambda}{\partial\lambda} = \zeta_{p+1}. \tag{7.170}$$

Now we are in sight of our goal. We know that like any other diagonal element G_{00}^λ can be expressed in terms of the associated moments μ_p^λ, albeit indirectly via its a_i and b_i. If we can just differentiate G_{00}^λ with respect to these moments we will be able to apply the chain rule of differentiation, inserting (7.170) to obtain:

$$\langle I\mu|\hat{G}|J\nu\rangle = \sum_{p=0}^{\infty}\frac{\partial G_{00}^\lambda}{\partial\mu_p^\lambda}\frac{\partial\mu_p^\lambda}{\partial\lambda} = \sum_{p=1}^{\infty}\frac{\partial G_{00}^c}{\partial\mu_p^c}\zeta_{p+1}. \tag{7.171}$$

The superscript c on G_{00}^c and its moments indicates that the complex starting orbital and the $\lambda = 0$ value have been used, as in eq. (7.127), in which the starting orbital is $|I\mu\rangle + \mathrm{i}|J\nu\rangle$. Its associated moments are just the moments of the average densities of states of the two orbitals.

Equation (7.171) is the expansion we want, and it is the main constituent of a BOP. The bond order itself is given from (1.210) by

$$\rho^{I\mu J\nu} = -\frac{2}{\pi}\int_{-\infty}^{\epsilon_F}\langle I\mu|\hat{G}|J\nu\rangle\,\mathrm{d}\epsilon, \tag{7.172}$$

where the orthogonality of the basis enables us to write $G^{I\mu J\nu} = G_{I\mu J\nu}$ in (1.210), and $G_{I\mu J\nu} \equiv \langle I\mu|\hat{G}|J\nu\rangle$ by definition.

Notice that the parameter λ has tactfully disappeared from the scene. We still have to evaluate the derivatives in (7.171), and this is best done by a second application of the chain rule, first differentiating G with respect to its recursion coefficients a_i and b_i^2:

$$\frac{\partial G_{00}^c}{\partial\mu_p^c} = \sum_{i=1}^{\mathcal{L}}\left\{\frac{\partial G_{00}^c}{\partial a_i}\frac{\partial a_i}{\partial\mu_p^c} + \frac{\partial G_{00}^c}{\partial b_i^2}\frac{\partial b_i^2}{\partial\mu_p^c}\right\}. \tag{7.173}$$

The reason why this is helpful is first that fairly simple expressions (7.112), are available for the continued fraction coefficients as functions of the moments, which one can differentiate, and second simple expressions for the a_i and b_i derivatives of a continued fraction can be obtained (e.g. Appendix B of Turchi and Ducastelle, 1985 and Pettifor and Aoki, 1991). Then the derivative $\partial G^c_{00}/\partial \mu^c_p$ can be expressed entirely in terms of the calculated coefficients a_i and b_i, for $i = 0 \cdots \mathcal{L}$ of G^c_{00}. Let us see how this works. We first obtain general results for the derivative of a continued fraction with respect to its a_i and b_i^2. At this point we can drop the superscript c, as the following results would apply for any starting orbital.

We can obtain an expression for $\partial G_{00}/\partial a_i$ by a recursive procedure, following the heirarchy defined by (7.110):

$$\begin{aligned}
\frac{\partial G^>_{00}}{\partial a_i} &= b_1^2 G^2_{00} \frac{\partial G^>_{11}}{\partial a_i}, \\
\frac{\partial G^>_{11}}{\partial a_i} &= b_2^2 G^{>\,2}_{11} \frac{\partial G^>_{22}}{\partial a_i}, \\
&\vdots \\
\frac{\partial G^>_{ii}}{\partial a_i} &= G^{>\,2}_{ii},
\end{aligned} \tag{7.174}$$

from which

$$\frac{\partial G_{00}}{\partial a_i} = b_1^2 b_2^2 \cdots b_i^2 G^2_{00} G^{>\,2}_{11} \cdots G^{>\,2}_{ii} = \frac{\mathcal{D}^2_{i+1}}{\mathcal{D}^2_0} b_1^2 b_2^2 \cdots b_i^2. \tag{7.175}$$

The polynomials $\mathcal{D}_i(\epsilon)$ were defined in (7.117). A similar differentiation down the heirarchy gives us

$$\frac{\partial G_{00}}{\partial b_i^2} = b_1^2 b_2^2 \cdots b_{i-1}^2 G^2_{00} G^{>\,2}_{11} \cdots G^>_{ii} = \frac{\mathcal{D}_i \mathcal{D}_{i+1}}{\mathcal{D}^2_0} b_1^2 b_2^2 \cdots b_{i-1}^2. \tag{7.176}$$

All the derivatives of the first four recursion coefficients with respect to the moments are given by:

$$\begin{aligned}
&\frac{\partial a_0}{\partial \mu_1} = 1, \quad \frac{\partial b_1^2}{\partial \mu_1} = -2a_0, \quad \frac{\partial b_1^2}{\partial \mu_2} = 1, \\
&\frac{\partial a_1}{\partial \mu_1} = \frac{-2b_1^2 + 2a_0 a_1 + a_0^2}{b_1^2}, \quad \frac{\partial a_1}{\partial \mu_2} = -\frac{2a_0 + a_1}{b_1^2}, \quad \frac{\partial a_1}{\partial \mu_3} = \frac{1}{b_1^2}, \\
&\frac{\partial b_2^2}{\partial \mu_1} = \frac{2(a_0 + a_1)(b_1^2 - a_0 a_1) + 2a_0^2 b_2^2}{b_1^2}, \quad \frac{\partial b_2^2}{\partial \mu_2} = \frac{a_0^3 + 4a_0 a_1 + a_1^2 - 2b_1^2 - 2b_2^2}{b_1^2}, \\
&\frac{\partial b_2^2}{\partial \mu_3} = \frac{-2a_1}{b_1^2}, \quad \frac{\partial b_2^2}{\partial \mu_4} = \frac{1}{b_1^2}.
\end{aligned} \tag{7.177}$$

It should be clear by now that b_i depends on the first $2i$ moments, and a_i depends on the first $2i+1$ moments. It is also worth pointing out that the a_i and b_i derivatives of G_{00} have a very physical interpretation when we integrate their imaginary parts up to the Fermi energy, which has to be done before they enter the expression for the bond energy. Think of the equivalent linear chain problem generated by the recursion method. These derivatives are describing the change in charge on the head of the chain with respect to variation in the potential at atoms or bonds along the chain. They are therefore response functions. To be more precise they must be the matrix elements of the response function χ_e for the chain problem.

From (7.171), inserting (7.175) and (7.176), we obtain the general expansion for $G_{I\mu J\nu}$. It is convenient to express this in terms of the quantities introduced by Pettifor and coworkers.

$$\delta(a_j) = \sum_{p=1}^{2j+1} \frac{\partial a_j}{\partial \mu_p} \zeta_{p+1}, \quad \delta(b_j^2) = \sum_{p=1}^{2j} \frac{\partial b_j^2}{\partial \mu_p} \zeta_{p+1}. \tag{7.178}$$

The resulting expression is:

$$\begin{aligned} G_{I\mu J\nu} = \frac{1}{\mathcal{D}_0^2} \Big\{ & \mathcal{D}_1^2 \zeta_2 + \mathcal{D}_1 \mathcal{D}_2 \delta(b_1^2) + \mathcal{D}_2^2 \delta(a_1) \\ & + \sum_{j=2}^{\mathcal{L}} \left[\mathcal{D}_j \mathcal{D}_{j+1} b_1^2 b_2^2 \cdots b_{j-1}^2 \delta(b_j^2) + \mathcal{D}_{j+1}^2 b_1^2 b_2^2 \cdots b_j^2 \delta(a_j) \right] \Big\}. \end{aligned} \tag{7.179}$$

Equation (7.179) is a central result of the theory of BOPs and the key formula for developing BOPs in the form of a multi-atom expansion. It takes in more and more of the environment of the bond the greater the value of $\mathcal{L}$, but becomes rapidly more difficult to evaluate.

There were many technical difficulties to be addressed before the formalism became a practical scheme. Indeed, the approach is still evolving, and for the details I refer the reader to the literature, in particular: Horsfield *et al.* (1996*a*,*b*), Aoki *et al.* (1997), Oleinik and Pettifor (1999), Ozaki *et al.* (2000), Pettifor and Oleinik (2000, 2002). As usual, it is desirable that the bond order derived from (7.179) produces an energy from which the forces can be derived by differentiating only the matrix elements of the Hamiltonian. The energy should also be rotationally invariant. And the energy and forces should be sufficiently converged with respect to $\mathcal{L}$. Block recursion was implemented to take care of the symmetry problem when it arose, for example in π-bonded systems. The convergence problem is trickier, since one would like to neglect some of the longer paths ζ_p. For example, if we are working to level $\mathcal{L} = 1$, we have

$$\mathcal{D}_2 = 1, \quad \mathcal{D}_1 = \epsilon - a_1, \quad \mathcal{D}_0 = (\epsilon - a_0)(\epsilon - a_1) - b_1^2. \tag{7.180}$$

From (7.179) and (7.178), this implies paths up to and including ζ_4 for getting $\delta(a_2)$, which are 2-, 3- and 4-body terms, namely all the paths of three steps linking I with J. While this is still manageable, already with $\mathcal{L} = 2$ it seems that we would need to count paths up to ζ_6, which is a more daunting prospect. Fortunately they are not all of equal importance, and there is a way to simplify the procedure that Pettifor and coworkers (Pettifor and Oleinik, 2002) have recently found to be useful. This is to exploit the requirement that all the matrix elements of $\hat{G}$ have at least a subset of the poles of G_{00}. It implies that $\mathcal{D}_0(\epsilon)$ must be a factor of the expression within the braces of (7.179), so that the resulting expression for $G_{I\mu J\nu}$ has the same denominator $\mathcal{D}_0$ as G_{00} does. The procedure is best illustrated by an example.

7.12.2 Example: Analytic Bond Orders in a p-Bonded Trimer

The main physical ideas of BOPs can be illustrated by studying a bond between atoms I and J in a p-bonded system. The extension to sp-bonded systems in the literature is more complicated, and involves some additional approximations (Pettifor and Oleinik, 2002), but the physical ideas are the same. Let us only consider the interaction of this bond with atom K, and suppose for simplicity that the bond $\mathbf{R}_{JK}$ makes an angle θ with $\mathbf{R}_{IJ}$. We also suppose there is no matrix element linking I with K. This would be the normal case in tetrahedrally bonded semiconductors. The configuration is illustrated in Fig. 7.6. Although there are 9 p-orbitals altogether in the minimum basis, on atoms I and K we only need consider the ones that point along the bonds, which we label $|I\rangle$ and $|K\rangle$. On atom J we must consider the two orbitals pointing in x and y directions, $|Jx\rangle$ and $|Jy\rangle$, since they each have components along the bond to K. None of the remaining 5 orbitals couple with this set if we neglect π-bonding integrals ($pp\pi$). This is a good approximation in bulk silicon. It is not adequate to describe the range of bonding

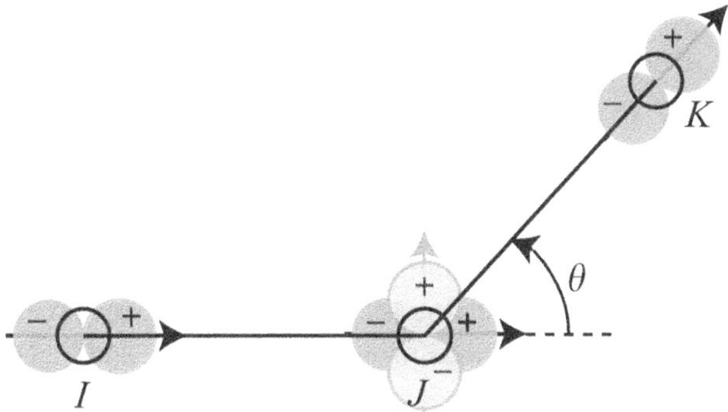

FIG. 7.6 The bond $I - J$ and neighbour K, showing the conventional sign and direction of the p-orbitals.

between different structural types, or between single, double and triple bonds in the carbon system for example, but it will suffice to illustrate the method and some of its further general features.

BOP theory introduces bond-angle dependence of the energy, which is absent in the familiar on-site second-moment model discussed previously. Taking a step beyond the second-moment approximation, Pettifor showed how the resulting BOP turns out to be somewhat similar in functional form to the empirical potential of Tersoff (1986).

The Hamiltonian in the space of orbitals we have selected is:

$$\begin{aligned}\hat{H} = \epsilon_p \sum_{I'\mu} |I'\mu\rangle\langle I'\mu| &+ pp\sigma\{|I\rangle\langle Jx| + |Jx\rangle\langle I|\} \\ &+ pp\sigma \cos\theta\{|Jx\rangle\langle K| + |K\rangle\langle Jx|\} \\ &+ pp\sigma \sin\theta\{|Jy\rangle\langle K| + |K\rangle\langle Jy|\}.\end{aligned} \tag{7.181}$$

Several of the features of this Hamiltonian are worth noticing. First, by choosing ϵ_p as the zero of energy we can eliminate the first term. Second, the p-orbitals on J can be resolve parallel and perpendicular to the bond $\mathbf{R}_{JK}$. Only their components parallel to $\mathbf{R}_{JK}$ are involved in the σ bond integrals with atom K, hence the factors $\cos\theta$ and $\sin\theta$ appear in the overlap of the $|Jx\rangle$ and $|Jy\rangle$ orbitals with the orbital $|K\rangle$. We have seen the bond angles entering the theory in an analogous way for d-orbitals in the fourth-moment model described previously, in which the angle factor was $P_2(\theta)$. In the present case when the bonds are collinear ($\theta = 0$) there is no coupling of I or K to the third orbital, $|Jy\rangle$. This coupling increases as $\sin\theta$ and is essential for a proper description of the energy of bond-bending. Finally, our choice of axes means that there is never coupling between $|I\rangle$ and $|Ky\rangle$.

This Hamiltonian, which refers to an isolated trimer, has the 4×4 matrix representation:

$$\mathbf{H} = pp\sigma \begin{pmatrix} 0 & 1 & 0 & 0 \\ 1 & 0 & 0 & \cos\theta \\ 0 & 0 & 0 & \sin\theta \\ 0 & \cos\theta & \sin\theta & 0 \end{pmatrix}. \tag{7.182}$$

First, let us review the solution obtained by conventional matrix diagonalisation, then we will revisit the problem using the multi-atom expansion.

We can easily solve the eigenvalue problem in this case, and find the eigenvalues:

$$\begin{aligned}\epsilon_1 &= -pp\sigma\sqrt{2}\cos\tfrac{1}{2}\theta \quad \epsilon_2 = -pp\sigma\sqrt{2}\sin\tfrac{1}{2}\theta, \\ \epsilon_3 &= +pp\sigma\sqrt{2}\sin\tfrac{1}{2}\theta \quad \epsilon_4 = +pp\sigma\sqrt{2}\cos\tfrac{1}{2}\theta,\end{aligned} \tag{7.183}$$

with associated eigenvectors:

$$\begin{aligned}
\psi_1 &= \tfrac{1}{2}(+1, -\sqrt{2}\cos\tfrac{1}{2}\theta, -\sqrt{2}\sin\tfrac{1}{2}\theta, +1),\\
\psi_2 &= \tfrac{1}{2}(+1, -\sqrt{2}\sin\tfrac{1}{2}\theta, +\sqrt{2}\cos\tfrac{1}{2}\theta, -1),\\
\psi_3 &= \tfrac{1}{2}(+1, +\sqrt{2}\sin\tfrac{1}{2}\theta, -\sqrt{2}\cos\tfrac{1}{2}\theta, -1),\\
\psi_1 &= \tfrac{1}{2}(+1, +\sqrt{2}\cos\tfrac{1}{2}\theta, +\sqrt{2}\sin\tfrac{1}{2}\theta, +1).
\end{aligned} \tag{7.184}$$

The entries for each eigenvector are the coefficients of $|I\rangle$, $|Jx\rangle$, $|Jy\rangle$ and $|K\rangle$ in its expansion. Hence the matrix elements of $\hat{G}$ in the atomic basis are given by inverting $\epsilon\mathbf{1} - \mathbf{H}$. This is a straightforward exercise for the reader.

Three of these matrix elements, the ones associated with atoms I, J and the energy of bond IJ, are:

$$\begin{aligned}
\langle I|\hat{G}|I\rangle &= \frac{1}{4}\left(\frac{1}{\epsilon-\epsilon_1} + \frac{1}{\epsilon-\epsilon_2} + \frac{1}{\epsilon-\epsilon_3} + \frac{1}{\epsilon-\epsilon_4}\right),\\
\langle Jx|\hat{G}|Jx\rangle &= \frac{1}{4}\left(\frac{2\cos^2\frac{1}{2}\theta}{\epsilon-\epsilon_1} + \frac{2\sin^2\frac{1}{2}\theta}{\epsilon-\epsilon_2} + \frac{2\sin^2\frac{1}{2}\theta}{\epsilon-\epsilon_3} + \frac{2\cos^2\frac{1}{2}\theta}{\epsilon-\epsilon_4}\right),\\
\langle I|\hat{G}|Jx\rangle &= \frac{1}{4}\left(-\frac{\sqrt{2}\cos\frac{1}{2}\theta}{\epsilon-\epsilon_1} - \frac{\sqrt{2}\sin\frac{1}{2}\theta}{\epsilon-\epsilon_2} + \frac{\sqrt{2}\sin\frac{1}{2}\theta}{\epsilon-\epsilon_3} + \frac{\sqrt{2}\cos\frac{1}{2}\theta}{\epsilon-\epsilon_4}\right).
\end{aligned} \tag{7.185}$$

The Green function matrix elements $\langle I|\hat{G}|Jy\rangle$ and $\langle I|\hat{G}|K\rangle$ do not affect the bond energy, in which they are multiplied by vanishing Hamiltonian matrix elements. Now, if there are just two electrons (or two holes) we know that the bond energies must sum to $-2\epsilon_1$, so let us check this. The bond order between I and J is

$$\rho_{IJx} = -\frac{2}{\pi}\mathrm{Im}\int_{-\infty}^{\epsilon_F} G_{IJx}\mathrm{d}\epsilon = -\tfrac{1}{2}\sqrt{2}\cos\tfrac{1}{2}\theta. \tag{7.186}$$

The energy of the bond IJ is:

$$\rho_{IJx}\, pp\sigma + \rho_{JxI}\, pp\sigma = -pp\sigma\sqrt{2}\cos\tfrac{1}{2}\theta = -\epsilon_1. \tag{7.187}$$

It is gratifying that this is half the total bond energy, since the two bonds are physically equivalent. We must have the same result for the total energy of bond JK:

$$(\rho_{JxK} + \rho_{KJx})pp\sigma\cos\theta + (\rho_{JyK} + \rho_{KJy})pp\sigma\sin\theta = -\epsilon_1. \tag{7.188}$$

The sum of the bond energies thus correctly gives the energy of the doubly occupied lowest eigenstate. There is a restoring force to bond-bending of magnitude

$$\frac{\mathrm{d}^2\epsilon_1}{\mathrm{d}\theta^2} = \tfrac{1}{4}\sqrt{2}pp\sigma. \tag{7.189}$$

Now let us approach the problem from the point of view of the multi-atom expansion (7.179). For it to be exact, we have to include contributions up to $\mathcal{D}_4 = 1$. In this case, because all the on-site elements of the Hamiltonian are zero, $a_i = 0$ for all i. The polynomials are given by:

$$\begin{aligned} \mathcal{D}_4 &= 1 & \mathcal{D}_3 &= \epsilon & \mathcal{D}_2 &= \epsilon^2 - b_3^2 \\ \mathcal{D}_1 &= \epsilon(\epsilon^2 - b_2^2 - b_3^2) & \mathcal{D}_0 &= \epsilon^2(\epsilon^2 - b_1^2 - b_2^2 - b_3^2) + b_1^2 b_3^2. \end{aligned} \tag{7.190}$$

Within the multi-atom expansion we have to consider the following $\delta(a_i)$, $\delta(b_i^2)$:

$$\begin{aligned} \delta(b_1^2) &= \zeta_3 \\ \delta(a_1) &= -2\zeta_2 + \frac{1}{b_1^2}\zeta_4 \\ \delta(b_2^2) &= -\frac{2b_1^2 + b_2^2}{b_1^2}\zeta_3 + \frac{1}{b_1^2}\zeta_5 \\ \delta(a_2) &= \frac{b_1^2 + 2b_2^2}{b_2^2}\zeta_2 - 2\frac{b_1^2 + b_2^2}{b_1^2 b_2^2}\zeta_4 + \frac{1}{b_1^2 b_2^2}\zeta_6 \end{aligned} \tag{7.191}$$

$$\begin{aligned} \delta(b_3^2) &= ? \\ \delta(a_3) &= ? \end{aligned} \tag{7.192}$$

I have not evaluated the last two, but note that in general we would expect $\delta(a_3)$ to include interference terms up to ζ_8, which in this case includes the path starting on I and visiting all the orbitals once before returning to I and making the seventh and final step to J. ζ_8 would be very cumbersome to compute in a general environment of atoms. There is a special simplification in the present case because of the absence of rings of atoms, namely ζ_p vanishes for odd p. To obtain a non-vanishing ζ_3 for example we would require a triangle of orbitals $I - K - J$ in which both $I - K$ and $J - K$ are linked by non-zero matrix elements of $\hat{H}$. No such triangle exists in our trimer, or indeed quite generally in a tetrahedrally bonded structure.

For our trimer (or any system without rings) the multi-atom expansion now becomes:

$$\begin{aligned} G_{IJx} = \frac{1}{\mathcal{D}_0^2}\{ & \mathcal{D}_1^2\zeta_2 + \mathcal{D}_2^2[-2b_1^2\zeta_2 + \zeta_4] \\ & + \mathcal{D}_3^2[b_1^2(b_1^2 + 2b_2^2)\zeta_2 - 2(b_1^2 + b_2^2)\zeta_4 + \zeta_6] \\ & + \mathcal{D}_3\mathcal{D}_4 b_1^2 b_2^2 \delta(b_3^2) + \mathcal{D}_4^2 b_1^2 b_2^2 b_3^2 \delta(a_3)\}. \end{aligned} \tag{7.193}$$

Now if we operate the recursion method on $|I\rangle + \mathrm{i}|Jx\rangle$ to level 2 we find:

$$b_1^2 = pp\sigma^2(1 + \tfrac{1}{2}\cos^2\theta),$$
$$b_2^2 = \frac{pp\sigma^2\cos^2\theta(1 - \frac{1}{4}\cos^2\theta)}{1 + \frac{1}{2}\cos^2\theta}. \quad (7.194)$$

You can also easily count the paths in this trimer to find:

$$\zeta_2 = pp\sigma,$$
$$\zeta_4 = pp\sigma^3(1 + \cos^2\theta), \quad (7.195)$$
$$\zeta_6 = pp\sigma^5(1 + 3\cos^2\theta).$$

We now sidestep the cumbersome tasks of obtaining $\delta(a_3)$ and $\delta(b_3^2)$, and avoid going to the next level of recursion to get b_3, by using the factorisation mentioned at the end of the last section. G_{IJx} cannot have more poles than G_{00} (although it might have less if the weight of one happens to vanish), so we write:

$$G_{IJx} = \frac{A\epsilon^2 + B\epsilon + C}{\mathcal{D}_0}. \quad (7.196)$$

Now we equate this to the previous expression (7.193) and determine the coefficients A, B and C by equating the coefficients of ϵ^6, ϵ^5 and ϵ^4 respectively in the identity:

$$(A\epsilon^2 + B\epsilon + C)\mathcal{D}_0 = \mathcal{D}_1^2\zeta_2 + \mathcal{D}_2^2[-2b_1^2\zeta_2 + \zeta_4]$$
$$+ \mathcal{D}_3^2[b_1^2(b_1^2 + 2b_2^2)\zeta_2 - 2(b_1^2 + b_2^2)\zeta_4 + \zeta_6]$$
$$+ \mathcal{D}_3\mathcal{D}_4 b_1^2 b_2^2\delta(b_3^2) + \mathcal{D}_4^2 b_1^2 b_2^2 b_3^2\delta(a_3). \quad (7.197)$$

This gives us

$$A = \zeta_2,$$
$$B = 0, \quad (7.198)$$
$$C = -(b_1^2 + b_2^2 + b_3^2)\zeta_2 + \zeta_4.$$

Notice that by working with (7.196) we now no longer need to know $\delta(b_3^2)$ or $\delta(a_3^2)$ although they could be determined easily enough by equating coefficients of ϵ and the constant $Cb_1^2b_3^2$. The result is quite general for systems with vanishing on-site energies and no rings:

$$G_{I\mu J\nu} = \frac{(\epsilon^2 - b_1^2 - b_2^2 - b_3^2)\zeta_2 + \zeta_4}{\epsilon^2(\epsilon^2 - b_1^2 - b_2^2 - b_3^2) + b_1^2 b_3^2}. \tag{7.199}$$

Finally we still have the coefficients of ϵ^2 to equate, which gives us an expression for b_3:

$$b_3^2 = \frac{\zeta_6 - (b_1^2 + b_2^2)\zeta_4}{\zeta_4 - b_1^2 \zeta_2}. \tag{7.200}$$

Given the values of ζ_6 and ζ_4 above, the expression for b_3 simplifies to

$$b_3^2 = \frac{pp\sigma^2 \sin^2\theta}{b_1^2}, \tag{7.201}$$

which is exact for the trimer problem we are considering. Physically, as $\theta \to 0$, $b_3 \to 0$, which corresponds to the decoupling of $|Jy\rangle$ from the other three orbitals. Equations like (7.199) and (7.200) could be derived for many kinds of systems. An analogous procedure to this example was used by Pettifor and Oleinik (2000, 2002) to develop a level 4 BOP for semiconductors, including the complications of s-orbitals and ring terms, and the addition of hydrogen. Once we go beyond the trimer here by embedding our bond in a full environment of other atoms, it becomes necessary to explore approximations to the most difficult terms. In the present case for example we might treat ζ_6 by representing it as a sum of products of topologically simpler paths (e.g. irreducible paths). The bond-order approach seems to be the only general and systematic way of including the topology of the bonding and the quantum nature of chemical bonds within an analytic model that goes beyond the level of second moments.

Exercises

1. Show that $\delta(b_3^2) = 0$ in the above example.

2. Show that at level 1, an intersite Green function matrix element is given by

$$G_{I\mu J\nu} = \frac{\zeta_2}{\epsilon^2 - b_1^2}, \tag{7.202}$$

and the single bond energy for half occupation is

$$(\rho_{I\mu J\nu} + \rho_{J\nu I\mu})\zeta_2 = -\frac{\zeta_2^2}{b_1}. \tag{7.203}$$

It follows that in a crystal in which the atoms are z-fold coordinated, $b_1 \propto \sqrt{z}$. The bond energy summed over all the bonds is therefore proportional to $\sqrt{z}$, just as in the second-moment model.

8

HYBRID SCHEMES

In this chapter, we will discuss two methods that do not fall into either the approach of perturbing a free electron gas or the approach of superimposing atomic charge densites and doing tight-binding. Rather they combine aspects of both. The first, known as generalized pseudopotential theory, offers a systematic and powerful route to models for metallic systems. It subsumes and generalizes the pair potentials described in Chapter 6, while systematically adding three- and four-body contributions to the energy. It has proved very useful in many kinds of simulation for transition metals (Moriarty *et al.*, 2002). The other approach I discuss is the effective medium theory (EMT), which after much simplification becomes identical in form to the widely used embedded atom method (EAM).

8.1 Generalized Pseudopotential Theory

This approach has its origins in the generalization of pseudopotential perturbation theory to include transition metals, besides the nearly free electron metals dealt with in Chapter 6. To do this it is necessary to include matrix elements of the Hamiltonian with atom-centred, localized d-orbitals besides plane waves. This original concept of Harrison (1969) was extended over the subsequent 20 years mainly by Moriarty and coworkers. A detailed description of the theory for transition metals with references to the most significant steps on the way is given by Moriarty (1988). It is a first-principles DFT theory known as generalized pseudopotential theory (GPT), the achievement of which is to express the total energy as a multi-atom expansion of the form:

$$E^{\text{tot}} = N_{\text{a}} E^{\text{vol}}(\Omega) + \frac{1}{2} \sum_{IJ}{}' v_2(\mathbf{R}_I, \mathbf{R}_J; \Omega) + \frac{1}{6} \sum_{IJK}{}' v_3(\mathbf{R}_I, \mathbf{R}_J, \mathbf{R}_K; \Omega)$$

$$+ \frac{1}{24} \sum_{IJKL}{}' v_4(\mathbf{R}_I, \mathbf{R}_J, \mathbf{R}_K, \mathbf{R}_L; \Omega) + \cdots . \tag{8.1}$$

The primes on the summations indicate that the atom indices must be different, that is on-site terms are excluded. The first two terms closely resemble the total energy of simple metals, v_2 being a volume-dependent pairwise potential of the kind I have previously denoted by V_{IJ}. The problem of defining Ω in a very inhomogeneous system is present, just as in the case of simple metals discussed previously. It can only be circumvented by further assumptions and approximations. This is necessary if such a model is to be applied to surfaces, as in fracture simulation. Moriarty and Phillips (1991) handled it by defining a convenient local average of the electron density.

Moriarty's approach has now reached the point of being a practical scheme for large scale atomistic simulation, as described in the review by Moriarty *et al.* (2002). Applications have included the study of binary and ternary alloys, equations of state at high pressure, point defects, grain boundary structure, dislocation core structure and mobility, and fracture. To reach this point some approximations within the *ab initio* approach are made and the resulting multi-atom potentials, which comprise two-, three- and four-body terms, are known as model generalized pseudopotential theory (MGPT) potentials.

We have already met most of the physical concepts which are needed to derive (8.1), but there are some more. As the derivation is rather long and complicated, I will confine myself here to describing the main additional concepts, but refer the reader to the literature, particularly (Moriarty, 1988, 1990; Moriarty and Phillips, 1991; Moriarty *et al.*, 2002) and references therein, for details of the derivation and a survey of applications.

There is a strong similarity to our derivation of the TBBM, in that the cancellation of terms between free atoms and the solid is exploited by focussing attention on $E^{\rm bond}$. However in the mixed basis set, the five d-orbitals per atom, denoted $\phi_d(\mathbf{r}-\mathbf{R}_I)$ for atom I, and the plane-waves $|\mathbf{k}\rangle$ require different treatment, which we now look at. The point is that pseudopotentials which were seen as weak by plane waves are not weak to d-orbitals. In analogous fashion, it is convenient to make a new division of the charge density, which explicitly separates a part originating from d-states and a part which is nearly free electron like.

The eigenvalue equation looks just like that in simple metals, with an effective potential $V_{\rm eff}$ including a pseudopotential that weakly scatters plane waves:

$$\hat{H}|n\rangle \equiv (-\tfrac{1}{2}\nabla^2 + \hat{V}_{\rm eff})|n\rangle = \epsilon_n|n\rangle. \tag{8.2}$$

This is eq. (33) of Moriarty (1988) expressed in the notation of previous chapters. The ϕ_d are chosen to be eigenfunctions of an atomic-like potential $\hat{v}_{\rm at}$, the exact form of which remains to be defined:

$$\big(-\tfrac{1}{2}\nabla^2 + \hat{v_{\rm at}}\big)|\phi_d\rangle = \epsilon_d^0|\phi_d\rangle. \tag{8.3}$$

The sum of the atomic-like potentials, call it $V_{\rm at}(\mathbf{r})$, differs from $V_{\rm eff}(\mathbf{r})$ by an amount denoted by $\delta V(\mathbf{r})$:

$$V_{\rm at}(\mathbf{r}) \equiv \sum_I v_{\rm at}(\mathbf{r}-\mathbf{R}_I) = V_{\rm eff}(\mathbf{r}) + \delta V(\mathbf{r}). \tag{8.4}$$

The potential δV is itself divided into a large on-site constant term and a part called the *hybridization potential* Δ:

$$\delta V(\mathbf{r}) = \langle\phi_d|\delta\hat{V}|\phi_d\rangle + \Delta(\mathbf{r}). \tag{8.5}$$

Since the main function of δV is to represent the part of the potential that has d-symmetry, most of it near the core of the atom is absorbed in the first term in (8.5), so that Δ is small there. The various matrix elements of $\hat{H}$ can now be written down in terms of shifted d-state energies

$$\epsilon_d = \epsilon_d^0 - \langle\phi_d|\delta\hat{V}|\phi_d\rangle, \tag{8.6}$$

small $d - d'$ overlap matrix elements

$$\begin{aligned} \Delta_{dd'}(\mathbf{R}_{IJ}) &= \int \phi_d(\mathbf{r}-\mathbf{R}_I)\Delta(\mathbf{r})\phi_{d'}(\mathbf{r}-\mathbf{R}_J)\,\mathrm{d}\mathbf{r}, \\ S_{dd'}(\mathbf{R}_{IJ}) &= \int \phi_d(\mathbf{r}-\mathbf{R}_I)\phi_{d'}(\mathbf{r}-\mathbf{R}_J)\,\mathrm{d}\mathbf{r}, \end{aligned} \tag{8.7}$$

and small hybridization matrix elements:

$$\begin{aligned} \Delta_{kd} &= \frac{1}{\sqrt{V}}\int \exp(-i\mathbf{k}\cdot\mathbf{r})\Delta(\mathbf{r})\phi_d(\mathbf{r})\,\mathrm{d}\mathbf{r}, \\ S_{kd} &= \frac{1}{\sqrt{V}}\int \exp(-i\mathbf{k}\cdot\mathbf{r})\phi_d(\mathbf{r})\,\mathrm{d}\mathbf{r}. \end{aligned} \tag{8.8}$$

Notice that $\Delta(\mathbf{r})$ itself may be quite large outside the region near atomic cores, in the bonds between atoms. However, it only enters the theory multiplying a d-state, and the product will be small everywhere. The result is a Hamiltonian with $d - d'$ and $\mathbf{k} - \mathbf{k}'$ blocks around the diagonals, and hybridization blocks off the diagonals:

$$\left(\begin{array}{c:c} \epsilon_d S_{dd'} - \Delta_{dd'} & \epsilon_d S_{dk} - \Delta_{dk} \\ \hdashline \epsilon_d S_{kd} - \Delta_{kd} & \langle\mathbf{k}|\hat{V}_{\text{eff}}|\mathbf{k}'\rangle \end{array}\right). \tag{8.9}$$

So much for the Hamiltonian used in this theory. Equally important is the decomposition of the charge density, which is written as:

$$\rho = \rho^{\text{val}} + \sum_I \rho_I. \tag{8.10}$$

Notice in contrast to pure TB theory, which works with a superposition of atomic-like charge densities, and nearly free electron theory which starts with a uniform charge density, we have here something in between. ρ_I is constructed from the occupied fraction of the d-orbitals together with the inner core electrons, while ρ^{val} contains all the remaining electron density and is fairly uniform outside the atomic cores. In fact ρ^{val} resembles the electron density in nearly free electron metals. This partitioning enables an expansion of all functions of ρ, such as V_{xc}, in terms of one-, two-, three- and four-body functions. In TB theory these terms were just represented by a pairwise potential. Rather than a Taylor expansion, such as we introduced in eq. (7.31), Moriarty made an ingenious cluster expansion, which avoids the need for explicit derivatives and with which three-body and higher order terms can be generated in a natural way. Since it is a rather general procedure, it is worth looking at briefly here. Taking V_{xc} as an example, let us first define a finite difference, for any $\Delta\rho$:

$$V^*_{\text{xc}}(\Delta\rho) = V_{\text{xc}}(\rho^{\text{val}} + \Delta\rho) - V_{\text{xc}}(\rho^{\text{val}}). \tag{8.11}$$

The cluster expansion is then:

$$\begin{aligned} V_{\text{xc}}(\rho) &= V_{\text{xc}}(\rho^{\text{val}}) + \sum_I V^*_{\text{xc}}(\rho_I) \\ &+ \frac{1}{2}\sum_{I,J}{}' \left[V^*_{\text{xc}}(\rho_I + \rho_J) - V^*_{\text{xc}}(\rho_I) - V^*_{\text{xc}}(\rho_J) \right] + \cdots . \end{aligned} \tag{8.12}$$

The first term is a constant, dependent on the volume but not the structure. The second term adds the contributions of the atoms in a linear way, like a first derivative. The third term is pairwise in character, calculating the non-linear additive contributions in the region of each bond, and then adding these linearly. It is the discrete analogue to the second-derivative term in (7.31). And so on. For a finite number of atoms, this expansion converges by construction to the exact result: the final term *is* just the exact result minus all the preceeding terms!

Central to the development of the GPT are the elements of the Green function $(\epsilon\mathbf{1} - \mathbf{H})^{-1}$, which are obtained in perturbation series to second order in the hybridization parameter and the pseudopotential. The $d - d'$ block of the Green function matrix corresponds to the electron density contributed by the localised distributions ρ_I. Furthermore, like the tight-binding model with d-orbitals only, the $d - d'$ block leads to multi-atom terms in the energy coming from its contribution to E^{bond}. In the MGPT, this is the origin of the three- and four-body terms. The nearly free electron part of the density of states is included only to the level of pairwise potentials, as in Chapter 6.

8.2 Effective Medium Theory

There is more than one EMT, but the most complete development is probably due to Jacobsen *et al.* (1987). The version I will describe is fairly close to theirs, which is based on HKS theory, although for reasons of consistency with the rest of the book my development here is not exactly the same. A similar but simpler theory, which leads to the embedded atom model (EAM) (Daw and Baskes, 1984; Daw, 1989), is based on a completely local density functional theory of the Thomas–Fermi type (see Section 2.3). The EMT has features in common with Moriarty's GPT, but for practical applications the latter has gone further by its derivation of explicit three- and four-atom interactions.

The basic idea is to estimate the total energy by summing *embedding energies* atom by atom. An *embedding energy* is the change in total energy when a single atom I is placed in a free electron gas of a certain density ρ_I^0. This idea was originally developed as a way of calculating the binding energy of impurities in metals (Nørskov and Lang, 1980; Stott and Zaremba, 1980; Nørskov, 1982), because the calculations that need to be done for a single atom in jellium are relatively straightforward. A key question is, what is the appropriate value of ρ_I^0? It cannot be the average electron density in the real material, because that would include the density of the atom which is being inserted. It must be lower, simulating the electron density in a vacancy that the inserted atom would fill, but that density varies from a small value at the centre of the vacancy to its bulk value nearby. Some kind of average should be appropriate, but what kind would be best? One task of EMT is to answer this question. A second task of the theory is as far as possible to estimate and correct for the residual error, due to neglecting any effect of the atomic structure of the material into which atoms have been embedded. I will outline here how these questions can be tackled. As usual I will mostly use the notation of previous chapters, rather than adopting that of the original literature.

Jacobsen *et al.* first discuss the self-consistent change $\Delta\rho_I$ induced in the charge density ρ_I^0 of a jellium by adding an atom of atomic number Z_I (or alternatively a pseudoatom of valency Z_I), as sketched in Fig. 8.1. $\Delta\rho_I$ is generated formally by minimizing the embedding energy functional of an arbitrary change $\Delta\rho$:

$$\Delta E_I^{\text{jel}}(\rho_I^0, [\Delta\rho]) = E_I^{\text{jel}}[\rho_I^0 + \Delta\rho, V_{\text{eff}}] - E^{\text{jel}}[\rho_I^0] - E_I^{\text{atom}}. \tag{8.13}$$

$E^{\text{jel}}[\rho_I^0]$ and E_I^{atom} are the total energies of the perfect jellium and of the free atom, respectively. The effective potential is given by

$$V_{\text{eff}}(\mathbf{r}) = -\frac{Z_I}{|\mathbf{r} - \mathbf{R}_I|} + \int \frac{\Delta\rho(\mathbf{r}')}{|\mathbf{r} - \mathbf{r}'|}\,d\mathbf{r}' + V_{\text{xc}}[\rho_I^0 + \Delta\rho]. \tag{8.14}$$

At the variational minimum of (8.13) with respect to the density variation $\Delta\rho$, suppose $\Delta\rho = \Delta\rho_I$ and $V_{\text{eff}} = V_{\text{eff}}^I$. The definition of the embedding energy,

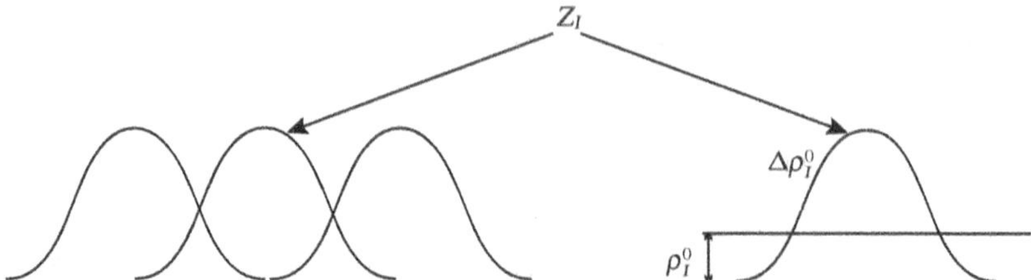

FIG. 8.1 Schematic diagram showing atom I in the solid (left) and in the effective medium (right), which is a jellium of density ρ_I^0.

which is a function only of ρ_I^0 and the atomic number of the embedded species Z_I, is now:

$$\Delta E_I^{\text{jel}}(\rho_I^0) = \Delta E_I^{\text{jel}}(\rho_I^0, [\Delta\rho_I]). \tag{8.15}$$

Notice that, from the theory of the generalization of the Harris–Foulkes functional (see Section 3.4.2), the embedding energy is stationary with respect to variations in both $\Delta\rho_I$ and V_{eff}^I. To a fairly good approximation one can simply take the free atom charge density as an approximation to $\Delta\rho_I$, although it lacks the wiggles in the tail (the Friedel oscillations) present in the charge density induced by an ion in jellium.

The total cohesive energy can be expressed within the approximation of the generalized Harris–Foulkes functional, eq. (3.62), in the form:

$$\begin{aligned} -E^{\text{coh}} &= E^{\text{HFG}} - \sum_I E_I^{\text{atom}} \\ &= \left(T_s[\rho^{\text{out}}] + \rho^{\text{out}} V_{\text{eff}}^{\text{in}}\right) - \rho^{\text{in}} V_{\text{xc}}^{\text{in}} + E_{\text{xc}}^{\text{in}} - E_{\text{H}}^{\text{in}} + E_{ZZ} - \sum_I E_I^{\text{atom}}. \end{aligned} \tag{8.16}$$

We now write this as the sum of the embedding energies plus a correction:

$$-E^{\text{coh}} = \sum_I E_I^{\text{jel}}(\rho_I^0) + \Delta E^{\text{EMT}}. \tag{8.17}$$

The first term, which is the sum of the embedding energies, allows the optimum jellium density to vary from site to site. For example we expect at a surface ρ_I^0 will be less than in the bulk. The second term is the correction to the simple effective medium embedding energy which we can now derive. We use the same decomposition of each of the three total energies E_I^{jel}, $E^{\text{jel}}[\rho_I^0]$ and E^{HFG} and cancelling the free atom energies, we can reverse (8.17) to obtain the expression for ΔE^{EMT}:

$$\begin{aligned}\Delta E^{\mathrm{EMT}} = &\left(T_s[\rho^{\mathrm{out}}] + \rho^{\mathrm{out}} V_{\mathrm{eff}}^{\mathrm{in}}\right) - \rho^{\mathrm{in}} V_{\mathrm{xc}}^{\mathrm{in}} + E_{\mathrm{xc}}^{\mathrm{in}} - E_{\mathrm{H}}^{\mathrm{in}} + E_{ZZ} \\ &+ \Big(\sum_I T_s[\rho_I^0] + \sum_I \rho_I^0 V_{\mathrm{xc}}[\rho_I^0]\Big) - \sum_I \rho_I^0 V_{\mathrm{xc}}[\rho_I^0] + \sum_I E_{\mathrm{xc}}[\rho_I^0] \\ &- \Big(\sum_I T_s[\rho_I^0 + \Delta\rho_I] + \sum_I (\rho_I^0 + \Delta\rho_I) V_{\mathrm{eff}}^I\Big) \\ &+ \sum_I (\rho_I^0 + \Delta\rho_I) V_{\mathrm{xc}}[\rho_I^0 + \Delta\rho_I] \\ &- \sum_I E_{\mathrm{xc}}[\rho_I^0 + \Delta\rho_I] + \sum_I \frac{1}{2} \Delta\rho_I V_{\mathrm{H}}[\Delta\rho_I]. \end{aligned} \tag{8.18}$$

The three sets of terms highlighted by including them in large parentheses are the sums of one-electron energies in the three cases: (i) complete solid, (ii) jellium, and (iii) atoms in jellium. This expression takes account of the cancellation of all the electrostatic terms in the embedding energy that involve the Hartree potential of the uniform charge distributions ρ_I^0, including their positive background.

A detailed examination of (8.18) reveals that there is a strong cancellation of terms, which justifies considering ΔE^{EMT} as a small correction to the sum of embedding energies, provided a sensible choice of ρ^{in}, $V_{\mathrm{eff}}^{\mathrm{in}}$ and ρ_I^0 is made. First, a natural choice of ρ^{in} would be

$$\rho^{\mathrm{in}} = \sum_I \Delta\rho_I, \tag{8.19}$$

although departures from this will only affect the result to second order, so it does not have to be calculated precisely. Thus $\Delta\rho_I$ is normally represented by an atomic charge density, which avoids any self-consistent calculations for the many atom system. The choice of $V_{\mathrm{eff}}^{\mathrm{in}}$, or equivalently $V_{\mathrm{xc}}^{\mathrm{in}}$, is not so obvious, neither is ρ_I^0. With this problem in mind, let us consider to what extent there is cancellation of the V_{xc} terms, with the above choice of ρ^{in}:

$$\sum_I \Delta\rho_I V_{\mathrm{xc}}^{\mathrm{in}} \stackrel{?}{=} \sum_I (\rho_I^0 + \Delta\rho_I) V_{\mathrm{xc}}[\rho_I^0 + \Delta\rho_I] - \sum_I \rho_I^0 V_{\mathrm{xc}}[\rho_I^0]. \tag{8.20}$$

The strategy adopted by Jacobsen *et al.* (1987) was to divide the material into neutral spheres Ω_I centred on each atom. A small error is thereby accepted, due to the fact that the spheres do not precisely fill space. This error can be corrected in close-packed systems, but it becomes more problematic to keep it under control when we are dealing with atoms at surfaces. It is an error the EMT just has to carry. With the division of the space into neutral spheres, ρ_I^0 is defined as the average over sphere I of the tails of all the $\Delta\rho_J$ that stick into Ω_I. If we denote the sum of these tails by $\Delta\rho_{-I}$, and assume the spheres fill space, (8.20) becomes:

$$\begin{aligned}\sum_I \int_{\Omega_I} (\Delta\rho_I + \Delta\rho_{-I}) V_{\text{xc}}^{\text{in}} \stackrel{?}{=} & \sum_I \int_{\Omega_I} (\rho_I^0 + \Delta\rho_I) V_{\text{xc}}[\rho_I^0 + \Delta\rho_I] \\ & + \sum_I \int_{V-\Omega_I} (\rho_I^0 + \Delta\rho_I) V_{\text{xc}}[\rho_I^0 + \Delta\rho_I] \\ & - \sum_I \int_V \rho_I^0 V_{\text{xc}}[\rho_I^0],\end{aligned} \tag{8.21}$$

where

$$\rho_I^0 = \frac{1}{\Omega_I} \int_{\Omega_I} \Delta\rho_{-I}. \tag{8.22}$$

Whereas the left-hand side is an integral over the overlapped charge densities that approximate the real solid, each term on the right-hand side refers to a single atom in its effective medium. The second integral, subscripted $V - \Omega_I$, is over the entire material outside sphere I. It is to be expected, and has been confirmed by calculations, that in the neighbourhood of atom I, the exact density closely resembles $\rho_I^0 + \Delta\rho_I$. The cancellation among terms referring to each site I is now easier to see. Within a sphere, $V_{\text{xc}}^{\text{in}}$ and $V_{\text{xc}}[\rho_I^0 + \Delta\rho_I]$ must resemble each other because they are both dominated by the exchange-correlation potential within an atom, and so a small adjustment of $V_{\text{xc}}^{\text{in}}$ (which we do not need to specify!) would make the left-hand side exactly cancel the first term on the right-hand side. As for two final integrals on the right-hand side, their integrands cancel exactly in the region far enough outside Ω_I for the tail of $\Delta\rho_I$ to have dropped to zero. The remainder is of second order in the amplitudes of the tails. This cancellation is helped by the constraint of charge conservation:

$$\int_{\Omega_I} \rho_I^0 V_{\text{xc}}[\rho_I^0] = \int_{V-\Omega_I} \Delta\rho_I V_{\text{xc}}[\rho_I^0]. \tag{8.23}$$

A similar argument shows that the three E_{xc} terms also cancel strongly, although this requires the local density approximation (2.47).

Now let us look at the remaining terms in (8.18). The easiest of all to deal with are the electrostatic terms, because they involve spherical charge distributions; these are the two last terms on the first line of (8.18) and the very last term, which taken altogether have the form of a pairwise sum of screened Coulomb repulsions:

$$-E_{\text{H}}^{\text{in}} + E_{ZZ} + \sum_I \frac{1}{2} \Delta\rho_I^0 V_{\text{H}}[\Delta\rho_I^0] \equiv \frac{1}{2} \sum_{I \neq J} V_{IJ}(R_{IJ}). \tag{8.24}$$

Notice that the electrostatic self-energy of the $\Delta\rho_I$, the summation on the left-hand side of (8.24), cancels out with the same contribution to E_{H}^{in}.

By these manoeuvres we have reduced the cohesive energy to an embedding energy, a pairwise term, and the term in the difference of one-electron energies:

$$-E^{\rm coh} = \sum_I \Delta E_I^{\rm jel}(\rho_I^0) + \frac{1}{2}\sum_{I \neq J} V_{IJ} + \Delta \sum \epsilon_n, \tag{8.25}$$

where

$$\Delta \sum \epsilon_n = \left(T_s[\rho^{\rm out}] + \rho^{\rm out} V_{\rm eff}^{\rm in}\right) - \Big\{\Big(\sum_I T_s[\rho_I^0 + \Delta\rho_I] + \sum_I (\rho_I^0 + \Delta\rho_I) V_{\rm eff}^I\Big) - \Big(\sum_I T_s[\rho_I^0] + \sum_I \rho_I^0 V_{\rm xc}[\rho_I^0]\Big)\Big\}. \tag{8.26}$$

The embedding energy and the pairwise energy together have the familiar form of the semi-empirical effective medium or embedded atom models. In these models, ρ_I^0 is parameterized in terms of the sum of a pairwise function over neighbouring atoms J. The embedding energy itself is parameterized, as is the pair potential, and all the parameters are fitted to bulk properties of the crystal. Examples of such potentials and their application abound in the literature, and I will not consider here any particular realizations of the EAM or such empirical effective medium models. When the functions are arbitrary in form, and parameters are simply fitted to bulk properties, there may be little difference in practice whether one is parameterizing an effective medium model or a second-moment tight-binding model, in which the embedding term becomes the bond energy. The square-root dependence on coordination of the bond energy, which is a feature of the second-moment approximation, is not unlike the concave dependence of the embedding energy on the jellium density.

Finally we must take a closer look at the term $\Delta \sum \epsilon_n$, which we have given no justification to neglect so far. Consider first the terms in braces $\{\cdots\}$ in (8.26). They represent the change in band energy in a jellium to which the atom I is added. Now think in terms of the local projection of the band energy, eq. (1.183). The effective potential $V_{\rm eff}^I$ will be screened within a few atomic distances and tends to its jellium value $V_{\rm xc}[\rho_I^0]$ far from the embedded atom. Similarly, the net change in band energy induced by embedding will be localized around atom I. This is seen in calculations of transition metals in jellium, where the atomic d levels broaden into a *d-resonance*, due to hybridization of the d-orbitals with free electron wavefunctions. The local density of states in the vicinity of the atom thus displays a peak on top of the jellium $\sqrt{\epsilon}$ background, gradually reverting to the jellium density of states at some distance from the atom. Within a tight-binding model on the other hand, a similar broadening of the d density of states is due to the effective overlap integrals between d-orbitals. As several authors have elucidated, there is no contradiction here, simply a transformation of basis functions.

The local band energies for each atom in the braces can be defined over the cells Ω_I, so that they become a sum of atom-projected band energies. Because of the similarity in the potentials within these cells for the embedded atom in jellium and for the same atom in the real solid, there will be a strong cancellation for each atom between the first term in (8.26) and the terms in braces. The cancellation of these terms shows us that the effect on the bonding energy attributed in tight-binding models to the *second moment* μ_2 of the density of states, seen for example in the width of the local density of d-states in transition metals, has already been taken care of in EMT within the embedding energy. This is reasonable enough, since the width of the density of states is measured by μ_2, which in tight binding is a measure of the local density of atoms, and in EMT the local density of atoms is captured by the value of ρ_I^0. The band energy difference $\Delta \sum \epsilon_n$ still accounts for all the features in the density of states of the material that depend on the topology of the bonding, which in a crystalline solid means all the features related to the particular crystal structure. Such are the structural features that determine the fourth moment of the d-density of states, which as discussed before, is responsible for the relative energies of BCC and FCC crystals. The inevitable conclusion is that for the purpose of structural predictions in transition metals, the kind of embedded atom method that uses only an embedding energy and a pairwise energy is inadequate, just as the second-moment approximation in tight-binding theory is inadequate.

An enhanced model might express $\Delta \sum \epsilon_n$ as a real-space potential, involving three- and perhaps four-body terms, which is effectively what is achieved in the GPT of Moriarty. There are also empirical potentials, 'modified' embedded atom models, with this aim, that include 'bond-bending' terms in the energy, however their functional forms are fairly arbitrary. In the noble metals and sp-bonded metals, the simplified EMT or EAM may be more reliable for structural predictions than they are in transition metals, but a cautious approach is always advisable, especially when applying such potentials at surfaces or similarly gross perturbations to the bulk density.

9

IONIC MODELS

In previous chapters, I have described several models for metals, and a large class of models derived from tight binding that have been applied over the last few decades to transition metals, semiconductors, and also organic and molecular systems. This class has recently been enlarged even further to include ionic materials, as described in Chapter 7. In this final chapter, I discuss in general terms the models that are appropriate for ionic materials, including more specific features of the tight-binding approach. It will be seen how the machinery of density functional theory and the second-order functional again provides a unified way of deriving ionic models that were apparently unrelated, including the Born model, shell models and variable charge transfer models.

9.1 Introduction

The scope of ionic models is very wide, and complementary to that of pair potential models. Although so far they have been much less developed and applied, self-consistent tight-binding models have potentially a similar scope, embracing the traditional text book materials NaCl, KF and all the other alkali halides, the insulating oxides with wide band-gaps such as MgO and Al_2O_3, perovskites such as $SrTiO_3$ as well as oxides which are thought of as less ionic, such as TiO_2 and ZrO_2. In fact it is likely that someone somewhere has attempted to model any of the insulating inorganic materials you can think of with an ionic model. We can distinguish models by their increasing degrees of sophistication, which are described briefly below. Following this introduction there is a more detailed discussion of how ionic models in general can be related to the theory of the second-order density functional.

The simplest ionic model is the *rigid ion model*. It is sometimes called the *Born model* after its pioneer Max Born and his coworkers in Edinburgh, who over 50 years ago developed most of the mathematical description of the ionic model that we still use; see for example the book by Born and Huang (1954). In the terms of the present book we can think of it as a first-order model, as we shall see in the following section. The ions interact by the Coulomb interaction, in addition to which there is a short range pairwise repulsion.

Ions are not rigid, and the *shell model* was first developed by Dick and Overhauser (1958) to include their polarizability. In its simplest version a spherical shell of charge $-q^s$ surrounds an ion of reduced charge $\Delta Z + q^s$ and is attached to it by a harmonic spring. The repulsion between ions is now modelled as a repulsion between the centres of these shells.

The compressible ion model is an extension of the shell model in which the radius of the shells is a variable. The energy of a compressible ion is a function of this radius, which adjusts according to the environment of the ion. This model is most strongly motivated by the case of oxygen, as discussed further below. The traditional models were mainly constructed by fitting their parameters to experimental data. This was a disincentive to make them too sophisticated, since the number of parameters would become so large that they could fit anything without any guarantee of being physically sound. The compressible ion model was held back for this reason, since in its early days, the parameters were determined by fitting them to experimental phonon dispersion curves. The work of Madden, Wilson, Pyper and others (see e.g. Wilson *et al.*, 1996*a* and references therein) attempted to reduce the empirical nature of the parameters by fitting them individually, rather than altogether. They were fitted to total energy data obtained by first principles calculation on molecules and crystals at chosen interatomic distances. This approach also made it feasible to introduce independent parameters to describe quadrupolar polarization. Even so, the most sophisticated ionic models take no explicit account of covalent effects, and this may be a significant shortcoming.

None of the previous ionic models allow for the fact that the charge associated with an ion also depends on its environment. This has been treated in an empirical way by the model introduced by Streitz and Mintmire (1994). A more recent approach is to use self-consistent tight-binding, which in principle can be regarded as an ionic model with a degree of covalency included, as described in Section 7.7. In fact other models can be derived from a tight-binding description by making specific approximations, perhaps the most drastic being the complete neglect of interatomic Hamiltonian matrix elements as in the variable charge-transfer model. The rigid ion model can be thought of as the simplest in this category.

9.2 The Rigid Ion Model Derived

Ionic models, like the other models in this book, involve the idea of an input charge density ρ^{in}. It is constructed, at least notionally, by superimposing spherical 'atomic' charge densities. The atomic charge density in the sense used here is not necessarily the same as the charge density of free atoms. A judicious choice of atomic charge density could make the input charge ρ^{in} closer to the exact, self-consistent charge density ρ^{ex}, improving the accuracy of a first-order model. The usual picture of the charge in an ionic crystal is of course a superposition of *ions*

		1 H	2 He	3 Li	4 Be	5 B	6 C		
7 N	8 O	9 F	10 Ne	11 Na	12 Mg	13 Al	14 Si		
15 P	16 S	17 Cl	18 Ar	19 K	20 Ca	21 Sc	22 Ti	23 V	24 Cr
33 As	34 Se	35 Br	36 Kr	37 Rb	38 Sr	39 Y	40 Zr	41 Nb	42 Mo
51 Sb	52 Te	53 I	54 Xe	55 Cs	56 Ba	57 La	58 Ce	59 Pr	60 Nd
83 Bi	84 Po	85 At	86 Rn	87 Fr	88 Ra	89 Ac	90 Th	91 Pa	92 U

FIG. 9.1 The periodic table centred on the noble gases.

rather than of atoms, and we might expect that these would be a better starting point.

A neat general recipe for the 'atomic charge density' is to take the electron density of the nearest noble gas in the same row of the periodic table (see Fig. 9.1 and Harrison, 1980). Thus oxygen, fluorine, sodium, magnesium and aluminium would all be represented by the charge density of neon atoms, chlorine and potassium by argon, and so forth.

This leads us to construct a first-order model for the total energy as follows. First, superimpose the noble gas atoms at the positions required in the material, keeping their charge densities rigid. Let us denote the superimposed charge density by $\rho_{\text{ng}}^{\text{in}}$, where the individual noble gas atom on site I has a density ρ_{ng}^{I}. Thus

$$\rho_{\text{ng}}^{\text{in}}(\mathbf{r}) = \sum_{I} \rho_{\text{ng}}^{I}(\mathbf{r} - \mathbf{R}_I). \tag{9.1}$$

Let us denote by $E^{\text{HKS}}[\rho_{\text{ng}}^{\text{in}}, V_{\text{ext}}]$ the first-order energy of this system, where:

$$E^{\text{HKS}}\left[\rho_{\text{ng}}^{\text{in}}, V_{\text{ext}}\right] = T_s\left[\rho_{\text{ng}}^{\text{in}}\right] + \rho_{\text{ng}}^{\text{in}} V_{\text{ext}} + E_{\text{H}}^{\text{in}} + E_{\text{xc}}^{\text{in}} + E_{ZZ}. \tag{9.2}$$

It is strongly repulsive, because the equilibrium spacing of noble gas atoms is considerably larger than that of the cations and anions in ionic materials. Indeed, because we are dealing with closed shell atoms, we expect the first-order charge density and energy are rather good approximations to the exact, self-consistent quantities we denote $\rho_{\rm ng}$ and $E_{\rm ng}$. We now 'switch on' the integer charge differences ΔZ_I on the nuclei which are required to create the nuclei of the actual cations and anions. We regard this as a perturbing external potential to the electrons, given by

$$\Delta V_{\rm ext}(\mathbf{r}) = -\sum_I \frac{\Delta Z_I}{|\mathbf{r} - \mathbf{R}_I|}. \tag{9.3}$$

To first-order this does not disturb the charge density but it introduces a strong electrostatic cohesive energy, due to the attraction between the unlike charges, which because of their alternating arrangement in space more than compensates for the repulsion of like charges. We will now examine how this works. The total change in the electrostatic interactions of the nuclei introduced by our 'transmutation' is:

$$\Delta E_{ZZ} = \sum_{I \neq J} \frac{Z_I \Delta Z_J}{|\mathbf{R}_I - \mathbf{R}_J|} + \frac{1}{2} \sum_{I \neq J} \frac{\Delta Z_I \Delta Z_J}{|\mathbf{R}_I - \mathbf{R}_J|}. \tag{9.4}$$

The second term is called the total *Madelung energy* of the particular structure, and we denote it here as $N_{\rm a} E_{\rm M}$, where $N_{\rm a}$ is the total number of atoms and $E_{\rm M}$ is the Madelung energy per atom:

$$E_{\rm M} = \frac{1}{2N_a} \sum_{I \neq J} \frac{\Delta Z_I \Delta Z_J}{|\mathbf{R}_I - \mathbf{R}_J|}. \tag{9.5}$$

The Madelung energy we are dealing with here closely resembles the quantity of that name discussed in Section 6.2.3, which was the electrostatic energy of like charges embedded in a compensating uniform background of charge. Let us consider in more detail the example of a simple binary crystal structure, in which a formula unit consists of n^C cations of charge ΔZ^C and n^A anions of charge ΔZ^A. These charges derived according to the above definition are sometimes called *formal* charges, to distinguish them from actual electronic charge transfers. Generally, for any material, summing over all sites, charge neutrality requires that:

$$\sum_I \Delta Z_I = 0. \tag{9.6}$$

For our binary crystal, the charge neutrality condition holds for each formula unit:

$$n^C \Delta Z^C + n^A \Delta Z^A = 0. \tag{9.7}$$

In this case, the Madelung energy per atom can be written as

$$E_{\rm M} = \alpha_{\rm M} \frac{\Delta Z^C \Delta Z^A}{2R_{\rm CA}}, \tag{9.8}$$

TABLE 9.1 Madelung constants for ionic crystals, after Johnson and Templeton (1961)

Structure	α_M
NaCl	1.75
CsCl	1.76
α-Al_2O_3	1.68
V_2O_5	1.49

where R_{CA} is the nearest-neighbour cation–anion distance and α_M is the Madelung constant, which, like the corresponding quantity for the structures of the metallic elements (Table 6.1), is tabulated in the literature for a number of crystal structures, e.g. Johnson and Templeton (1961), Ashcroft and Mermin (1976, Table 20.4). Examples are given in Table 9.1. For an arbitrary arrangement of ions, for example within a supercell which may comprise thousands of atoms, or without periodic boundary conditions, there are special strategies for calculating $N_a E_M$, such as the Ewald method for periodic systems, which are well described in the textbooks, e.g. Ziman (1972, p. 39).

Our first-order model for the total energy now follows directly from eq. (4.14), omitting all the second-order response terms, which include the deviation of both ρ_{ng} and of the final exact charge density ρ^{ex} from ρ^{in}_{ng}:

$$\begin{aligned} E^{(1)} &= E^{HKS}\left[\rho^{in}_{ng}, V_{ext}\right] + \rho^{in}_{ng}\Delta V_{ext} + \Delta E_{ZZ} \\ &= E^{HKS}\left[\rho^{in}_{ng}, V_{ext}\right] + N_a E_M - \sum_{I\neq J}\int \rho^{I}_{ng}(\mathbf{r}-\mathbf{R}_I)\frac{\Delta Z_J}{|\mathbf{r}-\mathbf{R}_J|}\,d\mathbf{r} \\ &\quad + \sum_{I\neq J}\frac{Z_I \Delta Z_J}{|\mathbf{R}_I-\mathbf{R}_J|} - \sum_{I}\int \rho^{I}_{ng}(\mathbf{r}-\mathbf{R}_I)\frac{\Delta Z_I}{|\mathbf{r}-\mathbf{R}_I|}\,d\mathbf{r}. \end{aligned} \tag{9.9}$$

The interaction of the atomic charge density with the perturbation on its own site has been separated out as the final term. This expression is dominated by the first two terms, namely $E^{HKS}[\rho^{in}_{ng}, V_{ext}]$ and the Madelung energy. It is a reasonable approximation to write $E^{HKS}[\rho^{in}_{ng}, V_{ext}]$ as a sum of pairwise interactions V_{IJ}, since we are dealing with closed-shell atoms. With this assumption we obtain the simplest form of the rigid ion model:

$$E_{rim} = \frac{1}{2}\sum_{I\neq J} V_{IJ} + N_a E_M. \tag{9.10}$$

The remaining first-order terms will be much smaller, as we now discuss. If all the atoms happen to be represented by the same noble gas, the final sum must vanish identically by the condition of charge neutrality. But in any case it is a structure independent term that is not relevant to interatomic forces. The third and fourth terms taken together represent the electrostatic interaction between the noble gas atomic densities, which are neutral objects, and the perturbations ΔZ_I, excluding the self-interactions. These are pairwise interactions that are short-ranged and weak compared to the electrostatic interactions contained in $E^{\rm HKS}[\rho^{\rm in}_{\rm ng}, V_{\rm ext}]$. They fall to zero by Gauss's theorem when the neighbour J is entirely outside the charge density $\rho^I_{\rm ng}$, which must be a good approximation beyond the nearest neighbours. Furthermore the nearest neighbours will usually have opposite signs of ΔZ_I and ΔZ_J, so the two contributions $\rho^I_{\rm ng}\Delta Z_J$ and $\rho^J_{\rm ng}\Delta Z_I$ tend to cancel. Indeed the cancellation is obviously exact if the noble gas is the same for I and J and if $\Delta Z_I = -\Delta Z_J$, such as in binary compounds like MgO or NaF. Now compare these terms with the pairwise electrostatic contributions to $E^{\rm HKS}[\rho^{\rm in}_{\rm ng}, V_{\rm ext}]$:

$$\int \frac{\rho^I_{\rm ng}(\mathbf{r})\rho^J_{\rm ng}(\mathbf{r}')}{|\mathbf{r}-\mathbf{r}'|}\,\mathrm{d}\mathbf{r}\,\mathrm{d}\mathbf{r}' + \frac{Z_I Z_J}{R_{IJ}} - \int \left\{ \frac{\rho^I_{\rm ng}(\mathbf{r})Z_J}{|\mathbf{r}-\mathbf{R}_J|} + \frac{\rho^J_{\rm ng}(\mathbf{r})Z_I}{|\mathbf{r}-\mathbf{R}_I|} \right\} \mathrm{d}\mathbf{r}. \quad (9.11)$$

This interaction will generally be much bigger. Notice that it involves the full nuclear charges Z rather than ΔZ, secondly, it is always repulsive, because it comprises the direct Coulomb interaction between nuclei partially screened by the electron distributions, and thirdly, it is of longer range, being non-zero when $\rho^I_{\rm ng}$ overlaps $\rho^J_{\rm ng}$, not merely Z_J. The terms we are considering are all strictly pairwise, and so can be exactly accounted for by a relatively small correction to the V_{IJ} we have introduced to represent $E^{\rm HKS}[\rho^{\rm in}_{\rm ng}, V_{\rm ext}]$. Since the form of V_{IJ} in practical applications has been parameterized, the extra terms could be absorbed without explicitly evaluating them.

Let us now return to the rigid ion model in the form (9.10). It was first derived on the basis of overlapping noble gas atomic charge densities by Gordon and Kim (1972) and Kim and Gordon (1974). They used a completely local DFT (including the local kinetic energy functional) to evaluate the pair potential between noble gas atoms, with atomic charge densities overlapped but unrelaxed, and they discovered that this could account fairly well for the lattice parameters and bulk moduli of ionic crystals. Their $E^{\rm HKS}[\rho^{\rm in}_{\rm ng}, V_{\rm ext}]$ could also be fairly well fitted by the Lennard–Jones potential (Gordon and Kim, 1972), the parameters of which I have used, together with the appropriate Madelung energy, to calculate bond lengths in a number of compounds, as illustrated in Fig. 9.2, where the results are compared with experimental bond lengths. Because some of the crystal structures are a bit

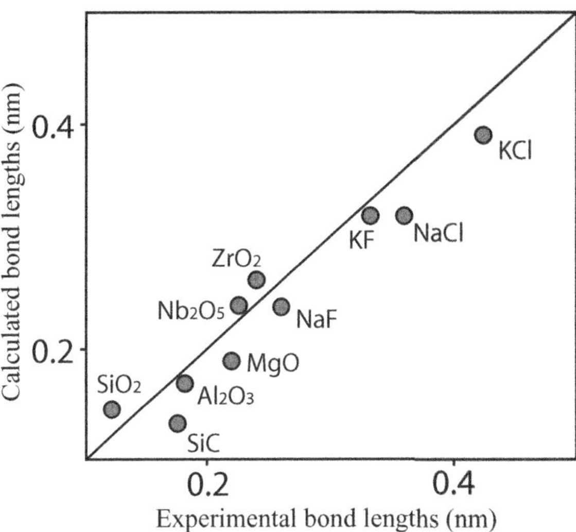

FIG. 9.2 Bond lengths calculated from Lennard–Jones and Madelung energies.

complicated, I have made convenient simplifications, so that the bond length can be predicted by minimizing the energy analytically with respect to R_{CA}. For example, I have only included nearest neighbour interactions in the Lennard-Jones potential. Then for Al_2O_3 I took as the 'experimental' bond length the sum of ionic radii for Al^{3+} and O^{2-}, namely 0.191 nm (Kittel, 1976), which is also the mean of the two close neighbour bond lengths observed in the actual corundum structure. Such ionic radii incidentally reproduce the experimental bond lengths of the other common oxides to an accuracy of 0.002 nm. Notice that in some cases a mixture of rare gases is involved, for example Nb_2O_5 has to be built out of Kr and Ne, because Nb in the 5+ state of ionisation resembles Kr. Since a value for the Madelung constant was not listed in this case by Johnson and Templeton, I assumed their V_2O_5 value would do for Nb_2O_5. The agreement of calculated and observed bond lengths in such diverse materials as ZrO_2 and KCl, with no fitting of parameters to any experimental data, is quite striking. Not surprisingly the worst candidate tested here is SiC, which is a classic covalent compound, but even in this case the crude ionic model does better than we might have expected, since the perturbations ΔZ_I are by no means small. Such is the power of the variational principle. I should add that the relatively good agreement of this model breaks down when it comes to calculating the second derivative of the energy, the bulk-modulus. However, that is easily repaired if one allows the pair potential to be empirically fitted. Predictions of phonon frequencies and the proporties of defects nevertheless generally require a higher quality model than rigid ions can provide, even when some degree of empirical fitting is allowed.

9.3 Beyond the Rigid Ion Model

9.3.1 The Basic Second-order Model

Several other models can be derived within the broad framework of the second-order functional, which is given by eq. (3.32) or (4.14). The second-order functional describing the energy of our ionic crystal can be written in terms of the reference density from the noble gas atoms as:

$$E^{(2)}[\rho] = E^{\mathrm{HKS}}[\rho_{\mathrm{ng}}^{\mathrm{in}}, V_{\mathrm{ext}}] + \rho_{\mathrm{ng}}^{\mathrm{in}} \Delta V_{\mathrm{ext}} + \Delta E_{ZZ} + (\rho - \rho_{\mathrm{ng}}^{\mathrm{in}}) \Delta V_{\mathrm{ext}} - \tfrac{1}{2}(\rho - \rho_{\mathrm{ng}}^{\mathrm{in}}) \chi_{\mathrm{el}}^{-1} (\rho - \rho_{\mathrm{ng}}^{\mathrm{in}}). \tag{9.12}$$

Recall that the first term represents the energy of the noble gas solid to first order, and the idea is that the above rigid ion model has taken care of this and the next two terms, that is all terms of first order in ΔV_{ext}. The task is then to approximate the two remaining terms that depend on $(\rho - \rho_{\mathrm{ng}}^{\mathrm{in}})$ and that are both of second order in ΔV_{ext}. The second order energy functional in this picture is therefore:

$$E^{(2)}[\rho] = \frac{1}{2} \sum_{I \neq J} V_{IJ} + N_{\mathrm{a}} E_{\mathrm{M}} + (\rho - \rho_{\mathrm{ng}}^{\mathrm{in}}) \Delta V_{\mathrm{ext}} - \frac{1}{2}(\rho - \rho_{\mathrm{ng}}^{\mathrm{in}}) \chi_{\mathrm{el}}^{-1} (\rho - \rho_{\mathrm{ng}}^{\mathrm{in}}). \tag{9.13}$$

The difference between this functional and the HKS second-order functional is that the one-electron energies are here subsumed into the classical pairwise potential, an approximation that is only at all justifiable for the closed-shell electronic structure of noble gas atoms. By this manoeuvre we have shifted the entire responsibility for the description of any covalent bonding effects, which are of second order in $\rho - \rho_{\mathrm{ng}}^{\mathrm{in}}$, onto χ_{el}. The response function $\chi_{\mathrm{el}}(\mathbf{r}, \mathbf{r}')$ refers to an electron density distribution $\rho_{\mathrm{ng}}^{\mathrm{in}}$, and we can only model it in rather crude ways. Whatever way we do this, and an example is given below, the model so generated is entirely classical in form, and requires $E^{(2)}[\rho]$ to be minimized directly with respect to ρ, without the solution of any Schrödinger equation. The resulting density must satisfy

$$\rho - \rho_{\mathrm{ng}}^{\mathrm{in}} = \chi_{\mathrm{el}} \Delta V_{\mathrm{ext}}. \tag{9.14}$$

9.3.2 Deformable Ions

The above formalism is a way to derive the shell model, which is the simplest extension of the rigid ion model to introduce deformable ions. Originally this model was entirely empirical, providing sufficient disposable parameters to fit phonon spectra much better than previous models. Now from the previous analysis we are in a better position to see how its parameters can be related to real physical quantities. The shell model is in effect a representation of the terms in $\rho - \rho_{\mathrm{ng}}^{\mathrm{in}}$:

$$\Delta E^{(2)} \equiv \left(\rho - \rho_{\mathrm{ng}}^{\mathrm{in}}\right)\Delta V_{\mathrm{ext}} - \tfrac{1}{2}\left(\rho - \rho_{\mathrm{ng}}^{\mathrm{in}}\right)\chi_{\mathrm{el}}^{-1}\left(\rho - \rho_{\mathrm{ng}}^{\mathrm{in}}\right). \qquad (9.15)$$

In this case, the density $\rho - \rho_{\mathrm{ng}}^{\mathrm{in}}$ of eq. (9.13) is modelled by the sum of displacements by $\mathbf{u}_I^s$ of rigid, spherical shells of negative charge $-\Delta q_I^s$ originally centred about each atom. Each shell is tethered to its nucleus by a spring. At equilibrium in a perfect crystal an undisturbed shell can be thought of as representing part of the density $\rho_{\mathrm{ng}}^{\mathrm{in}}$. The shells are assumed to be massless, since the inertia of electrons does not enter the dynamics of the system in the Born–Oppenheimer approximation, which is the framework for all the models discussed in this book. When the atoms are not in their equilibrium positions, the shells are subject to three kinds of forces, corresponding to three terms in the total energy. First, they are pulled towards their nuclei by the tethering spring, second, they are pulled towards other nuclei by the Coulomb attraction, and third, the shells repel each other. The first two kinds of forces are a representation of the term $(\rho - \rho_{\mathrm{ng}}^{\mathrm{in}})\Delta V_{\mathrm{ext}}$ in eq. (9.13). In the harmonic approximation, the change in energy when the shells are displaced by $\mathbf{u}_I^s$ from their nuclei has the form

$$\Delta E^{(2)} = \frac{1}{2}k u_I^{s\,2} + \sum_{J \neq I}\left\{-q_I^s \Delta Z_J\left[\frac{1}{|\mathbf{R}_I + \mathbf{u}_I^s - \mathbf{R}_J|} - \frac{1}{|\mathbf{R}_I - \mathbf{R}_J|}\right]\right.$$
$$\left. + \frac{1}{2}q_I^s\, q_J^s\left[\frac{1}{|\mathbf{R}_I + \mathbf{u}_I^s - \mathbf{R}_J - \mathbf{u}_J^s|} - \frac{1}{|\mathbf{R}_I - \mathbf{R}_J|}\right]\right\}. \qquad (9.16)$$

A number of points can be made about this model. First, it captures the long-ranged Coulomb interactions between induced dipoles. The on-site elements of χ_{el}^{-1} are represented within the first term $\frac{1}{2}k u_I^{s\,2}$, while the last term in the braces represents its inter-site elements. The Coulomb interactions in (9.16) could be expanded to first order in $\mathbf{u}_I^s$ to display the dipole–dipole interactions explicitly, and indeed the model could be formulated in terms of variable dipoles on the sites instead of explicit shell charges and displacements. These Coulomb interactions are an important part of the inverse response function we are trying to model, as shown in eq. (4.17). However, they are certainly not the whole story, since the true inverse response function contains terms in the kinetic energy and the exchange and correlation energy. A short-ranged empirical potential between the shells can be introduced to model these contributions.

In a more realistic model the shells should be allowed to change their radius in response to their environment. Such an effect can be accomplished by including a kind of compressibility of the ion as a further parameter, which adds a functional dependence on $(r^s - r_0^s)$ to the potential energy, where r^s and r_0^s are the current and relaxed values of the shell radius. This is the idea behind the 'breathing shell' model, or more recently the 'compressible ion' model. For recent developments in the compressible ion model I refer the reader to the paper by Marks *et al.* (2001) and references therein. The compressibility of an ion has a significant effect on properties, and is most noticeable in the case of oxygen, because the size of the O^{2-} ion is particularly sensitive to its environment. Indeed, in free space the O^{2-} ion is unstable, which can be thought of as the limiting case of the radius of the ion going to infinity as its coordination is reduced to zero. A noteworthy example of the importance of compressible oxygen is MgO. Within the rigid ion model the Cauchy relation between elastic constants, $C_{12} = C_{44}$, holds, since the potential energy is purely pairwise (see Section 5.3.2), yet experimentally C_{12} and C_{44} differ by more than 50%. The shell model does not help here, because by symmetry the elastic strains do not displace the shells from their lattice sites, therefore the Cauchy relation should still hold. The difference between C_{12} and C_{44} is only obtained when compressible oxygen ions are introduced. This is because the elastic constant C_{12} is associated with a change in volume, which induces compression or expansion of the ions if the model allows them this degree of freedom. It may help to appreciate this if you think of C_{12} as a linear combination of the bulk modulus $B = \frac{1}{3}(C_{11} + 2C_{12})$ and the shear modulus $C' = \frac{1}{2}(C_{11} - C_{12})$, that is $C_{12} = \frac{1}{3}(3B - 2C')$.

There are now more sophisticated models with polarizable ions, going beyond the notion of spherical shells and harmonic springs. Closer in spirit to our density functional framework is the approach of Ivanov and Maksimov (1996). These authors represent $(\rho - \rho_{\mathrm{ng}}^{\mathrm{in}})$ as a superposition of atomic charge densities ρ_{ng}^{I}, each of which is allowed to shift rigidly with respect to its nucleus, thereby modelling the dipole moments. The associated changes in energy they calculate with a completely local density functional along the lines of Gordon and Kim (1972). A natural extension also considered by Ivanov and Maksimov is to include parameters that describe distortions of the atomic charge densities, such as expansion and contraction or quadrupolar distortion. Besides the local density approximations for kinetic, exchange and correlation energies, the only limitation of this approach is the number of free parameters one is prepared to include to characterize the charge density.

Another line of attack starts as before with the assumption that the ions are polarizable as dipoles and quadrupoles, and that they are compressible. The strategy is then as far as possible to obtain the many parameters such a model requires as independently as possible, by fitting functions to an extensive set of first-principles calculations on ions in specially chosen potential wells or in clusters. It is a great improvement on the simple shell model, and can reproduce a range of strain energy data and phonon spectra. However, in spite of the extensive use of first principles calculations, it is impossible to eliminate some arbitrary choices regarding the functional forms used, and some pairwise approximations. A detailed example of this approach is given by Marks *et al.* (2001).

9.3.3 Variable Charge Transfer Models

In the models described so far in this chapter, a fixed charge transfer between ions is defined. The net charges on ions are fixed at the formal charges, which are specified by the electronic structure of noble gases. In reality, the occupancy of atomic orbitals varies with the local atomic environment, so it is worth thinking about how this effect can be modelled. In fact a model of ionic crystals that includes covalency, polarizable ions (dipoles and quadrupoles) and variable charge transfer was introduced in Section 7.7 under the heading of self-consistent tight-binding. Self-consistent tight-binding offers a way to treat pure metal atoms as well as their oxides, since, at least for sd-bonded metals, it models the nature of chemical bonding from the metallic to the ionic state.

Conceptually the application of self-consistent tight-binding to an ionic material is no different to its application to metals. However, an alternative input charge density might be considered. In the tight-binding models discussed previously, atomic charge densities were superimposed to create the notional input charge density, so that all the long-ranged electrostatic interactions appeared in the second order term. However, one could also formulate the tight-binding model starting from the superimposed atomic charge densities of noble gases, in which case the Coulomb interactions would appear in the pair potential, exactly as they do in the rigid ion model. A systematic study of the consequences of one or the other starting point has not yet been made.

In spite of advantages of self-consistent tight-binding in comparison to the models introduced earlier in this chapter, it has certain disadvantages that should be mentioned. These include the obvious one of greater computation time, but there may be others. For example, in current implementations the second order term is represented by Coulomb interactions between point multipoles. This appears to take no account of the effect that the *compressible ion model* is trying to capture. The logical way to describe a compressible oxygen ion within tight-binding would be to include more orbitals of s and p character in the basis set, so that the spread of the charge density on an anion could vary by transfer of electrons beween the orbitals on a site. The price to pay would be more parameters to determine and more computation time. The benefits over a first-principles tight-binding method would be diminished or eliminated.

On the other hand, notions of compressible ions and charge transfer are not absolute; a transfer of electrons from cations to anions might be rather well described in terms of the charge density by a contraction of anion densities accompanied by an expansion of cation charge densities, and vice versa. The tight-binding model predicts a violation of the Cauchy relations because of its non-central forces and covalency, and possibly charge transfer, whereas the compressible ion model predicts the Cauchy violation from the form of its classical non-pairwise potential energy. A one-to-one mapping of these properties such as charge transfer between the compressible ion model and the tight-binding model is not unique, so, at least at the time of writing, it would be unfair to dismiss one or the other description as wrong; they may both be grasping at the same physics from different points of view.

It may be noted that the classical variable charge model (9.13) is also the basis of a popular empirical model that includes charge transfers (Streitz and Mintmire, 1994). These authors consider the determination of ρ as a variational problem for which the solution equalizes a classical chemical potential of the electrons everywhere. The result is a model that has some similarities to that of Ivanov and Maksimov (1996). We can think of it as a second order ionic model with charge transfer, or as a tight-binding model in which the effect of covalency (intersite matrix elements of the Hamiltonian) is neglected. This is a model which enables large scale molecular dynamics simulations to be carried out, giving a broad-brush description that may be adequate if details such as the relative energies of different structures are not of interest.

BIBLIOGRAPHY

Ackland, G.J., Finnis, M.W. and Vitek, V. Validity of the second moment tight-binding model. *J. Phys. F: Metal Phys.* **18**, L153–L157 (1988).

Allan, G. and Lannoo, M. Vacancies in transition metals: formation energy and formation volume. *J. Phys. Chem. Solids* **37**, 699–709 (1976).

Andersen, O.K. Simple approach to the band-structure problem. *Solid State Commun.* **13**, 133–136 (1973).

Aoki, M. Rapidly convergent bond order expansion for atomistic simulations. *Phys. Rev. Lett.* **71**, 3842–3845 (1993).

Aoki, M., Horsfield, A.P. and Pettifor, D.G. Tight-binding bond order potential and forces for atomistic simulations. *J. Phase Equilib.* **18**, 614–623 (1997).

Aoki, M. and Pettifor, D.G. Angularly-dependent many-atom bond order potentials within tight-binding Huckel theory. *Int. J. Mod. Phys. B* **7**, 299–304 (1993). This is a volume of conference proceedings called *Physics of Transition Metals*, P. M. Oppeneer and J. Kübler (eds.), published by World Scientific Publishers, Singapore.

Ashcroft, N.W. Quantum-solid behaviour and the electronic structure of the light alkali metals. *Phys. Rev. B* **39**, 10552–10559 (1989).

Ashcroft, N.W. and Mermin, N.D. *Solid State Physics* (Harcourt Brace, Orlando, 1976).

Baroni, S., de Gironcoli, S. and DalCorso, A. Phonons and related crystal properties from density-functional perturbation theory. *Rev. Mod. Phys.* **73**, 515–562 (2001).

Bester, G. and Fähnle, M. Interpretation of *ab initio* total energy results in a chemical language: I. formalism and implementation into a mixed-basis pseudopotential code. *J. Phys.-Condens. Mat.* **13**, 11541–11550 (2001).

Born, M. and Huang, K. *Dynamical Theory of Crystal Lattices* (Clarendon Press, Oxford, 1954).

Börnsen, N., Meyer, B., Grotheer, O. and Fähnle, M. E_{cov}—a new tool for the analysis of electronic structure data in a chemical language. *J. Phys.: Condens. Mat.* **11**, L287–L293 (1999).

Bowler, D.R., Miyazaki, T. and Gillan, M.J. Recent progress in linear scaling *ab initio* electronic structure techniques. *J. Phys.: Condens. Mat.* **14**, 2781–2798 (2002).

Brovman, E.G. and Kagan, Yu. Long wavelength phonons in metals. *Zh. Eksp. Teor. Fiz.* **57**, 1329–1341 (1969). (*Sov. Phys.-JETP* **30**, 721 (1970)).

Brovman, E.G., Kagan, Yu. and Kholas, A. The compressibility problem and violation of the Cauchy relation in metals. *Zh. Eksp. Teor. Fiz.* **57**, 1635–1645 (1969). (*Sov. Phys.-JETP* **30**, 883 (1970)).

Brown, R.H. and Carlsson, A.E. Critical-evaluation of low-order moment expansions for the bonding energy of lattices and defects. *Phys. Rev. B* **32**, 6125–6130 (1985).

Carlsson, A.E. Beyond pair potentials in elemental transition metals and semiconductors. *Solid State Phys.* **43**, 1–91 (1990).

Carlsson, A.E. Angular forces in Group-VI transition metals: application to W(100). *Phys. Rev. B* **44**, 6590–6597 (1991).

Casimir, H.B.G. and Polder, D. The influence of retardation on the London-Van der Waals forces. *Phys. Rev.* **73**, 360–372 (1948).

Ceperley, D.M. and Alder, B.J. Ground state of the electron gas by a stochastic method. *Phys. Rev. Lett.* **45**, 566–569 (1980).

Chadi, D.J. and Cohen, M.L. Intrinsic (111) surface states of Ge, GaAs and ZnSe. *Phys. Rev. B* **11**, 732–737 (1975).

Chetty, N., Jacobsen, K.W. and Nørskov, J.K. Optimized and transferable densities from 1st-principles local density calculations. *J. Phys.: Condens. Mat.* **3**, 5437–5443 (1991).

Cohen, M.L., Heine, V. and Weaire, D. Volume 24 of *Solid State Physics* (Academic, New York, 1970).

Coulson, C.A. The electronic structure of some polyenes and aromatic molecules VII. Bonds of fractional order by the molecular orbital method. *Proc. R. Soc. Lond. A* **169**, 413–428 (1939).

Cyrot-Lackmann, F. Sur le calcul de la cohésion et de la tension superficielle des métaux de transition par une méthode de liasons forts. *J. Phys. Chem. Solids* **29**, 1235–1243 (1968).

Daw, M.S. Model of metallic cohesion—The embedded-atom method. *Phys. Rev. B* **39**, 7441–7452 (1989).

Daw, M.S. and Baskes, M.I. Embedded atom method: Derivation and application to impurities, surfaces and other defects in metals. *Phys. Rev. B* **29**, 6443–6453 (1984).

Dick, B.G. and Overhauser, A.W. Theory of the dielectric constants of alkali halide crystals. *Phys. Rev.* **112**, 90–103 (1958).

Dirac, P.A.M. Note on exchange phenomena in the Thomas atom. *Proc. Cambridge Philos. Soc.* **26**, 376–385 (1930).

Dreyssé, H. (ed.) *Electronic Structure and Physical Properties of Solids*. Lecture Notes in Physics. (Springer, Berlin, 2000).

Ducastelle, F. Modules élastiques des métaux de transition. *J. Phys.* **31**, 1055–1062 (1970).

Elstner, M., Porezag, D., Jungnickel, G., Elsner, J., Haugk, M., Frauenheim, Th., Suhai, S. and Seifert, G. Self-consistent-charge density-functional tight-binding method for simulations of complex materials properties. *Phys. Rev. B* **58**, 7260–7268 (1998).

Esfarjani, K. and Kawazoe, Y. Self-consistent tight-binding formalism for charged systems. *J. Phys.: Condens. Mat.* **10**, 8257–8267 (1998).

Evans, R. and Finnis, M.W. Vacancy formation energies and linear screening theory. *J. Phys. F: Metal Phys.* **6**, 483–497 (1976).

Evans, R. and Kumaravadivel, R. A thermodynamic perturbation theory for the surface tension and ion density profile of a liquid metal. *J. Phys. C: Solid State Phys.* **9**, 1891–1906 (1976).

Faber, T.E. *An Introduction to the Theory of Liquid Metals* (CUP, Cambridge, 1972).

Farid, B., Heine, V., Engel, G.E. and Robertson, I.J. Extremal properties of the Harris-Foulkes Functional and an improved screening calculation for the electron gas. *Phys. Rev. B* **48**, 11602–11621 (1993).

Fermi, E. Un metodo statistice per la determinazione di alcune proprieta dell'atomo. *Rend. Accad. Lincei* **6**, 602–607 (1927). Translation into English in March, 1975, 'Self Consistent Fields in Atoms'.

Feynman, R.P. Forces in molecules. *Phys. Rev.* **56**, 340–343 (1939).

Finnis, M.W. The energy and elastic constants of simple metals in terms of pairwise interactions. *J. Phys. F: Metal Phys.* **4**, 1645–1656 (1974).

Finnis, M.W. The Harris functional applied to surface and vacancy formation energies in aluminium. *J. Phys.: Condens. Mat.* **2**, 331–342 (1990).

Finnis, M.W., Kaschner, R., Kruse, C., Furthmüller, J. and Scheffler, M. The interaction of a point charge with a metal surface: theory and calculations for (111), (100) and (110) aluminium surfaces. *J. Phys.: Condens. Mat.* **7**, 2001–2019 (1995).

Finnis, M.W., Kear, K.L. and Pettifor, D.G. Interatomic forces and phonon anomalies in bcc 3d transition metals. *Phys. Rev. Lett.* **52**, 291–294 (1984*a*).

Finnis, M.W. and Sinclair, J.E.A. Simple Empirical N-Body Potential for Transition Metals. *Phil. Mag.* A **50**, 45–55 (1984b); Erratum: *Phil. Mag.* A **53**, 1, p. 161 (1986).

Finnis, M.W., Paxton, A.T., Methfessel, M. and van Schilfgaarde, M. The crystal structure of zirconia from first principles and self consistent tight binding. *Phys. Rev. Lett.* **81**, 5149–5152 (1998*a*).

Finnis, M.W., Paxton, A.T., Methfessel, M. and van Schilfgaarde, M. Self-consistent tight-binding approximation including polarisable ions. In P.E.A. Turchi, A. Gonis and L. Colombo, (eds.), *Proc. MRS Symposium: Tight-Binding Approaches to Computational Materials Science* (Materials Research Society of America, 1998*b*).

Finnis, M.W., Walker, A.B. and Gumbsch, P. Representations of the local atomic density. *J. Phys.: Condens. Mat.* **10**, 7983–7993 (1998*c*).

Finnis, M.W. and Pettifor, D.G. Response functions and interatomic forces. In D.G. Pettifor and D.L. Weaire (eds.), *The Recursion Method and its Applications*, pp. 120–131 (Springer Verlag, Berlin, 1985). Proc. Conf. London, England, 13–14 September 1984.

Finnis, M.W. and Sachdev, M. Vacancy formation volumes in simple metals. *J. Phys. F: Metal Phys.* **6**, 965–978 (1976).

Finnis, M.W. and Sinclair, J.E.A. Simple empirical N-body potential for transition metals. *Philos. Magazine A* **50**, 45–55 (1984b); Erratum *Philos. Magazine A* **53**, 161 (1986).

Foiles, S.M. Interatomic interactions for Mo and W based on the low-order moments of the density-of-states. *Phys. Rev. B* **48**, 4287–4298 (1993).

Foulkes, W.M.C. PhD Thesis, Cambridge (1987).

Foulkes, W.M.C. Accuracy of the chemical-pseudopotential method for tetrahedral semiconductors. *Phys. Rev. B* **48**, 14216–14225 (1993).

Friedel, J. On the possible impact of quantum mechanics on physical metallurgy. *Trans. Metallurg. Soc. AIME* 616–632 (1964).

Friedel, J. Transition metals. electronic structure of the d-band. its role in the crystalline and magnetic structures. In J.M. Ziman (ed.), *The Physics of Metals, Vol. I—Electrons*, pp. 340–408 (CUP, Cambridge, 1969).

Gaunt, J.A. The Triplets of Helium. *Trans. R. Soc. (Lond.) A* **228**, 151–196 (1929).

Gibson, J.B., Goland, A.N., Milgram, M. and Vineyard, G.H. Dynamics of radiation damage. *Phys. Rev.* **120**, 1229–1253 (1960).

Glanville, S., Paxton, A.T. and Finnis, M.W. A comparison of methods for calculating tight-binding bond energies. *J. Phys. F: Metal Phys.* **18**, 693–718 (1988).

Godin, T.J. and Haydock, R. The block recursion library—accurate calculation of resolvent submatrices using the block recursion method. *Comp. Phys. Commun.* **64**, 123–130 (1991).

Goedecker, S. Integral representation of the fermi distribution and its applications in electronic-structure calculations. *Phys. Rev. B* **48**, 17573–17575 (1993).

Goodwin, L., Skinner, A.J. and Pettifor, D.G. Generating transferable tight-binding parameters—application to silicon. *Europhys. Lett.* **9**, 701–706 (1989).

Gordon, R.G. and Kim, Y.S. Theory for the forces between closed-shell atoms and molecules. *J. Chem. Phys.* **56**, 3122–3133 (1972).

Gumbsch, P. and Gao, H. Dislocations faster than the speed of sound. *Science* **283**, 965–968 (1999).

Hafner, J. *From Hamiltonians to Phase Diagrams* (Springer, Berlin, 1987).

Hafner, J. and Heine, V. Theory of the atomic interactions in (s,p)-bonded metals. *J. Phys. F: Metal Phys.* **16**, 1429–1458 (1986).

Hagen, M. and Finnis, M.W. Point defects and chemical potentials in ordered alloys. *Philos. Magazine A* **77**, 447–464 (1998).

Harris, J. Simplified method for calculating the energy of weakly interacting fragments. *Phys. Rev. B* **31**, 1770–1779 (1985).

Harrison, W.A. *Pseudopotentials in the Theory of Metals* (Benjamin, New York, 1966).

Harrison, W.A. Transition metal pseudopotentials. *Phys. Rev.* **181**, 1036–1053 (1969).

Harrison, W.A. *Electronic Structure and the Properties of Solids* (W.H. Freeman, San Francisco, 1980).

Hartford, J., Hansen, L.B. and Lundqvist, B.I. Harris functional densities: From solid to atom. *J. Phys.: Condens. Mat.* **8**, 7379–7391 (1996).

Hartmann, W.M. and Milbrodt, T.O. Model-potential calculations of phonon energies in aluminium. *Phys. Rev. B* **3**, 4133–4143 (1971).

Haydock, R. Efficient electronic energy functionals for tight-binding. In P.E.A. Turchi, A. Gonis and L. Colombo (eds.), *Materials Society Research Symposium*, Vol. 491, pp. 35–43 (Boston, 1998). MRS. Symposium on Tight-Binding Approach to Computational Materials Science at the 1997 MRS Fall Meeting.

Haydock, R., Heine, V. and Kelly, M.J. Electronic structure based on the local atomic environment for tight-binding bands. *J. Phys. C: Solid State Phys.* **5**, 2845–2858 (1972).

Haydock, R., Heine, V. and Kelly, M.J. Electronic structure based on the local atomic environment for tight-binding bands: II. *J. Phys. C: Solid State Phys.* **8**, 2591–2604 (1975).

Haydock, R. and Nex, C.M.M. Comparison of quadrature and termination for estimating the density of states within the recursion method. *J. Phys. C: Solid State Phys.* **17**, 4783–4789 (1984).

Haydock, R. and Nex, C.M.M. A general terminator for the recursion method. *J. Phys. C: Solid State Phys.* **18**, 2235–2248 (1985).

Heine, V. In H. Ehrenreich, F. Seitz and D. Turnbull (eds.), *Solid State Physics*, Vol. 35, pp 1–127 (Academic, New York, 1980). D.W. Bullett, *ibid.* pp. 129–214; R. Haydock, *ibid.* pp. 215–294, M.J. Kelly, *ibid.* pp. 295–383.

Heine, V. and Weaire, D. In H. Ehrenreich, F. Seitz and D. Turnbull (eds.), *Solid State Physics*, Vol. 24, pp. 249–463 (Academic, New York, 1970).

Hellmann, H. *Einführung in die Quantenchemie* (Deuticke, Leipzig and Vienna, 1937).

Hohenberg, P. and Kohn, W. Inhomogeneous electron gas. *Phys. Rev.* **136**, B864–B871 (1964).

Horsfield, A.P., Bratkovsky, A.M., Fearn, M., Pettifor, D.G. and Aoki, M. Bond-order potentials: Theory and implementation. *Phys. Rev. B* **53**, 12694–12712 (1996*a*).

Horsfield, A.P., Bratkovsky, A.M., Pettifor, D.G. and Aoki, M. Bond-order potential and cluster recursion for the description of chemical bonds: efficient real-space methods for tight-binding molecular dynamics. *Phys. Rev. B* **53**, 1656–1666 (1996*b*).

Hubbard, J. The description of collective motions in terms of many-body perturbation theory II. The correlation energy of a free-electron gas. *Proc. R. Soc. (Lond.) A* **243**, 336–352 (1958).

Ichimaru, S. Strongly coupled plasmas–high-density classical plasmas and degenerate electron liquids. *Rev. Mod. Phys.* **54**, 1017–1059 (1982).

Ichimaru, S. and Utsumi, K. Analytic-expression for the dielectric screening function of strongly coupled electron liquids at metallic and lower densities. *Phys. Rev. B* **24**, 7385–7388 (1981).

Inkson, J.C. *Many-Body Theory of Solids—An Introduction* (Kluwer Academic, 1984).

Inoue, J. and Ohta, Y. Orbital symmetrization of the recursion method. *J. Phys. C: Solid State Phys.* **20**, 1947–1964 (1987).

Ivanov, O.V. and Maksimov, E.G. Generalized variational approach to Kim–Gordon electron gas theory for ionic crystals. *Solid State Commun.* **97**, 163–167 (1996).

Jackson, J.D. *Classical Electrodynamics*, 2nd edn. (John Wiley and Sons, New York, 1975).

Jacobsen, K.W., Nørskov, J.K., and Puska, M.J. Interatomic interactions in the effective medium theory. *Phys. Rev. B* **35**, 7423–7442 (1987).

Johnson, Q.C. and Templeton, D.H. Madelung constants for several structures. *J. Chem. Phys.* **34**, 2004–2007 (1961).

Jones, R. and Lewis, M.W. Electronic charge-densities and the recursion method. *Philos. Magazine B* **49**, 95–100 (1984).

Jones, R.O. and Gunnarsson, O. Density functional formalism, its applications and prospects. *Rev. Modern Phys.* **61**, 689–746 (1989).

Kim, Y.S. and Gordon, R.G. Theory of binding of ionic crystals: application to alkali-halide and alkali-earth-dihalide crystals. *Phys. Rev. B* **9**, 3548–3554 (1974).

Kittel, C. *Introduction to Solid State Physics*, 5th edn. (Wiley, New York, 1976).

Koch, W. and Holthauser, M.C. *A Chemist's Guide to Density Functional Theory*, 2nd edn. (Wiley-VCH, 2001).

Kohn, W. and Sham, L.J. Self-consistent equations including exchange and correlation effects. *Phys. Rev.* **140**, A1133–A1138 (1965).

Kress, J.D. and Voter, A.F. Low-order moment expansions to tight binding for interatomic potentials: successes and failures. *Phys. Rev. B* **52**, 8766–8775 (1995).

Landau, L.D. and Lifshitz, E.M. *Quantum Mechanics*. (Pergamon Press, Oxford, 1965).

Lindan, P.J.D. First-principles simulation: ideas, illustrations and the CASTEP code. *J. Phys.: Condens. Mat.* **14**, 2717–2744 (2002).

Lindholm, E. and Lundqvist, S. Semiempirical MO methods, deduced from density functional theory. *Phys. Scrip.* **32**, 220–224 (1985).

Luchini, M.U. and Nex, C.M.M. A new procedure for appending terminators in the recursion method. *J. Phys. C: Solid State Phys.* **20**, 3125–3130 (1987).

Majewski, J.A. and Vogl, P. Crystal stability and structural transition pressures of sp-bonded solids. *Phys. Rev. Lett.* **57**, 1366–1369 (1986).

Marks, N.A., Finnis, M.W., Harding, J.H. and Pyper, N.C. A physically transparent and transferable compressible ion model for oxides. *J. Chem. Phys.* **114**, 4406–4414 (2001).

Mayer, I. Charge, bond order and valence in the *ab initio* SCF theory. *Chem. Phys. Lett.* **97**, 270–274 (1983).

Mead, L.R. and Papanicolaou, N. Maximum-entropy in the problem of moments. *J. Math. Phys.* **25**, 2404–2417 (1984).

Methfessel, M. Independent variation of the density and the potential in density-functional methods. *Phys. Rev. B* **52**, 8074–8081 (1995).

Methfessel, M.S. *Multipole Green Functions for Electronic Structure Calculation.* Ph.D. Thesis (Katholieke Universiteit te Nijmegen, 1986). Printed by Springelkamp, Groningen.

Monkhorst, H.J. and Pack, J.D. Special points for Brillouin-zone integrations. *Phys. Rev. B* **13**, 5188–5192 (1976).

Moriarty, J.A. Density-functional formulation of the generalized pseudopotential theory. III. Transition-metal interatomic potentials. *Phys. Rev. B* **38**, 3199–3231 (1988).

Moriarty, J.A. Analytic representation of multi-ion interatomic potentials in transition metals. *Phys. Rev. B* **42**, 1609–1628 (1990).

Moriarty, J.A., Belak, J.F., Rudd, R.E., Söderlind, P., Streitz, F.H. and Yang, L.H. Quantum-based atomistic simulation of materials properties in transition metals. *J. Phys.: Condens. Mat.* **14**, 2825–2858 (2002).

Moriarty, J.A. and Phillips, R. First-principles interatomic potentials for transition-metal surfaces. *Phys. Rev. Lett.* **66**, 3036–3039 (1991).

Nex, C.M.M. The recursion method—processing the continued-fraction. *Comp. Phys. Commun.* **34**, 101–122 (1984).

Nex, C.M.M. The block Lanczos-algorithm and the calculation of matrix resolvents. *Comp. Phys. Commun.* **53**, 141–146 (1989).

Nielsen, O.H. and Martin, R.M. Stresses in semiconductors: *ab initio* calculations on Si, Ge, and GaAs. *Phys. Rev. B* **32**, 3792–3805 (1985).

Nørskov, J.K. Covalent effects in the effective-medium theory of chemical binding: hydrogen heats of solution in the 3d metals. *Phys. Rev. B* **26**, 2875–2885 (1982).

Nørskov, J.K. and Lang, N.D. Effective-medium theory of chemical binding: application to chemisorption. *Phys. Rev. B* **21**, 213–2136 (1980).

Nye, J.F. *Physical Properties of Crystals* (Clarendon Press, Oxford, 1985).

Oleinik, I.I. and Pettifor, D.G. Analytic bond-order potentials beyond Tersoff-Brenner. II. Application to the hydrocarbons. *Phys. Rev. B* **59**, 8500–8507 (1999).

Ozaki, T., Aoki, M. and Pettifor, D.G. Block bond-order potential as a convergent moments-based method. *Phys. Rev. B* **61**, 7972–7988 (2000).

Parr, R.G. and Yang, W. *Density-Functional Theory of Atoms and Molecules* (OUP, 1989).

Paxton, A.T., Methfessel, M. and Polatoglou, H.M. Structural energy-volume relations in 1st-row transition-metals. *Phys. Rev. B* **41**, 8127–8138 (1990).

Paxton, A.T., van Schilfgaarde, M., MacKenzie, M. and Craven, A.J. The near-edge structure in energy-loss spectroscopy: many-electron and magnetic effects in transition metal nitrides and carbides. *J. Phys.: Condens. Mat.* **12**, 729–750 (2000).

Perdew, J.P. and Zunger, A. Self-interaction correction to density-functional approximations for many electron systems at finite temperatures. *Phys. Rev. B* **23**, 5048–5079 (1981).

Pettifor, D.G. The structures of binary compounds: I. Phenomenological structure maps. *J. Phys. C: Solid State Phys.* **19**, 285–313 (1986).

Pettifor, D.G. A quantum-mechanical critique of the Miedema rules for alloy formation. *Solid State Phys.* **40**, 43–92 (1987).

Pettifor, D.G. New many-body potential for the bond order. *Phys. Rev. Lett.* **63**, 2480–2483 (1989).

Pettifor, D.G. *Bonding and Structure in Molecules and Solids* (Clarendon Press, Oxford, 1995).

Pettifor, D.G. and Aoki, M. Bonding and structure of intermetallics: a new bond order potential. *Phil. Trans. R. Soc. Lond. A* **334**, 439–449 (1991).

Pettifor, D.G. and Oleinik, I.I. Analytic bond-order potentials beyond Tersoff–Brenner. I. Theory. *Phys. Rev. B* **59**, 8487–8499 (1999).

Pettifor, D.G. and Oleinik, I.I. Bounded analytic bond-order potentials for sigma and pi bonds. *Phys. Rev. Lett.* **84**, 4124–4127 (2000).

Pettifor, D.G. and Oleinik, I.I. Analytic bond-order potential for open and close-packed phases. *Phys. Rev. B*, **65**, 172103 (2002).

Pettifor, D.G. and Podloucky, R. The structure of binary compounds: II. Theory of pd-bonded AB compounds. *J. Phys. C: Solid State Phys.* **19**, 315–330 (1986).

Pettifor, D.G. and Ward, M.A. An analytic pair potential for simple metals. *Solid State Commun.* **49**, 291–294 (1984).

Pettifor, D.G. and Weaire, D.L. (eds.) *The Recursion Method and its Applications*, Springer Series in Solid-State Sciences 58 (Springer Verlag, Berlin, 1985). Proc. Conf. London, England, 13–14 September 1984.

Pick, R.M., Cohen, M.H. and Martin, R.M. Microscopic theory of force constants in the adiabatic approximation. *Phys. Rev. B* **1**, 910–920 (1970).

Pines, D. and Nozieres, P. *The Theory of Quantum Liquids* (Benjamin, New York, 1966).

Polatoglou, H.M. and Methfessel, M. Cohesive properties of solids calculated with the simplified total-energy functional of Harris. *Phys. Rev. B* **37**, 10403–10406 (1988).

Polatoglou, H.M. and Methfessel, M. Comparison of the Harris and the Hohenberg-Kohn-Sham functionals for calculation of structural and vibrational properties of solids. *Phys. Rev. B* **41**, 5898–5903 (1990).

Pulay, P. *Ab initio* calculations of force constants and equilibrium geometries in polyatomic molecules. I theory. *Molec. Phys.* **17**, 197–204 (1969).

Quong, A.A. and Klein, B.M. Self-consistent-screening calculation of interatomic force constants and phonon dispersion curves from first principles: Application to aluminium. *Phys. Rev. B* **46**, 10734–10737 (1992).

Rasolt, M. and Taylor, R. Charge densities and interionic potentials in simple metals: nonlinear effects, I. *Phys. Rev. B* **11**, 2717–2725 (1975).

Robertson, I.J. and Farid, B. Does the Harris Energy Functional posess a local maximum at the ground-state density? *Phys. Rev. Lett.* **66**, 3265–3268 (1991).

Rosenfeld, A.M. and Stott, M.J. Density-dependent pair potentials and the compressibility problem. *J. Phys. F: Metal Phys.* **17**, 605–627 (1987).

Sankey, O.F., Adams, G.B., Weng, X., Dow, J.D., Huang, Y.-M., Spence, J.C.E., Drabold, D.A., Bu, W.-M., Wang, R.P., Klemm, S. and Fedders, P.A. First-principles electronic structure calculations with molecular dynamics made easy. *Superlattices Microstruct.* **10**, 407–414 (1991).

Schelling, P.K., Yu, N. and Halley, J.W. Self-consistent tight-binding atomic-relaxation model of titanium dioxide. *Phys. Rev. B* **58**, 1279–1293 (1998).

Shohat, J.A. and Tamarkin, J.D. *The Problem of Moments.* Math. Surv. I rev. edn. (American Mathematical Society, Providence Rhode Island, 1950).

Slater, J.C. and Koster, G.F. Simplified LCAO method for the periodic potential problem. *Phys. Rev. B* **94**, 1498–1524 (1954).

Soler, J.M., Artacho, E., Gale, J.D., García, A.I., Junquera, J., Ordejón, P. and Sánchez-Portal, D. The SIESTA method for *ab initio* order-N materials simulation. *J. Phys.: Condens. Mat.* **14**, 2745–2780 (2002).

Stokbro, K., Chetty, N., Jacobsen, K.W. and Nørskov, J.K. Construction of transferable spherically-averaged electron potentials. *J. Phys.: Condens. Mat.* **6**, 5415–5421 (1994).

Stone, A.J. *The Theory of Intermolecular Forces* (Clarendon Press, Oxford, 1996).

Stott, M.J. and Zaremba, E. Quasiatoms: An approach to atoms in nonuniform electronic systems. *Phys. Rev. B* **22**, 1564–1583 (1980).

Streitz, F.H. and Mintmire, J.W. Electrostatic potentials for metal-oxide surfaces and interfaces. *Phys. Rev. B* **50**, 11996–12003 (1994).

Sutton, A.P. *Electronic Structure of Materials* (Clarendon Press, Oxford, 1993).

Sutton, A.P., Finnis, M.W., Pettifor, D.G. and Ohta, Y. The tight-binding bond model. *J. Phys. C: Solid State Phys.* **21**, 35–66 (1988).

Tersoff, J. New empirical-model for the structural properties of silicon. *Phys. Rev. Lett.* **56**, 632–635 (1986).

Thomas, L.H. The calculation of atomic fields. *Proc. Cambridge Philos. Soc.* **23**, 542–548 (1927).

Thompson, W.J. *Angular Momentum* (Wiley, New York, 1994).

Turchi, P. and Ducastelle, F. Continued fractions and perturbation theory: application to tight binding systems. In D.G. Pettifor and D.L. Weaire, (eds.) *The Recursion Method and its Applications*, pp. 104–119 (Springer Verlag, Berlin, 1985). Proc. Conf. London, England, 13–14 September 1984.

Vanderbilt, D. and Louis, S.G. Total energies of diamond (111) surface reconstructions by a linear combination of atomic orbitals method. *Phys. Rev. B* **30**, 6118–6130 (1984).

Vosko, S.H., Wilk, L. and Nusair, M. Accurate spin-density liquid correlation energies for local spin density calculations: a critical analysis. *Can. J. Phys.* **58**, 1200–1211 (1980).

Walker, A.B. and Taylor, R. Density-dependent potentials for simple metals. *J. Phys.: Condens. Matt.* **2**, 9481–9499 (1990).

Wallace, D.C. *Thermodynamics of Crystals* (Wiley, New York, 1972).

Whittaker, E.T. and Watson, G.N. *A Course of Modern Analysis*, 4th edn. (CUP, Cambridge, 1927).

Wilkinson, J.H. *Algebraic Eigenvalue Problems* (Clarendon, Oxford, 1965).

Wilson, M., Huang, Y.M., Exner, M. and Finnis, M.W. Transferable model for the atomistic simulation of Al_2O_3. *Phys. Rev. B* **54**, 15683–15689 (1996*a*).

Wilson, M., Schönberger, U. and Finnis, M.W. Transferable atomistic model to describe the energetics of zirconia. *Phys. Rev. B* **54**, 9147–9161 (1996*b*). Erratum: Using the data reported, the rutile structure turns out to have the lowest energy, in contradiction to the figure.

Zaremba, E. Extremal properties of the Harris Energy Functional. *J. Phys.: Condens. Mat.* **2**, 2479–2486 (1990).

Ziman, J.M. *Principles of the Theory of Solids*, 2nd edn. (CUP, Cambridge, 1972).

INDEX

The manufacturer's authorised representative in the EU for product
safety is Oxford University Press España S.A. of El Parque Empresarial
San Fernando de Henares, Avenida de Castilla, 2 - 28830 Madrid
(www.oup.es/en or product.safety@oup.com). OUP España S.A. also acts
as importer into Spain of products made by the manufacturer.
Printed and bound by CPI Group (UK) Ltd, Croydon, CR0 4YY
06/07/2026
02157632-0011